T0331712

ASTRONOMY FOR THE DEVELOPING WORLD

PROCEEDINGS OF SPECIAL SESSION NO. 5

COVER ILLUSTRATION:

Star-watching at Mt John observatory. Photograph courtesy of Fraser Gunn, Lake Tekapo.

IAU SYMPOSIUM PROCEEDINGS SERIES

2006 EDITORIAL BOARD

Chairman

K.A. VAN DER HUCHT, IAU Assistant General Secretary
SRON Netherlands Institute for Space Research
Sorbonnelaan 2, NL-3584 CA Utrecht, The Netherlands
K.A.van.der.Hucht@sron.nl

Advisors

O. ENGVOLD, IAU General Secretary, *Institute of Theoretical Astrophysics, University of Oslo, Norway*
E.J. DE GEUS, *Netherlands Foundation for Research in Astronomy, Dwingeloo, The Netherlands*
M.C. STOREY, *Australia Telescope National Facility, Australia*
P.A. WHITELOCK, *South African Astronomical Observatory, South Africa*

Members

IAUS233
V. BOTHMER, *Universitäts Sternwarte, Georg-August-Universität, Göttingen, B.R. Deutschland*
IAUS234
M.J. BARLOW, *Department of Physics and Astronomy, University College London, London, UK*
IAUS235
F. COMBES, *LERMA, Observatoire de Paris, Paris, France*
IAUS236
A. MILANI, *Dipartimento di Matematica, Università di Pisa, Pisa, Italia*
IAUS237
B.G. ELMEGREEN, *IBM Research Division, T.J. Watson Research Center, Yorktown Heights, NY, USA*
IAUS238
V. KARAS, *Astronomical Institute, Academy of Sciences of the Czech Republic, Praha, Czech Republic*
IAUS239
F. KUPKA, *Max-Planck-Institut für Astrophysik, Garching-bei München, B.R. Deutschland*
IAUS240
W.I. HARTKOPF, *U.S. Naval Observatory, Washington D.C., USA*
IAUS241
A. VAZDEKIS, *Instituto de Astrofísica de Canarias, La Laguna, Tenerife, Canary Islands, Spain*

INTERNATIONAL ASTRONOMICAL UNION
UNION ASTRONOMIQUE INTERNATIONALE

ASTRONOMY FOR
THE DEVELOPING WORLD

PROCEEDINGS OF SPECIAL SESSION NO. 5
OF THE 26TH IAU GENERAL ASSEMBLY
HELD IN PRAGUE, THE CZECH REPUBLIC
21 AND 22 AUGUST 2006

Edited by

JOHN HEARNSHAW
University of Canterbury, New Zealand

and

PETER MARTINEZ
South African Astronomical Observatory
Cape Town, South Africa

CAMBRIDGE
UNIVERSITY PRESS

University Printing House, Cambridge CB2 8BS, United Kingdom

One Liberty Plaza, 20th Floor, New York, NY 10006, USA

477 Williamstown Road, Port Melbourne, VIC 3207, Australia

314-321, 3rd Floor, Plot 3, Splendor Forum, Jasola District Centre, New Delhi - 110025, India

103 Penang Road, #05-06/07, Visioncrest Commercial, Singapore 238467

Cambridge University Press is part of the University of Cambridge.

It furthers the University's mission by disseminating knowledge in the pursuit of education, learning and research at the highest international levels of excellence.

www.cambridge.org
Information on this title: www.cambridge.org/9780521876575

© International Astronomical Union 2007

This publication is in copyright. Subject to statutory exception and to the provisions of relevant collective licensing agreements, no reproduction of any part may take place without the written permission of Cambridge University Press.

First published 2007

A catalogue record for this publication is available from the British Library

ISBN 978-0-521-87657-5 Hardback
ISSN 1743-9213

Cambridge University Press has no responsibility for the persistence or accuracy of URLs for external or third-party internet websites referred to in this publication, and does not guarantee that any content on such websites is, or will remain, accurate or appropriate.

Table of Contents

Special session no. 5: ASTRONOMY FOR THE DEVELOPING WORLD

Section 1. Overview to astronomy in the developing world
Chair: J. Pasachoff

Section 2: Astronomy in Latin America and in the Caribbean
Chair: J. Hearnshaw

Section 3: Astronomy in Africa
Chair: J. Fierro

Section 4: Astronomy in Eastern Asia and the Pacific
Chair: J. Fierro

Section 5: Astronomy in the Middle East and central Asia
Chair: P. Martinez

Section 6: Astronomy in eastern Europe
Chair: J. Percy

Section 7: Astronomy education in developing countries
Chairs: J. Percy and M. Gerbaldi

Section 8: Promoting astronomy in developing countries through the UN, the IHY and COSPAR
Chair: B. Jones

Section 9: The virtual observatory and developing countries
Chair: H. Haubold

Preface

The International Astronomical Union has a strong commitment to the development of astronomical education and research throughout the world, especially in those countries developing economically. This commitment is in part through the work of IAU Commission 46 for Astronomy Education and Development. Within that commission, the Program Group for the Worldwide Development of Astronomy (PGWWDA) coordinates many of these activities, promoting the development of astronomy in developing countries.

Six years ago, at the time of the 24th IAU General Assembly in Manchester, Alan Batten, who was then chair of the PGWWDA, organized a special session on 'Astronomy for developing countries' (Batten 2001). The success of that meeting led Commission 46 to propose a second special session, this time at the 26th IAU General Assembly in Prague. This volume presents the papers from that two-day session, known as Special Session 5, 'Astronomy for the developing world'.

A key theme proposed for SPS5 was a survey of the development of astronomy in different geographical regions of the world, such as Latin America, eastern Europe, Africa, central Asia and the Far East. There are contributions here from all these places. In addition SPS5 strived to bring together several other programmes promoting astronomy and space science in the developing world, from agencies outside the IAU. These include the United Nations Office for Outer Space Affairs (UNOOSA), the International Heliophysical Year (IHY) programme for 2007 and the Committee on Space Research (COSPAR), amongst others.

Moreover, SPS5 had as one of its aims to promote the concept of establishing a Third-world Astronomy Institute or Network (TWAI/TWAN) – an idea championed by Professor Jayant Narlikar and presented in the introductory session. Moves are already underway to give the Inter-University Centre for Astronomy and Astrophysics in Pune, India, an international dimension. This will be a step towards realizing this dream, to do for astronomy what the International Centre for Theoretical Physics in Trieste is already doing for physics.

The interest in SPS5 'Astronomy for the developing world' was much greater than expected. In these pages, 56 papers are presented, although 61 abstracts were originally accepted. At the conference, there were 16 invited papers, 26 contributed oral talks and the rest were poster papers. The first authors came from 37 different countries. What is more, about 280 astronomers from 61 different countries registered their interest in participating in the SPS5 session; it was a truly multinational gathering.

All authors whose submitted abstracts were accepted were informed that acceptance of their paper at SPS5 entailed an obligation to send the full text for publication in this volume. In spite of this clear requirement, sadly some authors, in spite of numerous and repeated entreaties, failed to honour it. These included two of the invited authors, David Webb and Athem Alsabti. Certainly this volume is the poorer for not having their full contributions.

The editors recognize that this is the symptom of a much wider malaise in scientific conference publishing in general. Editors go to great lengths to produce the best scientific record possible of any conference, efforts which are partially thwarted by those few who, for whatever reason, fail to submit their papers. Conference participation has never been so popular; submitting conference papers is much less so. Even those who receive financial support to participate are sometimes reluctant to submit papers for publication. Perhaps more editors should consider making full paper submission in advance of any meeting a criterion for paper acceptance.

Reference:
Batten, A. (ed.), 2001, Astronomy for developing countries, IAU Special session at Manchester GA, publ. by Astron. Soc. Pacific, pp. xvii+376

John Hearnshaw (University of Canterbury, New Zealand) and Peter Martinez (South African Astronomical Observatory)
Co-editors of Special Session #5,
Christchurch, NZ, and Cape Town, SA,
March, 2007

Scientific organizing committee of SPS5

The following people served on the SOC for the special session, SPS5:
- John Hearnshaw (New Zealand, chair)
- Athem Alsabti (U.K./Iraq)
- Julieta Fierro (Mexico)
- Michele Gerbaldi (France)
- Hans Haubold (Germany, UNOOSA)
- Barrie Jones (U.K.)
- Ajit Kembhavi (India)
- Hugo Levato (Argentina)
- Peter Martinez (South Africa)
- Jayant Narlikar (India)
- Jay Pasachoff (USA)
- John Percy (Canada)
- Boonrucksar Soonthornthum (Thailand)
- Peter Willmore (U.K., COSPAR)
- Jay White (USA)

Participants at Special Session no. 5

Special session no. 5 editors, John Hearnshaw (left) and Peter Martinez (right)

Section 1:

Overview to astronomy in the developing world

Astronomy for the developing world
IAU Special Session no. 5, 2006
J.B. Hearnshaw and P. Martinez, eds.

© 2007 International Astronomical Union
doi:10.1017/S1743921307006606

TWAN: A way of networking third-world astronomers

Jayant V. Narlikar

Inter-University Centre for Astronomy and Astrophysics, Ganeshkhind, Pune 411007, India
email: jayant@iucaa.ernet.in

Abstract. This talk describes a proposal to set up a series of international institutions in different parts of the world to serve as nodes in a network that links astronomers from the developing nations worldwide. This network, along with its nodes is visualized as an economic way of upgrading the facilities for teaching, research and development of astronomy in the Third World countries. By way of illustration, the *modus-operandi* of the Inter-University Centre for Astronomy and Astrophysics in Pune, India is described. A network of this kind is suggested as a cost-efficient way of sharing limited resources.

Keywords. Third World Astronomy Network (TWAN); IUCAA, India

1. Introduction

The idea of a Third World Astronomy Network (TWAN) was aired first at the special session of the IAU General Assembly held in Manchester in 2000 (*see* Narlikar 2001a). I will first briefly review the basic idea as proposed then.

The aim of TWAN is to bring together astronomers from third world countries together in a network that is empowered to provide them with assistance to improve their research, teaching and developmental facilities. The network concept helps share limited resources in a cost-effective way, using the benefits of the emerging communications technology.

The basic structure of TWAN can be visualized as a group of nations in a region which are linked together and serviced by a 'node', the nodes themselves being connected to one another in a worldwide network. A typical node is an international centre or institute that carries out certain mandated activities to serve the above-mentioned aim of TWAN.

By way of illustration, I mention two institutions that carry out activities somewhat similar to what is intended for the TWAN nodes. These are the Abdus Salam International Centre for Theoretical Physics (AS-ICTP) in Trieste, Italy and the Inter-University Centre for Astronomy and Astrophysics (IUCAA) in Pune, India.

The AS-ICTP was set up more than four decades ago by Abdus Salam, a distinguished theoretical physicist who hailed from Pakistan. From his own personal experience Salam had felt the need to create an international resource centre which would help physicists from the Third World nations to fulfill their aspirations of carrying out top class research. To this end the ICTP was created in Trieste in 1964 and it has served as a place where Third world scientists can visit to use the Centre's library and other resources, attend schools and workshops and meet scientists from other countries to forge new collaborations. Since its inception, the ICTP has expanded its sphere of subjects beyond theoretical physics. The centre is supported largely by the Italian Government as well as by UNESCO and the International Atomic Energy Agency. It was named after its founder Professor Salam, after he passed away in 1996. For details of how the centre came about, *see* the interesting article by Professor Salam (1990).

The IUCAA was set up in 1988 as a centre of excellence by the University Grants Commission of the Government of India, with the mandate that the centre shall provide facilities and guidance to astronomers and astrophysicists from Indian universities in their research, teaching and developmental activities. Since the IUCAA was specifically aimed at astronomers and astrophysicists, it comes close to what is being proposed for a TWAN node. I will therefore spend some time on describing how the IUCAA functions.

2. What does IUCAA do?

For those interested in the details of what led to the setting of IUCAA and the impact it has had on university education and research, I may refer to my article published a few years ago (Narlikar 2001b). Briefly, I may add that the creation of IUCAA was motivated by the growing gap in the standards of academic functioning of universities on the one hand and autonomous research institutes (ARIs) on the other. The latter had been created largely since Indian independence in 1947, were generally limited to their specific subject areas and had scientists carrying out research and developmental activities unburdened by teaching graduates and undergraduates. Compared to universities the ARIs were much better endowed, had better facilities and better working conditions.

Although the universities also proliferated in number, their quality steadily went down. The academics at the universities were required to spend most of their time on lecturing, leaving very little scope for research and development. The funds for the latter also shrank so that the balance of research began to tilt heavily towards the ARIs.

To redress this balance the University Grants Commission, the parent funding agency for universities in India, came up with the idea of setting up inter-university centres (IUCs). In a typical IUC a subject that is not very well funded in a typical university is taken up. For example, the IUCAA was created to support astronomy and astrophysics (A&A). Given that there are 200+ universities, it would be prohibitively expensive to provide all A&A-related facilities to each and every university. Moreover, if the number of research workers (teachers and students included) in a typical university were no more than 20-30, these facilities would remain largely under-utilized. So a way had to be found whereby precious resources would be maximally used. The IUC mode is such a way: it is endowed liberally with the necessary resources and it ensures that they are optimally used by the university academics. How is this done?

Let me briefly describe the in-house facilities at IUCAA. It contains an up to date A&A library which also contains books of peripheral interest to astronomers, like physics, mathematics, computer science, etc. It has a good stock of A&A and physics journals and electronic subscription to many. It has mirror images of leading data centres and is currently playing a lead role in the Indian contributions to the international Virtual Observatory programme. Then there is a computer centre with state of the art software for use by astronomers as well as by theoreticians. To help development of instruments, there is a laboratory that guides users in making various astronomical instruments. Recently, a new-technology telescope with 2m aperture has been set up not far from Pune which will provide observing experience to astronomers from universities.

I should also mention the circumstance that influenced the siting of IUCAA in Pune. Its location is next door to the National Centre for Radio Astrophysics (NCRA) of the Tata Institute of Fundamental Research. The NCRA has been operating the Giant Metrewave Radio Telescope (GMRT) situated around 90 km from Pune. The nearness to this national centre also provides IUCAA and its university visitors with additional facilities and guidance.

How do university astronomers use these facilities? To facilitate this, IUCAA has introduced an associateship programme along the lines of AS-ICTP. This programme identifies active and promising university scientists as prospective users of IUCAA by making them visiting associates. A visiting associate can visit IUCAA frequently and for extended periods for using its facilities for his/her research. A senior research worker from a university who is a visiting associate can also bring his student with him if he feels that the student would benefit by such a visit. All costs for such visits are borne by IUCAA.

Following the AS-ICTP, the IUCAA has also an annual calendar of pedagogical activities. It runs a graduate school jointly with NCRA, to which university students are also invited. There are introductory schools or workshops in A&A as well as advanced research level meetings. Resource persons for such meetings are drawn from all over India and also sometimes from abroad.

Given a year-round influx of visiting scientists and students at IUCAA, an important facility of course is the hostel and guest house. These are attached to the main IUCAA buildings so that the visitors need spend very little time commuting to work. Long term visitors are accommodated in guest apartments. Having such in-house facility greatly helps the efficiency of the visiting scientist.

Another aspect of IUCAA deals with encouraging astronomical observing at the professional level. University scientists have been encouraged to go out of their campuses to carry out observations that relate to their research. Not only national facilities like the GMRT, the Vainu Bappu Telescope in the south or the ARIES Observatory in the north: the IUCAA encourages going beyond the national boundary to participate in guest observing programmes on major telescopes abroad. This is done through collaborations forged with scientists abroad. IUCAA helps and advises the university scientists in practical matters such as writing a proposal for observing.

To be able to carry out these tasks, IUCAA has provision for a resident faculty of active research workers who also carry out pedagogical activities. In order that IUCAA's guidance commands credibility, the faculty members have to be seen to belong to the top echelons in their respective subject areas. This faculty also has research students to guide and is supported by a vigorous post-doctoral programme. Indeed, from early days IUCAA's post-doctoral programme has had an international flavour with post-docs from Japan, China, France, Italy, the USA, the UK, Iran, etc.

Last, but not the least, IUCAA runs various programmes of public outreach, such as public lectures, night sky viewing, special viewing programmes at eclipses or transits, workshops for making telescopes for amateur observing, etc. Knowing the importance of attracting and motivating students towards science and astronomy in particular, IUCAA has an activity centre for school students around which many programmes are built. On the second and fourth Saturdays for example, there are morning lectures for students from secondary and higher secondary levels. Well illustrated and with demonstrations to accompany them, these lectures are very popular and the 500-seater auditorium is often overflowing for such events.

During the summer vacations IUCAA invites interested school students to do projects with the IUCAA staff. In a week long period a student (or a group of students) complete a project which teaches them some science. A special occasion is the *National Science Day*, when IUCAA stages an open house for the general public, which turns up in thousands. On this day we also have science quiz and competitions for school children.

3. The impact of IUCAA

What has been the impact of IUCAA? Several indicators suggest that the experiment has been remarkably successful. One ready reckoner is the list of publications of university academics. The number of papers published in standard refereed journals have multiplied

Figure 1. IUCAA's symbol displayed on the wall at the main entrance emphasizes its networking with the universities. It also suggests that like the loop without ends, astronomers are engaged in an unending cosmic search.

several times. The number of university departments offering A&A at post graduate level has also trebled since before IUCAA started.

The guest-observing programmes have brought international level observing to many university campuses. One of the speakers at this meeting (S.K. Pandey) started off as a theoretical solar physicist in a small university; but after collaborations with the IUCAA faculty, has blossomed into an observational astronomer with experience of observing galaxies, using many telescopes abroad.

The associateship programme brings faculty members from many different universities to the IUCAA campus at the same time. This circumstance, arising because there are common vacation windows for many universities, has led to collaborations between scientists who otherwise would not have met.

Instrumentation had long remained a neglected area in universities and most university scientists were forced by the lack of experimental facilities to be theorists. With the guidance provided by IUCAA's instrumentation laboratory, the culture of instrumentation has been growing.

IUCAA's public outreach programme has made the institution well known in the city of Pune and the state of Maharashtra. Its school-student oriented programmes have created a strong base of young amateur astronomers in the state and has inspired students to perform very well in the astronomy Olympiads.

Perhaps, it is necessary to recognize that IUCAA was the first Indian scientific institution to have instituted a scientific advisory committee with an international flavour, to oversee its programmes and monitor its performance. The SAC-IUCAA has provided valuable guidance that has helped IUCAA progress towards its objectives in more efficient ways.

4. A TWAN node

I have gone into details in describing IUCAA, because I feel the model of this institution comes very close to what we expect from a node of TWAN, and can be easily adapted to its requirements. Read 'third world countries' for 'universities' in the above scenario, and expand the 'catchment area' of the node to cover several countries in the region. Thus a node in India can very well serve the Asia-Pacific region, which, of course may require an additional node because of its geographical spread.

A typical node should have at least 6-10 experienced and productive astronomers on its staff carrying front-line research, who would guide the node's academic programmes. To support them, the node would be expected to have the following in-house academic facilities:

1. A very good library
2. An advanced computer centre
3. A data-mining facility
4. An R&D centre for instrumentation
5. Access to a good observing facility
6. A guest house for visitors

What will be the academic programmes that a node should be mandated to offer? Going by the example of IUCAA, the following suggest themselves:

1. *Associateship programme*: An associateship programme along the lines of IUCAA or AS-ICTP should be introduced. This should provide for, say, one visit to the node in two years over a period of 6 years for the associate. The duration of a visit should be at least for one month and may extend up to six months at a time. Associates may be selected from the region under the node and the node will take care of travel and subsistence of the associate for this visit.

2. *Visits by third world academics*: A scheme whereby a 1-3 month visit to the node by an active senior academic from the region may be encouraged. Unlike the associateship programme, this would be a one-off visit, with its academic objective well defined before-hand. The node may have an arrangement with the Third World Academy of Sciences which supports visits of this kind.

3. *Schools for young astronomers*: The node may take up the responsibility of organizing international schools for young astronomers coming from the region. This will be similar to the ISYA of the IAU, but limited to the region under the node.

4. *Advanced workshops*: To encourage research workers in a specific area of A&A, advanced schools may be arranged by the node. A workshop can be of 3-4 week duration and the lecture notes based on it may be published/or widely circulated through node's website.

5. *Guest observer programme*: The faculty of the node should actively encourage participation of the region's astronomers in guest observing in international observing facilities. The faculty may also advise the user from the region on data-mining and usage of virtual observatory facilities.

6. *Instrumentation programme*: The node should provide guidance in making instruments for astronomical observing. The role of the node staff should be catalytic in the sense that the actual instrument building would be done by the astronomers of the region.

The node should be governed by an Executive Council having representatives of the funding agencies, some user nations of the region and a few experienced astronomers. It should also have a scientific advisory committee of distinguished astronomers who are sympathetic to the objectives of TWAN. An ideal funding arrangement suggested here (along the lines of the AS-ICTP) is through the International Astronomical Union, UNESCO and the country in which the node is sited.

5. Budgetary aspects

How much will such a centre cost to set up? How much will be its running cost? These matters will need to be looked at in detail. What are presented here are tentative figures based on my IUCAA experience. If such a centre were set up in India, I expect that the cost of buildings and the in-house facilities will come up to around US\$ 10 million. It is assumed that the Government of India would provide land at no cost or at a nominal lease rent (as is often done). The Government may also be requested to bear the cost of the building. So far as the running cost is concerned, at rates prevalent in India, we arrive at the following figures for annual running costs in US\$:

10 Faculty Members	250 000
20 Infrastructural staff (10 employees + 10 on contract)	250 000
Maintenance, utilities, etc.	100 000
Academic facilities (as described in section 4)	200 000
Academic programmes (as described in section 4)	200 000
Total annual budget	1 000 000

If IUCAA were entrusted to carry out the functions of a node until an independent one were set up, the annual cost may work out close to US\$ 200 000. This is because IUCAA already has most of the infrastructure in place and will need a supplementary grant to take on the additional responsibility of a TWAN node.

6. Concluding remarks

The worldwide network envisaged under the auspices of TWAN will require more such nodes to take care of other regions. I expect there may be as many as 6 nodes to cover the developing countries in Asia, Africa and Latin America. The costs, compared to what is needed to create a major observing facility are modest, but the dividend in terms of upgrading the output of third world astronomers will be very significant.

I hope that this idea is taken up by the International Astronomical Union as a worth-while activity to support. An ideal time for launching a TWAN node is the year 2009, the proposed International Year for Astronomy.

Acknowledgements

I thank the IAU for giving me the opportunity for airing my vision of TWAN. At a personal level I thank Alan Batten and John Hearnshaw for their backing to the project.

References

Salam, A. 1990, in *Proceedings of the 25th anniversary conference 'Frontiers in Physics, High Technology and Mathematics'*, eds. Cerdeira, H.A. and Lundqvist, S.O. (World Scientific), 74–81

Narlikar, J.V. 2001a, in *Proceedings of the IAU Special Session*, ed. Alan H. Batten (Astronomical Society of the Pacific), 324–328

Narlikar, J.V. 2001b, in *Organizations and Strategies in Astronomy*, Vol 2, ed. Andre Heck (Kluwer, Dordrecht), 29–45

Astronomy for the developing world
IAU Special Session no. 5, 2006
J.B. Hearnshaw and P. Martinez, eds.

© 2007 International Astronomical Union
doi:10.1017/S1743921307006618

A survey of published astronomical outputs of countries from 1976 to 2005 and the dependence of output on population, number of IAU members and gross domestic product

John Hearnshaw[1]

[1]Department of Physics and Astronomy, University of Canterbury, Christchurch, New Zealand
email: john.hearnshaw@canterbury.ac.nz

Abstract. In this paper I report the results of a survey of the astronomical outputs of all 63 IAU member countries as well as several non-member countries, based on an analysis of the affiliations of the authors given for nearly 900 thousand astronomical papers appearing in ADS between the years 1976 and 2005. The results show a roughly three-fold increase in the number of published papers per year over this 30-year interval. This increase is seen both in developed and also in most developing countries. The number of publications per IAU member correlates strongly with gross domestic product. It is over 2 papers per IAU member per year in the countries with the strongest economies but less than 0.5 in the countries with low GDP per capita. Since 2001 there has been a dramatic increase in the number of multi-author multinational papers published. This increase is especially noticeable for authors in developing countries, indicating that astronomers in these countries are increasingly participating in international collaborations for their research activities.

Keywords. Developing countries, published outputs, IAU members, gross domestic product

1. Introduction

In an earlier paper (Hearnshaw 2001) I presented some statistical data on the astronomical activity in various developed and developing countries, and correlated this activity against the wealth of the country measured by IMF quota per capita. The parameter to measure astronomical activity was simply the number of IAU individual members per million of population. However, that is a fairly crude measure of the astronomical activity in any country, especially as the percentage of professional astronomers joining the IAU differs widely from one country to another. In this paper the analysis is improved and extended, by using published astronomical outputs as recorded in the ADS (Astrophysics Data System) database as another and probably more reliable measure of astronomical activity. In addition, this paper adopts the more widely understood gross domestic product per capita (in $US/capita) as a measure of a nation's intrinsic wealth.

2. Use of the ADS affiliation field to count astronomical publications by country

The main task of this paper is to record and then analyse the number of astronomical papers published by astronomers in different countries as given by their affiliations in ADS. In addition, the growth in astronomical activity in each country, as given by ADS over 30 years between 1976 and 2005 is also explored. Although ADS has an affiliation field for many of the items in the database, and although the affiliation field of each author nearly always lists the country or countries where each author is working, this information is by no means always given. Moreover, even when the country is listed, it is often done so in a variety of different formats: thus an astronomer in Cambridge, England might list his or her country as England, U.K., United Kingdom or Great Britain. For many papers published in the United States, the country is often not mentioned at all, and ADS might only give the state, and on occasions only the institution. It became obvious that the ADS affiliation fields represent fairly noisy data and the counts of papers based on country affiliations need to be used with caution.

In spite of this caution, there is still a large body of relatively good statistical data that can be extracted from the affiliation field in ADS which indicates in broad terms where the astronomical research and activity in the world is taking place. It also shows the growth in astronomical activity with time in the different countries.

The tool used in ADS for interrogation of the affiliation fields can be found at adswww.harvard.edu/affil/ and a discussion of the use of this tool can be found in the ADS frequently asked questions (FAQ). The sophistication of the affiliation interrogation tool has improved in recent years, and now enables mass statistical data to be gathered for most countries fairly readily. The USA and the countries that made up the former Soviet Union posed by far the greatest difficulties; in the former case, the country name is often omitted, and hence a search for all the individual state names and their abbreviations was necessary. (Puerto Rico was included in the publication count of the USA). As for countries of the former Soviet Union, the name of each republic is generally cited in each affiliation even before the Soviet break-up. If not, then city names were also useful for locating the country of origin.

In using the affiliation interrogation tool, the count of published papers was limited to refereed papers. Unrefereed papers have a much higher incidence of no country name in the affiliation field. They include items such as book reviews, published reports and popular articles in astronomy magazines for amateur astronomers. Unless otherwise stated, 'published papers' will henceforth refer to refereed published papers.

The tool was used with a date filter so as to count the number of published papers from each country selected in the time period January 1976 to December 2005. Within this 30-year span, the count was taken in 5-year intervals 1976–80, 1981–85, etc.

Results of the refereed paper counts by country and within 5-year time intervals from 1976 to 2005 are given in Table 1. The table lists results for 62 IAU members (based on membership in early 2006 before the 26th IAU General Assembly). Data for the Czech Republic and Slovakia are, however, combined into a single entry.

In addition, data for 24 non-IAU member countries are given in Table 2. These 24 countries comprise all 23 non-IAU member countries which have at least one IAU individual astronomer; in addition Mongolia (with no individual members) is included in this table. (In August 2006 Mongolia, Thailand and Lebanon all became member countries of the IAU, but their accession to IAU membership came after this analysis was completed.)

In Tables 1 and 2 the columns are:

• Col. 2–7. The number of papers published in 5-year intervals from 1976 to 2005, and where at least one author lists an affiliation to an institution in the country listed.

• Col. 8. The total number of papers 1976–2005.

It is noted that the paper counts in Tables 1 and 2 will count many papers more than once, if there are two or more countries in the affiliation field in ADS. This will be the case where a single author has affiliations to institutions in two or more countries, or multiple authors from different countries have collaborated. Hence the sum of papers counted over all countries may exceed the actual number of papers published. On the

Table 1. Refereed papers published, 1976–2005: IAU member countries

country	1976–80	1981–85	1986–90	1991–95	1996–2000	2001–05	total 1976–2005
Argentina	73	158	261	193	407	822	1914
Armenia	195	244	168	71	88	287	1053
Australia	1077	1346	1271	1088	1447	2967	9196
Austria	115	208	246	296	464	789	2118
Belgium	338	479	442	329	501	1121	3210
Bolivia	1	5	2	7	4	10	29
Brazil	211	359	486	493	847	1421	3817
Bulgaria	97	248	225	136	145	238	1089
Canada	1168	1609	2156	2040	1654	3617	12244
Chile	33	83	146	145	262	912	1581
China (P.R.C.)	153	1050	1092	935	950	2592	6772
China Taipei	10	17	24	73	188	577	889
Croatia	5	10	17	26	38	113	209
Cuba	0	1	2	5	6	10	24
Czech Rep + Slovakia	339	406	535	383	420	800	2883
Denmark	157	307	323	352	560	901	2600
Egypt	19	65	113	83	54	94	428
Estonia	46	59	68	58	72	100	403
Finland	103	200	288	314	443	884	2232
France	1829	3205	3735	3295	4665	8943	25672
Germany	2445	4240	4742	4163	6169	11594	33353
Greece	140	276	334	283	354	682	2069
Hungary	70	150	202	151	233	518	1324
Iceland	4	5	10	14	28	60	121
India	558	1190	1161	996	920	1408	6233
Indonesia	5	17	18	27	32	34	133
Iran	35	12	13	12	42	137	251
Ireland	49	74	96	112	179	403	913
Israel	288	285	405	377	469	967	2791
Italy	1228	2191	2513	2377	3474	7700	19483
Japan	804	1623	2266	2602	3196	5858	16349
Korea	26	59	112	276	538	1112	2123
Latvia	13	24	30	20	21	24	132
Lithuania	11	28	19	16	79	97	250
Malaysia	6	6	7	0	0	6	25
Mexico	97	189	271	296	802	1797	3452
Morocco	0	1	1	16	19	24	61
Netherlands	840	1422	1617	1222	1911	3549	10561
New Zealand	104	111	147	133	117	249	861
Nigeria	6	26	21	30	20	12	115
Norway	130	155	198	200	176	423	1282
Peru	9	5	7	8	5	31	65
Philippines	3	0	0	1	6	5	15
Poland	423	459	513	474	696	1338	3903
Portugal	5	21	24	40	191	541	822
Romania	31	27	8	126	103	149	444
Russia	3923	3892	3568	2668	2475	4035	20561
Saudi Arabia	1	10	19	16	15	38	99
Serbia, Montenegro	19	55	78	59	93	199	503
South Africa	216	276	394	275	401	583	2145
Spain	124	361	950	1009	2193	4337	8974
Sweden	310	529	604	480	668	1368	3959
Switzerland	438	495	584	540	895	1938	4890
Tajikistan	75	101	99	50	17	13	355
Turkey	52	82	122	90	102	281	729
Ukraine	486	566	505	418	690	832	3497
UK	3147	4232	4530	2916	1387	3270	19482
Uruguay	1	2	6	11	25	41	86
USA	10890	7800	7322	10103	16294	40745	93154
Vatican	6	20	40	27	52	94	239
Venezuela	18	38	47	59	86	147	395
totals	33005	41114	45203	43015	58388	123837	344562

Table 2. Refereed papers published, 1976–2005: IAU non-member countries

country	1976–80	1981–85	1986–90	1991–95	1996–2000	2001–05	total 1976–2005
Albania	0	0	0	0	0	0	0
Algeria	0	0	0	0	0	0	0
Azerbaijan	0	0	0	2	3	19	24
Columbia	0	1	3	2	0	1	7
Ecuador	0	0	0	0	2	0	2
Ethiopia	0	0	0	0	0	1	1
Georgia (Republic)	0	0	0	36	48	91	175
Honduras	0	0	0	0	1	0	1
Iraq	2	10	19	3	0	0	34
Kazakhstan	0	0	0	3	5	12	20
Korea DPR	0	0	0	0	0	0	0
Lebanon	1	4	5	2	4	14	30
Macedonia	0	0	0	1	1	1	3
Malta	0	0	0	1	0	1	2
Mauritius	0	0	0	1	1	12	14
Mongolia	2	2	5	1	2	4	16
Sri Lanka	2	2	1	0	1	1	7
Singapore	13	4	10	2	6	21	56
Pakistan	3	7	8	11	7	21	57
Slovenia	0	3	5	4	19	78	109
Thailand	0	0	0	3	4	54	61
United Arab Emirates	1	1	0	0	0	5	7
Uzbekistan	0	0	0	35	44	63	142
Vietnam	1	1	1	2	7	9	21
totals	25	35	57	109	155	408	789

other hand, it may also be less than the actual number of papers published, given that not all papers have countries in the affiliation field.

3. Country statistical data

The following data have been collected for the 86 countries in Tables 3 and 4:

• Col. 2. The number of IAU individual members in each country as at February 2006. These are professional astronomers resident and working in that country, according to IAU records. They are not necessarily citizens of the country in which they reside and work.

• Col. 3. The total number of refereed papers published 1976–2005 (from Tables 1 and 2).

• Col. 5. Gross domestic product per capita in \$US/capita. The data have been taken from

http://www.photius.com/rankings/economy/gdp_per_capita_2006_1.html and refer to data for the year 2006.

• Col. 6. Population in millions taken from

http://www.photius.com/rankings/population/population_2006_1.html. These figures also refer to 2006.

From these data the following statistics have been calculated for each country:

• Col. 4. Published papers per IAU member per year. This is a measure of the productivity of the astronomers in a country. It is recognized that many publishing astronomers may not be IAU members. Hence the actual average published output per individual will in general be less than the figure given here. However, the parameter still allows comparisons between countries to be made, and it should be regarded as a relative productivity index.

- Col. 7. IAU individual members/million of population. This is a measure of the support for astronomy in a given country
- Col. 8. Papers published per year per billion US dollars of GDP. This is a measure of the fraction of a country's financial resources that has gone into astronomical research outputs. The mean papers per year between 1976 and 2005 has been used. The GDP (GDP/capita times population) is for 2006.

These parameters are listed in Table 3 for IAU member countries, and in Table 4 for non-member countries having individual astronomers.

Table 5 summarizes some data pertaining to the statistics of IAU individual and national members. It shows that 112 out of 197 countries (almost 57% of all countries) have no IAU members and only 62 out of 197 (31%) adhere to the IAU. However, the non-member countries often have small populations. It is noted that only two countries (out of 11) with populations that exceed 10^8 do not belong to the IAU (these are Pakistan and Bangladesh).

In fact, 76% (4.9 billion out of 6.5 billion) of the world's total population lives in an IAU member country. Only 14% of the world's population (937 million people) live in a country with no professional astronomical activity. Likewise, almost 99% (8867 out of 8977) of the world's IAU individual members live in a country that adheres to the IAU. These numbers show that the IAU has been very effective in reaching a majority of the countries where professional astronomers operate.

Taken globally, the ratio of IAU individual members to the world's total population is 1.39 IAU astronomers/million of population. For the 62 IAU member countries, this ratio is 1.81, and for 135 non-member countries it is just 0.07 astronomers/million.

4. Global statistics of published papers in astronomy

Table 6 gives global statistics of published papers in astronomy and astrophysics listed in ADS for the 30 years 1976–2005. Over this period nearly 900 000 published items appeared in print, or an average of nearly 30 000 per year. Of these, 43% were refereed papers.

The table shows that unrefereed papers have grown from about 8000 per year in 1976–80 to nearly 30 000 annually in 2001–05, an increase of 3.7 times in 25 years. For refereed papers, the growth is from 9000 a year in the late 1970s to 17 500 annually now, a growth by a factor of 1.9. Unrefereed papers have therefore grown the more strongly. Refereed papers accounted for over half of all papers a quarter of a century ago, but they are little more than a third of all papers today.

5. Comparison of publications from IAU and non-IAU member countries

The data of Tables 1 and 2 allow a comparison of the number of refereed papers from IAU member countries (Table 1) with the number from non-IAU countries (Table 2) as a function of time for the years 1976–2005. The sums of the numbers of papers for IAU member countries, and for non-IAU members is given in Table 7. Over this period only 789 refereed papers came from non-IAU member countries, compared with over a third of a million from IAU member countries. The ratio of IAU individual members in non-member to member countries is 110:8860 (in February 2006) or 0.0124 – just over 1 per cent. But the ratio of refereed papers is 2.29×10^{-3}, indicating the roughly five

Table 3. Country statistical data: IAU member countries

country	2006 data # IAU members	papers 1976–2005 total	papers/ IAU memb. per yr.	2006 data $US GDP/cap.	2006 data millions popul.	IAU members per mill.	papers/yr /GDP (GDP $bill.)
Argentina	105	1914	0.61	13700	39.9	2.63	0.117
Armenia	25	1053	1.40	5300	3.0	8.33	2.208
Australia	225	9196	1.36	32000	20.3	11.08	0.472
Austria	35	2118	2.02	32900	8.2	4.27	0.262
Belgium	99	3210	1.08	31900	10.4	9.52	0.323
Bolivia	0	29		2700	9.0	0.00	0.040
Brazil	142	3817	0.90	8400	188.1	0.75	0.081
Bulgaria	48	1089	0.76	9000	7.4	6.49	0.545
Canada	220	12244	1.86	32900	33.1	6.65	0.375
Chile	55	1581	0.96	11300	16.1	3.42	0.290
China (P.R.C.)	282	6772	0.80	6300	1314.0	0.21	0.027
China Taipei	24	889	1.23	26700	23.0	1.04	0.048
Croatia	13	209	0.54	11600	4.5	2.89	0.133
Cuba	5	24	0.16	3300	11.4	0.44	0.021
Czech Rep+Slovakia	107	2883	0.90	16950	15.7	6.82	0.361
Denmark	60	2600	1.44	33400	5.5	10.91	0.472
Egypt	57	428	0.25	4400	78.9	0.72	0.041
Estonia	23	403	0.58	16400	1.3	17.69	0.630
Finland	53	2232	1.40	30600	5.2	10.19	0.468
France	631	25672	1.36	30000	60.9	10.36	0.468
Germany	486	33353	2.29	29800	82.4	5.90	0.453
Greece	97	2069	0.71	22800	10.7	9.07	0.283
Hungary	44	1324	1.00	16100	10.0	4.40	0.274
Iceland	4	121	1.01	34900	0.3	13.33	0.385
India	209	6233	0.99	3400	1095.4	0.19	0.056
Indonesia	19	133	0.23	3700	245.5	0.08	0.005
Iran	18	251	0.46	8100	68.7	0.26	0.015
Ireland	30	913	1.01	34100	4.1	7.32	0.218
Israel	63	2791	1.48	22300	6.4	9.84	0.652
Italy	448	19483	1.45	28400	58.1	7.71	0.394
Japan	503	16349	1.08	30700	127.5	3.95	0.139
Korea	67	2123	1.06	20400	48.8	1.37	0.071
Latvia	15	132	0.29	13000	2.3	6.52	0.147
Lithuania	16	250	0.52	13900	3.6	4.44	0.167
Malaysia	5	25	0.17	10400	24.4	0.20	0.003
Mexico	92	3452	1.25	10100	107.4	0.86	0.106
Morocco	7	61	0.29	4300	33.2	0.21	0.014
Netherlands	190	10561	1.85	30600	16.5	11.52	0.697
New Zealand	27	861	1.06	24200	4.1	6.59	0.289
Nigeria	4	115	0.96	1000	131.9	0.03	0.029
Norway	21	1282	2.03	42400	4.6	4.57	0.219
Peru	4	65	0.54	6100	28.3	0.14	0.013
Philippines	3	15	0.17	5100	89.5	0.03	0.001
Poland	129	3903	1.01	12700	38.5	3.35	0.266
Portugal	35	822	0.78	18600	10.6	3.30	0.139
Romania	37	444	0.40	8400	22.3	1.66	0.079
Russia	366	20561	1.87	10700	142.9	2.56	0.448
Saudi Arabia	11	99	0.30	12900	27.0	0.41	0.009
Serbia, Montenegro	23	503	0.73	2700	10.8	2.13	0.575
South Africa	73	2145	0.98	12100	44.2	1.65	0.134
Spain	239	8974	1.25	25200	40.4	5.92	0.294
Sweden	108	3959	1.22	29800	9.0	12.00	0.492
Switzerland	92	4890	1.77	35300	7.5	12.27	0.616
Tajikistan	6	355	1.97	1200	7.3	0.82	1.351
Turkey	45	729	0.54	7900	70.4	0.64	0.044
Ukraine	160	3497	0.73	6800	46.7	3.43	0.367
UK	589	19482	1.10	30900	60.6	9.72	0.347
Uruguay	4	86	0.72	16000	3.4	1.18	0.053
USA	2347	93154	1.32	42000	298.4	7.87	0.248
Vatican	5	239	1.59		0.0		
Venezuela	17	395	0.77	6500	25.7	0.66	0.079
totals/mean value	8867	344501	1.30		4925.3		

Table 4. Country statistical data: IAU non-member countries

country	2006 data # IAU members	papers 1976–2005 total	papers/ IAU memb. per yr.	2006 data $US GDP/cap.	2006 data millions popul.	IAU members per mill.	papers/yr /GDP (GDP $bill.)
Albania	1	0	0.00	4900	3.6	0.28	0.000
Algeria	3	0	0.00	7200	32.9	0.09	0.000
Azerbaijan	8	24	0.10	4700	8.0	1.00	0.021
Columbia	2	7	0.12	7100	43.6	0.05	0.001
Ecuador	1	2	0.07	3900	13.5	0.07	0.001
Ethiopia	1	1	0.03	800	74.8	0.01	0.001
Georgia (Republic)	18	175	0.32	3300	4.7	3.83	0.376
Honduras	2	1	0.02	2800	7.3	0.27	0.002
Iraq	6	34	0.19	3400	26.8	0.22	0.012
Kazakhstan	9	20	0.07	8800	15.2	0.59	0.005
Korea DPR	20	0	0.00	1800	23.1	0.87	0.000
Lebanon	2	30	0.50	5300	3.9	0.51	0.048
Macedonia	1	3	0.10	7600	2.1	0.48	0.006
Malta	1	2	0.07	19000	0.4	2.50	0.009
Mauritius	2	14	0.23	13200	1.2	1.67	0.029
Mongolia	0	16		2200	2.8	0.00	0.087
Sri Lanka	3	7	0.08	4300	20.2	0.15	0.003
Singapore	3	56	0.62	29900	4.5	0.67	0.014
Pakistan	1	57	1.90	2400	165.8	0.01	0.005
Slovenia	6	109	0.61	21000	2.0	3.00	0.087
Thailand	3	61	0.68	8300	64.6	0.05	0.004
United Arab Emirates	3	7	0.08	29100	2.6	1.15	0.003
Uzbekistan	11	142	0.43	2000	27.3	0.40	0.087
Vietnam	3	21	0.23	3000	84.4	0.04	0.003
totals	110	789			635.3		

Table 5. Statistics of astronomers in countries, 2006 data

	Number of countries	Population (millions)	Number of IAU individual members
Countries in world	197	6465	8977
IAU member countries	62	4892	8867
Non-IAU member countries but with individual members	23	635	110
Non-IAU member countries with no astronomers	112	937	0

Table 6. Global statistics of published papers in astronomy, 1976–2005

years	unrefereed	refereed	total	per cent ref.
1976–1980	40 157	45 430	85 587	53.1
1981–1985	60 673	52 091	112 764	46.2
1986–1990	73 257	58 823	132 080	44.5
1991–1995	80 362	65 192	145 554	44.8
1996–2000	109 013	74 671	183 684	40.7
2001–2005	148 657	88 239	236 896	37.2
total	512 119	384 446	896 565	42.9

Table 7. Statistics of refereed papers in astronomy, 1976–2005, comparing papers from IAU member and non-member countries

years	papers from IAU member countries	papers from non-IAU member countries	sum of IAU plus non-IAU member countries	ratio non-member to member countries
1976–1980	33005	25	33030	7.6×10^{-4}
1981–1985	41114	35	41149	8.5×10^{-4}
1986–1990	45203	57	45260	1.26×10^{-3}
1991–1995	43015	109	43124	2.53×10^{-3}
1996–2000	58388	155	58543	2.65×10^{-3}
2001–2005	123837	408	124245	3.29×10^{-3}
total	344562	789	345351	2.29×10^{-3}

times lower productivity of astronomers in non-IAU member countries. However, the proportion of papers coming from non-IAU member countries has been steadily rising over the last 30 years, as is shown in Table 7.

6. Analysis of the data by country

Fig. 1 shows a plot of IAU members/million versus GDP/capita in US dollars. This is a measure of the relative amount of astronomical activity in a given country. Both IAU member countries (Table 3) and non-members (Table 4)are included.

The results show a strong correlation between these parameters. The wealthier countries have more astronomers per million of population. For most countries the approximate relation IAU members/million = (1/3500)(GDP/capita)applies. The IAU members/million value varies between zero and about 14 for nearly all countries. All IAU non-member countries have relatively few astronomers (less than or equal to 3.0 per million) except for Georgia (3.8/million).

However, a few countries stand out from the overall trend. Estonia with 17.7 astronomers/million has easily the highest number of astronomers per capita, and Armenia (8.3 astronomers/million) is unusually high for its GDP/capita.

Among IAU member countries, South Korea, China Taipei, Japan, Austria and Norway have relatively few astronomers for their GDP/capita values. So do the United Arab Emirates and Singapore for non-members. All these are developed countries.

Fig. 2 shows a plot of the papers/IAU member/year versus GDP/capita, for both IAU member countries and for non-members (Tables 1 and 2). The papers/IAU member/yr applies to total papers over the interval 1976–2005 as discussed above, while the number of IAU members was taken in February 2006. Papers/IAU member/year is a relative indicator of the productivity of astronomers in a given country. It correlates strongly with GDP/capita, as shown in the figure. For most countries, the approximate relation papers/IAU member/year = (1/20 000)GDP/capita applies. For most countries the papers/IAU member/yr varies between zero and 2.0.

A few countries stand out from the trend, for apparently having an unusually high productivity for their GDP/capita values. These are Nigeria, Tajikistan, Serbia, India, Armenia, Mexico, Russia and Germany. Undoubtedly the first two of these have so few astronomers that there is no reliable statistical sample. Others have large populations that do not fully participate in economic production, thereby lowering GDP/capita rather than boosting astronomical productivity. Only in Germany can one say that the astronomers are unusually productive.

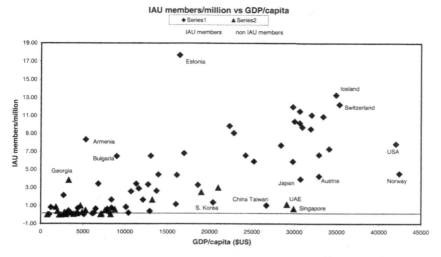

Figure 1. IAU individual members per million of population plotted against the wealth of a country given by GDP/capita. All figures are in $US. Solid squares: IAU member countries; solid triangles: IAU non-members

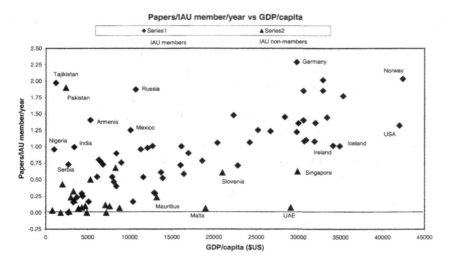

Figure 2. Papers/IAU member/yr plotted against the wealth of a country given by GDP/capita. All figures are in $US. Solid squares: IAU member countries; solid triangles: IAU non-members

Likewise, some countries stand out for astronomers of low productivity. These are Malta, United Arab Emirates and Singapore (all non-members of the IAU). Of the IAU member countries Ireland and Iceland have slightly low values of papers/IAU members/yr.

Finally, the parameter papers/year/billion dollars of GDP is plotted versus GDP/capita in Fig. 3. The ordinate (y) measures the fraction of a country's resources that have gone into astronomical research outputs, whereas the abscissa (x) is the intrinsic per capita wealth of a country. The figure shows that for IAU member countries there is a weak correlation between these parameters, as the wealthier countries tend to allocate a greater fraction of their GDP to astronomy. Approximately $y = (1./90\,000)x$. For

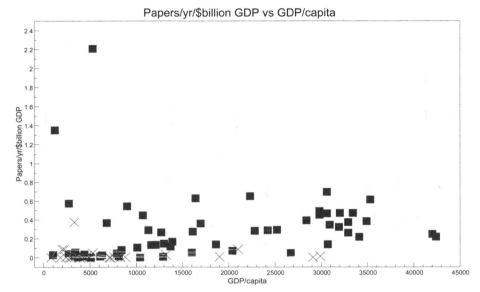

Figure 3. Papers/yr/billion dollars of GDP plotted against the wealth of a country given by GDP/capita. All figures are in $US. Solid squares: IAU member countries; crosses: IAU non-members

most IAU member countries, the values of the y parameter range between 0.0 and 0.7 papers/yr/billion dollars of GDP; however, two member countries stand out. These are Armenia ($y = 2.21$ papers/yr/billion dollars of GDP) and Tajikistan ($y = 1.35$). Apparently after the break-up of the Soviet Union, certain states such as these two were left with a strong tradition of support for astronomy, which was no doubt funded through the central Soviet economy. This strength has persisted through to recent times.

The mean value of papers/yr/billion dollars of GDP is $\bar{y} = 0.2925$ for 60 IAU member countries (excluding the Vatican and combining data for Czech Republic and Slovakia); excluding Armenia and Tajikistan gives $\bar{y} = 0.241$.

When we look at non-IAU member countries, the papers/year/billion dollars of GDP is dramatically lower. The mean value for 24 countries in Table 4 is $\bar{y} = 0.033$, which is about ten times less than for IAU members. Only Georgia ($y = 0.376$) has a value exceeding 0.1. The conclusion is that non-IAU member countries are generally applying a very small fraction of their resources to astronomy, whether or not they have strong economies (as measured by GDP/capita).

7. A comment on multi-author multi-country papers

A comparison of the numbers of refereed papers published in the period 1976–2005 in Tables 6 and 7 shows a discrepancy in the numbers reported. This is because Table 6 is the true number of individual papers published globally in astronomy, as obtained directly from ADS. On the other hand, Table 7 sums the papers published in 86 countries listed in Tables 1 and 2. (It is assumed that the papers published in countries not listed are negligible in number.) Bearing in mind the comments at the end of section 2, these two counts can differ. If the global paper count is larger than the sum of papers over all countries, this is because of the incompleteness of the country affiliations in ADS.

On the other hand, multi-author multi-country papers are counted more than once in Table 7, so the figure here can exceed the actual number of refereed papers in any year.

It is noted that between 1976 and 2000 the sum of refereed papers over countries was on average about 75 per cent of the total count. This is therefore an estimate of the lack of completeness of country affiliations in ADS.

On the other hand, for the time interval 2001–05, the sum of papers over countries exceeded the actual number by about 41 per cent. This represents a huge increase in the refereed paper count over countries in the last five years. The only simple explanation is that in these years each paper is counted something like twice, because there must be on average authors from two countries for each published paper. The actual number of countries per paper would depend on the completeness of the country affiliations in any time interval. The number two countries per paper comes from assuming that the completeness of country affiliations remains constant at 75 per cent.

8. Conclusions

In this survey, a number of key conclusions were reached. The main ones were as follows:

- There is a strong correlation between IAU individual members/million and GDP/capita.
- There is also a strong correlation between papers/IAU member/year and GDP/capita.
- The number of refereed papers published per year in astronomy has nearly doubled between 1976 and 2005.
- Just over one per cent of IAU individual members come from countries that do not adhere to the IAU. However, only about 0.3 per cent of papers are published by astronomers in non-IAU member countries. Although the proportion of papers from non-IAU member countries is very small, it is slowly rising.
- Non-IAU member countries allocate about a five times smaller proportion of their GDP to astronomy than do countries that adhere to the IAU.
- There has been a steep rise of multi-author multi-country papers published since 2001.

This survey represents an initial attempt to extract statistical information on country affiliations from the ADS data base. The uncertainties in the results obtained are acknowledged. Nevertheless much useful information on published paper statistics can be obtained.

It may be that more refined surveys can be carried out. In particular, a survey of citations to papers in different countries may be a better indicator of astronomical activity than simply counting the papers published from each country.

References

Hearnshaw, J.B. 2001. In *Astron. Soc. Pacific Conf. Series*, Special session on Astronomy in developing countries, ed. A.H. Batten, pp. 23–27.

Jayant Narlikar

John Hearnshaw

Astronomy for the developing world
IAU Special Session no. 5, 2006
J.B. Hearnshaw and P. Martinez, eds.
© 2007 International Astronomical Union
doi:10.1017/S174392130700662X

Promoting astronomy in developing countries: an historical perspective

Rajesh Kochhar[1]

[1]National Institute of Science, Technology and Development Studies, NewDelhi 110012, India
email: rkochhar2000@yahoo.com

Abstract. Any international effort to promote astronomy world wide today must necessarily take into account its cultural and historical component. The past few decades have ushered in an age, which we may call the Age of Cultural Copernicanism. In analogy with the cosmological principle that the universe has no preferred location or direction, Cultural Copernicanism would imply that no cultural or geographical area, or ethnic or social group, can be deemed to constitute a superior entity or a benchmark for judging or evaluating others.

In this framework, astronomy (as well as science in general) is perceived as a multi-stage civilizational cumulus where each stage builds on the knowledge gained in the previous stages and in turn leads to the next. This framework however is a recent development. The 19th century historiography consciously projected modern science as a characteristic product of the Western civilization decoupled from and superior to its antecedents, with the implication that all material and ideological benefits arising from modern science were reserved for the West.

As a reaction to this, the orientalized East has often tended to view modern science as "their" science, distance itself from its intellectual aspects, and seek to defend, protect and reinvent "our" science and the alleged (anti-science) Eastern mode of thought. This defensive mind-set works against the propagation of modern astronomy in most of the non-Western countries. There is thus a need to construct a history of world astronomy that is truly universal and unselfconscious.

Similarly, the planetarium programs, for use the world over, should be culturally sensitive. The IAU can help produce cultural-specific modules. Equipped with this paradigmatic background, we can now address the question of actual means to be adopted for the task at hand. Astronomical activity requires a certain minimum level of industrial activity support. Long-term maintenance of astronomical equipment is not a trivial task. There are any number of examples of an expensive facility falling victim to AIDS: Astronomical Instrument Deficiency Syndrome. The facilities planned in different parts of the world should be commensurate with the absorbing power of the acceptor rather than the level of the gifter.

Keywords. History and philosophy of astronomy, "Cultural Copernicanism", astronomy and civilization, astronomy in relation to Eastern and Western thought and culture

1. Introduction

Astronomy today is at the cutting edge of intellectual enquiry, and, at its most glamorous, a child of high technology. But it is more than a branch of modern science. It is a symbol of the collectivity and continuity of humankind's cultural heritage. This mixture of science and culture is astronomy's strength as well as dilemma. Strength, because support for astronomy transcends all boundaries: dilemma, because this support transcends science also.

For promoting astronomy in developing countries with memories of past contributions to science, scientific astronomy and cultural astronomy would need to be placed in a composite context, even though the question of judicious jettisoning of part of the cultural baggage is not going to be easy to address. Even more importantly, modern (post-Copernican) astronomy, or modern science in general, would need to be repositioned in a more extended evolutionary sequence.

2. Cultural Copernicanism

The past few decades have ushered in a new age, which we may call the Age of Cultural Copernicanism (Kochhar 1999). Analogous to the cosmological principle that the universe has no preferred location or direction, Cultural Copernicanism would imply that no cultural or geographical area, or ethnic or social group, can be deemed to constitute a superior entity or a benchmark for judging or evaluating others. In this framework, astronomy (as well as science in general) is perceived as a multi-stage civilizational cumulus where each stage builds on the knowledge gained in the previous stages and in turn leads to the next.

This framework is, however, a recent development. The 19th century historiography consciously projected modern science as a characteristic product of Western civilization, decoupled from and superior to its antecedents, with the implication that all material and ideological benefits arising from modern science (and technology) were reserved for the West.

As a reaction to this, the orientalized East has often tended to view modern science as Western science, distancing itself from its intellectual aspects, and seeking to defend, protect and reinvent their "own" science and the allegedly "anti-science" Eastern mode of thought. This defensive mindset works against the propagation of modern astronomy in most non-Western countries. There is thus the need to construct a history of world astronomy that is truly universal and unselfconscious.

It is customary for history books and modern astronomical texts to periodize astronomical developments in terms of Babylonian astronomy , Greek astronomy, Hindu astronomy, Arab astronomy, and modern astronomy. This approach can be faulted on a number of counts. Thus presented, modern astronomy becomes a synonym for European or Christian astronomy, which owes nothing to anybody else. The term Arab astronomy is insensitive and factually incorrect, because most of the astronomers in this phase were of non-Arabic ethnicity. It would be appropriate to use the term Zij astronomy, since the main concern in this phase was the construction of astronomical tables based on observations. Similarly a better term than Hindu astronomy is Siddhantic astronomy, because the main focus was mathematical calculations on the basis of algorithmic texts known as Siddhantas.

Interestingly, while terms like ancient China, ancient Egypt and ancient India are commonplace, ancient Iraq is not used. By using obscure phrases such as the Mesopotamian, Sumerian, Babylonian and Chaldean, we are denying the present inhabitants of these lands connectivity to and pride in their own remote past, forcing them to seek their identity from later events.

The philosophy behind this periodization mars the historical analysis also. Indians were told that their millennia-long astro-mathematical tradition was a derivative of Greek scholarship and lacked originality, although from about AD 500 till Kepler's laws, Indian astronomers were the only ones anywhere in the world who could calculate eclipses with a reasonable accuracy. As a reaction, some Indian historians have defensively asserted that the Greek inputs came into astrology but not into astronomy. Making a distinction between astronomy and astrology in older time-frames is an exercise in anachronism. Quite obviously bad scholarship does not encourage corrective research but begets more of its own kind.

In the same spirit, Arabs were dismissively told that their role had been no more than as librarians and archivists for preserving Greek science till Europe was in a position to take its heritage back. (Copernicus' own use of Al Tusi couple (Saliba 1995) was assigned to a dusty endnote in scholarly papers; it never made to the mainstream texts.) And yet when the Indians proudly pointed out that the Buddhists had worked extensively on health-related chemistry, they were told with a straight face that when their ancient texts mentioned the Buddhists, they probably meant Muslims!

Antecedents of western astronomy have been tunnelled backwards into a hypothetical unresolved monolithic period of classical Greek antiquity extending from the 6th century BC (Thales) to 2nd century AD (Ptolemy). In reality, Alexander represents an intellectually significant transition within this period. His conquests brought Greeks to Egypt and Mesopotamia, both of which had an older civilization, bigger economy and geography, as well as higher levels of practical knowledge and technological developments. The combination of these with the classical Greek tradition gave rise to a hybrid culture that produced science as we know now (Russo 2004). Thus the vastness of Egypt made possible the celebrated experiment by Eratosthenes to measure the circumference of the Earth. (Just as, centuries later, the vastness of British India permitted the measurement of the great meridional arc under George Everest.)

The standard reviews of astronomical history very often do not make any effort to synthesize into the main narrative the contributions from the Americas. It is mandatory for modern astrophysical texts to include a chapter on historical background. Since the book writers are not experts in this area they normally take recourse to copy-and-paste. Thus it has been stated that since Muslims must pray towards Mecca this directional constraint encouraged them to take to observations. It is difficult to say how this "insight" developed. But an obvious counter-example may be noted. The most spectacular Muslim empire anywhere was the Mughal in India. And yet in its entire golden age which lasted from the mid-16th century till the closing decades of the 17th, no observatory was ever built.

Insensitivities and casual remarks in the popular text books may leave most western students untouched, but they strengthen feelings of alienation elsewhere and thus work against the promotion of astronomy worldwide. Many former European colonies house old telescopes. They however do not view these telescopes as part of their scientific heritage. They are tolerated as signs of foreign presence.

In short, there is a pressing need for constructing a universal history of astronomy which emphasizes continuity and evolution, so that the whole world can relate to modern astronomy and participate in its furtherance. The International Astronomical Union, UNESCO and other agencies can join hands in initiating a campaign for creating an on-line source for a universal history of astronomy. Planetariums can play an important role in creating awareness about astronomy. Unfortunately, the tendency these days is to use pre-supplied programmes, which since they are produced for a diversity of audiences are bereft of cultural content. There is a need to produce cultural-specific modules to supplement reports on modern developments. The International Astronomical Union and other international agencies can help in this direction.

Combining cultural and modern scientific aspects of astronomy is a non-trivial task. Care needs to be taken to interpret past accomplishments and notions in a framework relevant for the period. There is a prevalent tendency to seek modern validation for old concepts. Indiscriminate use of modern-day judgmental terminology while discussing the past can only complicate matters further.

3. Modern astronomy and the non-West

So far we have focused on a conceptual framework which astronomy of today requires. There are certain difficulties in actual practice also. It is the Brahiminical aspects of astronomy that lend glamour to it. It is often forgotten that behind these results lies a very high level of coordinated artisanal activity. Indeed historically astronomy and industrialization have grown hand in hand in Europe. Thus while in the industrial economies, astronomical pursuits are an extension of military -industrial activity, it is difficult for semi-industrialized or non-industrial countries to maintain astronomical equipment. If sophisticated instruments are installed by a donor that are far above the maintenance capabilities of the host country, the experiment will fail, no matter how well intentioned the exercise was. As the contribution of service sector in national economies increases, interest in nut-and-bold activities is going down, shifting interest from observations and experimentation to applied mathematics.

If even modest observational facilities are to be sustained in developing countries, a certain minimum amount of industrial activity has to be ensured. Astronomy cannot be practised as a purely cultural activity. Historically, astronomy has benefited from industrial and strategic developments (the Dutch invention of the telescope, the silver-on-glass technique, pre-Hubble telescopic satellites, the CCD, etc.). It has grown in conjunction with the nut-and-bolt economic activity. Can this trend be reversed? In other words can the initiation of astronomical activity induce shop-floor culture? The question is not easy to answer, but the effort is worth making.

Unlike in laboratory sciences where a new facility renders the older ones redundant, the field science of astronomy requires cooperation and collaboration from facilities old and new. No matter how powerful or rich or industrially advanced a nation or how well-equipped its observatories, it still is permitted a view of only half of the celestial sphere. Understanding the celestial environment has been humankind's passion since times immemorial. This common heritage needs to be emphasized and strengthened so that the depth of our understanding about the cosmos, both literally and figuratively, can be enhanced.

References

Kochhar, R. 1999, Education and training in basic space science and technology. In: Space Benefits for Humanity in the Twenty-First Century (Vienna: United Nations) p. 245

Russo, L. 2004, The Forgotten Revolution: How science was born in 300 BC and why it had to be reborn (Berlin: Springer)

Saliba, G. 1995, A History of Arabic Astronomy: Planetary Theories During the Golden Age of Islam (New York University Press)

Section 2:

Astronomy in Latin America
and in the Caribbean

Astronomy for the developing world
IAU Special Session no. 5, 2006
J.B. Hearnshaw and P. Martinez, eds.

© 2007 International Astronomical Union
doi:10.1017/S1743921307006643

Formal education in astronomy in Latin America

Hugo Levato[1]

[1]Complejo Astronómico El Leoncito, CONICET, Argentina
email: hlevato@casleo.gov.ar

Abstract. We review the present situation of the formal education in astronomy in the Latin American countries. We have concluded that we can divide the countries into three categories according with the different development of the astronomical careers in astronomy.

Keywords. Education, astronomy, Latin American countries

1. Introduction

This paper has the purpose of identifying the present situation of formal education in Astronomy in Latin American countries. Previous work related to the subject may be found in Batten 2001 and Hearnshaw. In this publication of the Special Session on Astronomy for Developing Countries that took place in the XXIV General Assembly there were several papers that may help to understand the situation of the subject at the end of the century.

2. Information available

We have searched the WEB for those universities in the different Latin American countries that offer astronomy courses at undergraduate and graduate levels. We have found the following:

- **Argentina**: Three national universities offer astronomy courses. Also another two offer astrophysics courses at graduate and undergraduate level. There are four different PhD programs in astronomy and astrophysics among the national universities. We have found around 120 PhD in astronomy in Argentina.
- **Brazil**: The IAG at the University of Sao Paulo offers undergraduate and PhD programs. The Physics Institute at the University of Rio Grande do Sul offers an astronomy career (the only one in Brazil). Brazil has more the 200 PhDs in Astronomy working in the country.
- **Chile**: The University of Chile offers undergraduates studies in science and also a PhD program in Astronomy which takes advantage of an agreement with Yale university. The Pontificia Universidad Católica de Chile offers undergraduates studies in astronomy and physics and a Ph.D. program
- **Mexico**: The Instituto de Astronomía of UNAM (Universidad Nacional autńoma de México) offers master and PhD programs in sciences with specialization in astronomy The Universidad de Guanajuato offers undergraduates courses in physics and a PhD program in astrophysics. Also the Instituto Nacional de Astrofísica, Óptica y Electrónica offers PhD programs in astrophysics, astronomical instrumentation and computational astronomy.

• **Venezuela**: The Universidad Simon Bolivar and the Universidad de Los Andes offer undergraduate studies in physics and also PhD programs. Additional studies in astronomy are generally taken at the instituto de Investigaciones en Astronomía (CIDA)

These five countries have closed the cycle for the production of astronomers at PhD level. They can provide their own PhD programs and they have a good number of PhD scientists and professors to keep the system running.

• **Uruguay**: the Universidad de la República offers studies in physics with specialization in astronomy, a small PhD program in physics is also offers.

• **Colombia**: This country offers a Magister study in Sciences-Astronomy.

• **Honduras and Central America**: The Universidad Autónoma de Honduras through the Observatorio Astronómico Centro Americano de SUYAPA offers Maestría (Masters) in astronomy and astrophysics to graduate students in physics or mathematics.

3. Social analysis

We have tried to relate the above results with certain social and economic parameters of the different countries. We have used for that purpose the concepts of: Gross Domestic Product (GDP) that is the value of all final goods and services produced within a nation in a given year and the Gross Domestic Product per-capita at purchasing power parity (PPP) that is the sum value of all goods and services produced in the country valued at prices prevailing in the United States and divided by the population. This is the measure most economists prefer when looking at per-capita welfare and when comparing living conditions or use of resources across countries. The data was taken from www.photius.com/rankings/index.html. From the same database we have taken the population for each country.

Finally we have used the quality of life index as defined at www.economist.com and the numbers of astronomers members of the International Astronomical Union as obtained from www.iau.org.fr.

Figure 1 shows that there is a good relation between the number of astronomers members of the International Astronomical Union and the logarithm of the GDP expressed

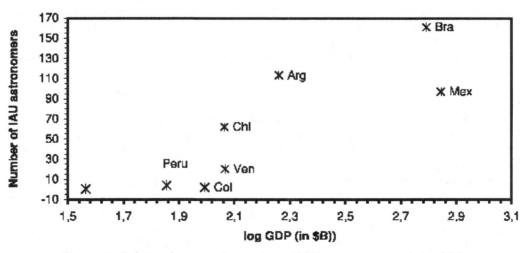

Figure 1. Relation between the number of IAU astronomers and the GDP

in billions of US dollars. From this figure one may conclude that probably Mexico will require more astronomers members of the IAU according to the level of its GDP.

Astronomers/million

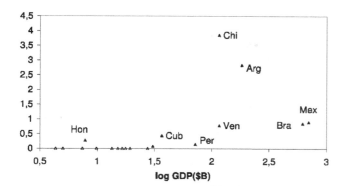

Figure 2. Number of IAU astronomers per millon habitants against the logarithm of the GDP in billions of US dollars

However if one takes into account the whole population in the statistics we see in Figure 2 the number of IAU astronomers per million inhabitants, against the logarithm of the GDP in billions of US dollars and in this case it seems that Chile and Argentina have more astronomers per million inhabitants than the general trend or probably one should say that Mexico and Brazil should increase their numbers of IAU astronomers.

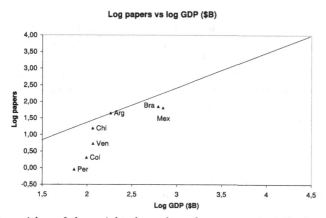

Figure 3. Logarithm of the weighted number of papers against the log of the GDP

Figure 3 shows the number of papers published in recognized scientific journals weighted by the impact of the journal against the GDP. There is a clear trend. The straight line corresponds to the line that adjusts the data for countries with more than 4 000 US dollars per capita as GDP. The data was taken from Abt 2004. It is clear that Peru, Colombia and Venezuela should increase their number of papers in recognized journals to keep the trend according to their GDP.

Finally we have checked the production of papers with the quality of life index because we think that probably a better life condition among the population favours the devotion

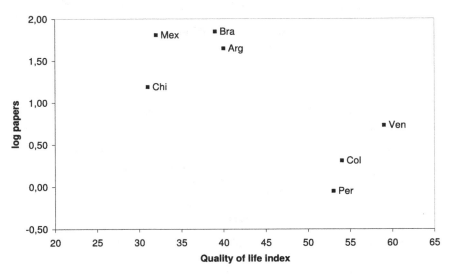

Figure 4. Log of the weighted number of papers against the quality of life index

to science for a larger portion of the population. Figure 4 shows clearly that the countries identified in the previous section as those that have closed the cycle of astronomers and PhD in astronomy production have better quality of life indexes than those who still need help to improve astronomy research and studies in their countries.

4. Conclusions

We conclude that only 5 countries in Latin America are in a position of producing their astronomers and PhDs in astronomy. Three of them (Brazil, Argentina and Mexico concentrate almost 90% of the PhDs in astronomy working in the region. We believe that Uruguay requires a little effort to boost professional education in astronomy. Colombia, Peru and Cuba will require moderate efforts jointly with Honduras for its Central America project. The rest of the countries will require moderate to strong efforts from the international community.

References

Abt, H.A. 2003 *National Astronomical Productivities* (Private Communication)

Batten, A.H. 2001 in: A.H. Batten (ed), *Astronomy for Developing Countries* (Publ. Astron. Soc. Pacific), p. 3

Hearnshaw, J.B. 2001 in: A.H. Batten (ed), *Astronomy for Developing Countries* (Publ. Astron. Soc. Pacific), p. 15

Astronomy for the developing world
IAU Special Session no. 5, 2006
J.B. Hearnshaw and P. Martinez, eds.

© 2007 International Astronomical Union
doi:10.1017/S1743921307006655

Astronomy for teachers in Mexico

Julieta Fierro[1]

[1]Instituto de Astronomía, UNAM, Mexico City, Mexico
email: julieta@astroscu.unam.mx

Abstract. In this paper I shall present ways in which professional astronomers in developing nations can aid basic education by using a few things that have been done in Mexico.

Recently the compulsory education in Mexico was increased from 6 to 12 years. An optional subject on astronomy was included in high school; so there are several ways in which one can contribute to the national education system.

I shall give a practical example on ways to teach and mention the importance of magazines and books dedicated to teachers. I will also address the way we have implemented brief courses for educators and organized conferences for science teachers.

It is important to emphasize that we must find ways to train people in a more effective way.

Keywords. Astronomy teaching, Mexico

1. How to teach

There is educational research going on in several nations, with many interesting results, unfortunately it is difficult for practical information to get to teachers. I encourage researchers to post their educational materials on the World Wide Web and to write for educators.

In Mexico we have several magazines for teachers, for instance "El Correo del Maestro" includes astronomy articles and edits 17 000 copies, which is a large number for a developing nation and a small one compared to the amount of elementary school teachers we have: over a million.

I believe we can extract from other species the basics of good education by observing how primates teach their young. If we analyse how a mother aids her child to use a stick to break nuts we can observe how she uses the following steps:

(*a*) She exposes him to interesting surroundings (she takes him to a site where other large apes are using tools to break nuts).

(*b*) She lets him experiment on his own (she watches while he tries to break nuts with an improper tool, does not succeed and even gets slightly injured).

(*c*) She acts at the precise moment, when the infant is about to give up, and teaches him how to grab the correct tool.

(*d*) She allows her child to practise till he masters nut breaking (the mother follows closely and only helps in case of necessity). In other words, the student must learn on his own and the teacher should go way beyond lecturing, set an example and become a facilitator.

These are the same items can be used for good teaching: Expose, experiment, facilitate, practice and succeed. We should of course add another step: creation. We want our students to go beyond current knowledge. Creation occurs when items that seemed disconnected are joined. For instance, painters or composers put together colour, shapes, sounds and rhythms in ways nobody has done before. So we must expose our students to an interesting and varied set of ideas and allow them to ponder about them and give challenging assignments.

2. Pre-school in Mexico

Pre-school in Mexico has recently been increased from one to three years. This means many new teachers must be trained. In Mexico education in mainly state run.

Luckily a group or elementary school educators contacted me with a list of 600 science questions that students had asked and they did not how to answer. So a group of educators answered them and five books were published with the answers and related topics. The group of original teachers sold the books personally; to make sure they got to the teachers that were interested in them.

Pre schoolteachers have several problems in answering science topics. Twenty one per cent of households have only one provider, women. So mothers have less time to teach their children how to talk and teachers are assuming this responsibility. Instructors have to understand science questions in a poor language, translate in a way a scientist can understand it, and translate the answer back in a way the child can comprehend it.

In order to contribute to the national effort I wrote a wordbook. The words are placed in alphabetical order, each has to do with science, I tried to make them difficult to pronounce, for instance they have three syllables or sounds such as "rui" that are hard for small children to pronounce. Each page has the word, its definition and a small sentence where it is used, written by a poet (Alberto Vital). The book is to aid both mothers and teachers: a word a day keeps poor learning skills away. As we all know words are necessary to build ideas.

I also wrote an astronomy book in small format. Then I realized teachers wanted large format books, so the children could see the pictures. So large format books were printed for teachers and small format books for students. Experience showed that students also wanted large format books. So I began a series of science books for pre school students in large format. The topics include light, sound and time, all with an astronomical component. I am preparing one on water and will continue with the sky.

3. Middle school

Three years of middle school, ages 12 to 15 have been added to compulsory education in Mexico.

I believe subjects that are meaningful for children are those that should be taught.

Part of the present curricula seems to me extremely boring it includes little astronomy in spite of it being attractive for both teachers and students.

I have written several books for that age level that has been included in the classroom library. This library is furnished with books chosen by the teachers. They can select about ten books every year. One of the books I wrote was on scatology to make a point about reading; students read if they are interested. Students read it in groups, laugh and talk about the book, that are qualities all books should have. This book has been on the top place in the list of most bought books on Mexico's most popular bookstore.

I am presently writing more books for this age level. Since Astronomy has been excluded in wont be taught in a boring way, due to its multidisciplinary nature it is ideal for informal reading.

4. High school teacher training

For many years there was an optional course on positional astronomy in the geography department of Mexican high schools. It was changed to a general astronomy course in the physics department.

A teachers' 180-hour-long course was designed for their training. At first professional astronomers lectured, and little by little the high school teachers have learned to train each other. Some of them have written materials and placed them with sets of images for lectures on line.

I believe on learns when one teaches, and having high school instructors train themselves have given them more confidence in what they teach.

5. Symposia

The Mexican Academy for Science Teachers is an association whose members include all level teachers, from Pre School to graduate. Its main activity is a bi-annual conference where about 1 500 teachers attend. Lectures, workshops, demonstrations, books on astronomy are always present. (www.ampcn.org.mx)

In developing nations such meetings are hard to organize, teachers are not used to presenting posters, it is not always easy to convince authorities to allow them and sponsor their participation and there is not a tradition of fund raising.

Nevertheless teachers benefit immensely by going to meetings. Multilevel bilingual teachers in remote areas can share experiences with their pears. Teachers feel proud of what they are achieving when they compare their work with the efforts made elsewhere. They can purchase books, magazines and teaching materials.

6. Books

I believe the importance of books cannot be over stated. Not only books should be edited but available.

Mexico has done three things to promote reading:
(*a*) Small libraries in school rooms
(*b*) Book fares
(*c*) A new major library that includes scholars on the board.

Classroom libraries hold books that have been selected by teachers from a group of 700. They can pick 10 per year. The 700 titles comes from a previous selection among the 13 000 that are printed in Spanish in different countries and are presented to the selection committee by the editors. Some examples of books in Spanish by this author that are available in schools are listed as references at the end of this paper.

A new large library was built in Mexico City in a section that has little cultural activities. There are several scholars on the board that not only have contributed by suggesting titles but to other activities such as workshops, lectures, use of the www, etc.

7. Conclusion

In developing nations good quality science education for a large portion of the population is needed. Curricula reforms are only a very small portion of what is needed. Children must be well nourished and healthy in order to learn.

In Mexico astronomy is not taught as such at any level, due to its appeal written materials must be available for teachers and students.

If one wants to improve the quality of education one must take time to work with teachers in order to understand their needs and difficulties.

References

Delgado, H. and Fierro, J., Volcanes y Temblores en México. Editorial SITESA, 2004

Domínguez, H. and Fierro, J., Albert Einstein: Un cientfico de nuestro tiempo. ISBN 970-32-1108-9, Editorial Lectorum, 2005

Domínguez, H. and Fierro, J., La luz de las estrellas Correo del Maestro, ediciones La Vasija, 2006-06-22 ISBN 970 756 095 9

Domínguez, H. and Fierro, J., Galileo La Vasija, en Prensa, 2007

Fierro, J. and Vital, A., Palabras para conocer el mundo, 2006, Editorial Santillana.

Fierro, J., Lo grandioso del sonido, Gran paseo por la ciencia, Editorial Nuevo México ISBN 970-677-180-8, 2006

Fierro, J., Lo grandioso de la luz, Gran paseo por la ciencia, Editorial Nuevo México ISBN 970-677-181-6, 2006

Fierro, J., Lo grandioso del tiempo, Gran paseo por la ciencia, Editorial Nuevo México ISBN 970-677-179-4, 2006

Fierro, J. and Sánchez, V.A. Cartas Astrales, Un romance cientfico del tercer tipo Editorial Alfaguara 2006 ISBN 968-19-1175-X

Fierro, J., El Sol, la Luna y las Estrellas DGDC, Colección Ciencia para Maestros, ISBN 970-32-1108-9, 2004

Fierro, J., La astronoma de México, ISBN 968-5270-55-4 Lectorum, 2001

Spadaccini, J., Fierro, J., Paglierani, R., Hawkins, I. and Cline, E.U. Úuchbenil le K'inó, Tradiciones del Sol, Traditions of the Sun Libro trilinge: Maya, Español, Inglés NASA, 2006

Tonda, J. and Fierro, J., El libro de las cochinadas, Ilustraciones de José Luis Perujo ADN Editores, 2005

Julieta Fierro and her monkey tricks

Astronomy for the developing world
IAU Special Session no. 5, 2006
J.B. Hearnshaw and P. Martinez, eds.

© 2007 International Astronomical Union
doi:10.1017/S1743921307006667

Astronomy–the Caribbean view from the ground up

Shirin Haque

Department of Physics, University of the West Indies, St. Augustine, Trinidad, West Indies.
email: shirin@tstt.net.tt

Abstract. The historical development of astronomy in the Caribbean is reviewed within its cultural and environmental framework. The present status of astronomy in education, research and at the popular level is presented also with the focus being on its development in the island of Trinidad and Tobago in particular. We review what works in small developing islands versus larger developed or developing countries and the peculiar trials and tribulations of our circumstances as well as the rewards of such efforts. The critical role of students and volunteer effort will be highlighted. The psychological and cultural aspect and its role in the development of astronomy in the Caribbean is also explored. The outlook for the next decade will be highlighted with a brief proposal of having a node for TWAN (Third World Astronomy Network) in Trinidad in the Caribbean.

Keywords. Astronomy in the Caribbean, Trinidad and Tobago

1. Introduction

The Caribbean islands and in particular Trinidad and Tobago have been known on the world scene for their carnival, music, and Miss Universes. The islands have produced Nobel laureates and made it to the World Cup in football but little is known if anything regarding astronomy in the Caribbean. This paper hopes to alleviate the misconception that there is no astronomy activity in the Caribbean. Astronomy exists in the Caribbean in all facets of education, research and popular outreach.

The history of astronomy is closely linked with the history of the University of the West Indies and the islands. The history of the Caribbean is one of slavery and indentured labour. With such a history, there was suppression of creativity and independent thought for hundreds of years in this region. Independence came to the Caribbean islands around the 1960s and with it the emergence of the University of the West Indies.

2. Astronomy at the University of the West Indies

The University of the West Indies (UWI) is one of the few regional universities in the world. Its three campuses are located respectively on the islands of Jamaica (Mona), Barbados (Cavehill) and Trinidad (St Augustine). Physics department was established at first at the Mona campus in Jamaica with areas of research in photometry and radio astronomy. Professional astronomy existed and continues to exist at St Augustine and at Mona campuses. Astronomy was established at Mona in the 1970s and at St Augustine from 1978 onwards. At any given time since, there have been 1–2 astronomers at the Mona campus and one at the St Augustine. At Cavehill in Barbados, occasionally there is an astronomer but there is no regular faculty position.

Mona and St Augustine have had taught programmes and research activity. At both campuses, the taught programme is relegated to one course in the undergraduate programme counting towards the physics major at the University. At both campuses, there has been an average of 1–2 postgraduate students pursuing astronomy over a ten-year period.

Observational astronomy was present at the Mona campus where they used a 21-inch telescope located at Stony Hill. However, this was damaged during some hurricanes and subsequently, due to increased crime problems and a downturn in the economy, it was never repaired or replaced. Astronomy at the campus there has been on a downslide due to the above-mentioned problems and deaths and migration of faculty reflecting also a general downturn in the physics department as a whole.

At the St Augustine campus, there was only research in theoretical astronomy at the start in areas of cosmology and quasars in the early 1980s. Up to the present, there are several research projects that undergraduate students perform each year as part of their degree programme. Lately, the size of the astronomy class has grown to almost 70 students from about 30 students in the past, in the advanced part of the programme.

However, around 2000, the St Augustine campus launched on an observational programme in conjunction with Tuorla Observatory, University of Turku in Finland. UWI now has access to 40-cm Meade telescope located at the University campus at St Augustine. This facility is known as the SATU (St AugustineTuorla) observatory. The primary project it is dedicated to is the monitoring of the variability of quasar OJ287. Our efforts have yielded a couple of publications in the Astrophysical Journal already.

Presently astronomy at the St Augustine campus is thriving and is on a growth path with three postgraduate students currently pursuing research in the areas of theoretical and observational astronomy, and astrobiology. In small countries, it is important to exploit all natural resources and Trinidad is home to one of the few natural pitch lake sites in the world. This is literally a hydrocarbon pool and we are examining it as an analogous site to the hydrocarbon lakes on Titan and the possibilities of the existence of extremophiles in such a condition. The astrobiology project is in conjunction with collaborators from Washington State University, Villanova University and University of British Columbia.

3. Education

Astronomy forms a small part of the primary school curriculum and is no longer in the high school curriculum since the Caribbean moved away from the Cambridge A-level exams and has its own equivalent examination. However, several schools have astronomy clubs that our groups assist with. As indicated in the previous section, it is part of the degree programme in Physics at the University of the West Indies and there are students doing postgraduate work in astronomy.

4. Amateur astronomy

It is interesting to note that the astronomy societies in the Caribbean exist on the three campuses where the University campus is located and in every case, the Astronomy Society predates the professional astronomy in the island. The Trinidad and Tobago Astronomical Society was formed in 1965 and the Barbadian Astronomical Society celebrated its 50th anniversary in 2006. The societies meet monthly and have viewing sessions and lectures. They own some 12-inch and smaller telescopes. We have conducted several nationwide surveys to gauge the interest in astronomy in schools. The results consistently show a high level of astronomy interest among high school students, but the teachers lack confidence in teaching in this subject area.

In 2002, there was the emergence of CARINA (Caribbean Institute of Astronomy) based in Trinidad, whose mandate is to unite and promote the growth of astronomy in the Caribbean as a whole. CARINA has been involved in running workshops and outreach programs in Trinidad and in the other islands. In addition, on the island of Tobago, there is a privately owned observatory SEAS (Sea, Earth And Sky) by an eye specialist and astronomy enthusiast, Dr Bruno Mitchell. CARINA is involved in running and overseeing its operation and the University has access to it as well. It currently houses a 12-inch telescope but plans are afoot for a 20-inch telescope to be put there.

5. Outreach

One of the popular annual events in Trinidad has been the star party run by CARINA. This event is sold out every year to capacity with over 400 persons attending the event this year. It is interesting to contrast the attendance at star parties versus attendance at astronomy seminars in the Department of Physics, which are held in comfortable environments, at convenient times with refreshments, versus star parties at distant locations and demanding hours where an entrance fee is charged and no seating is provided. The author suggests that the success of star parties versus seminars directly reflects basic innate human characteristics and a certain need-satisfaction mechanism. As part of the outreach programme, seminars and open public lectures and viewing sessions in astronomy are regularly held. This is often in conjunction with the National Science Center located in Trinidad that also has an astronomy popularization programme.

6. Cultural astronomy

Cultural astronomy deals with the disseminating and educating the population on aspects of astronomy of religious relevance to them in determining their festivals. The three main religious denominations in the Caribbean are Christianity, Islam and Hinduism. All rely on some aspect of astronomy in determining their festivals. Towards this end, a massive effort has been made in particular with the Muslim community that uses a lunar calendar and the sighting of the new moon to determine their festivals. This drive has been very successful in educating the nation to astronomical aspects of the lunar calendar.

7. Challenges in the Caribbean

Some of the problems on the ground are cultural, for example, email culture is not fully developed, although the technology is fully here. There is a serious lack of human resources and funding is often an issue as astronomy is not necessarily seen as a priority in the needs of the Caribbean people by decision makers. Those that are active in the field invariably become jack-of-all-trades as there is lack of infrastructure and to get any project underway one must play a multidisciplinary role. For such reasons, things generally take much longer to get done in the Caribbean than would take place in a developed nation. What could take a few months could easily take a year or two to pull together in the Caribbean due to lack of infrastructure and human resources. But the challenges can be overcome to an extent as we see in the discussion section.

8. Discussion

The small island plight is different from that of a larger developing nation. For us it is a numbers game. The population of Trinidad and Tobago is around 1.2 million, and Jamaica is around 2.7 million, with the other islands being in the order of 100 000s. With such numbers, the absolute number of professional astronomers will always be small and therefore systems in place will be small, which becomes a vicious cycle arresting the development of astronomy in such regions. However, today it is possible to overcome such limitations to an extent, thanks to the internet age, where our colleagues may not be there in real space but are just one click away for consultation. Collaboration is critical for small islands with institutions in other countries. Volunteer student effort is also critical in such situations for the successful progress and growth of astronomy. It is humanly impossible for one astronomer to do it all. It is thus important for students to gain exposure to the international community and towards this end, our undergraduate students have participated in astronomy research in Germany and La Palma, Canary Islands. Three of our students took part in the ISYA summer school in 2005 when it was held in Mexico.

Our thesis for the growth of astronomy in the Caribbean is one of mixing business with pleasure. And it comes from the answer to the question of why seminars are so poorly attended while star parties are heavily supported. The latter evokes the adventurous spirit and having fun and intellectual excitement – innate needs in persons. Seminars may provide intellectual stimulation but there is no fun and adventure in it in general. Tourism forms a big part of the economy of the Caribbean for its natural sun and sand, ecotourism and delightful cultures. This we think can be a motivator to attract astronomers to this part of the world. The greatest number of academic visitors to our department of Physics has been in the area of astronomy in the last few years. And visiting Astronomers initiate collaboration and give public lectures.

There is much discussion about the idea of the development of TWAN (Third world Astronomy Network). We would like to suggest, that although it may sound farfetched since the astronomy group seems so small in the Caribbean that, the Caribbean region and in particular Trinidad, be considered as a possible node for TWAN. The presence of the University campus provides the infrastructure and support that such a node would require and it would attract visitors which would act as a major impetus to the growth of astronomy in the Caribbean. The favourable aspects are that Trinidad is at latitude of about 10 degrees north, assuring a very large amount of skies. It is a country having an oil based economy that is doing well and Trinidad is listed as one of the safest places in the Caribbean during the hurricane season. In addition, English is the language spoken and there is total connectivity here. Also, the location of Trinidad is quite central to about three- quarters of the world - North America, South and Central America, Europe and Africa.

We hope in the next ten years or so, that astronomy grows and has a wider presence across the Caribbean in the areas of outreach and research in the non campus islands as well. The growth of online education can further facilitate this process. We hope to implement a course on astronomy for non science majors at the University of the West Indies. A feasibility study is currently underway to establish a radio telescope in Trinidad. We hope that the possibility of the Caribbean as a TWAN node may be considered.

9. Conclusion

Small developing islands are different from large developing countries. We have difficulties and opportunities that are different to the other developing nations and a lot of it is very much the numbers game - smaller populations by its very nature have reduced number of professionals and expertise. Some of the lessons learnt from our experience are that small developing islands must do small manageable projects. In astronomy, everything needs doing – It is not as important as what you are doing - but that what you do, you should do it well, do it right and achieve results. It takes a lot of dedicated effort but with the assistance of colleagues in developed nations, volunteer student effort, no task is impossible once the endeavour is passion driven. At the present time, in the Caribbean, often, the direct fruits of our labour are intangible and simply satiate the higher calling of the love of the discipline of astronomy, and that is reward enough.

Acknowledgements

The author would like to thank Dr Dipak Basu, former astronomy faculty at St Augustine, UWI and Dr Maura Imbert, President of the Trinidad and Tobago Astronomical Society for information

Hugo Levato

Patricia Rosenzweig

Ramón Rodríguez

Astronomy for the developing world
IAU Special Session no. 5, 2006
J.B. Hearnshaw and P. Martinez, eds.

© 2007 International Astronomical Union
doi:10.1017/S1743921307006679

Encounters with Science at ULA, Venezuela: an incentive for learning

Patricia Rosenzweig[1]†, O. Escalona[1], E. Guzmán[1], P. Bocaranda[2], R. Echeverría[3] and O. Naranjo[1]

[1]Grupo de Astrofísica Teórica, [2]Centro de Semiconductores, [3]Centro de Óptica,
Depto. de Física, Facultad de Ciencias, Núcleo La Hechicera, Universidad de Los Andes,
Mérida, Venezuela
email: patricia@ula.ve

Abstract. In the School of Science of the Universidad de Los Andes (ULA), in Mérida, Venezuela, a very successful event focused on elementary and high school students, was founded in 2000. The name of this event is "Encounter with Physics, Chemistry, Mathematics, and Biology" (hereinafter "Encounters with Science"), and it integrates these disciplines, as well as Astronomy. Its main purpose is that young minds can become familiar with the methods of science inquiry and reasoning, and can understand the concepts and processes of the sciences through thoroughly prepared experiences.

"Encounters with Science" has grown each year in an exponential way. As a matter of fact, in its sixth edition (2005), the number of elementary and high school students coming from all over the country, has reached the outstanding number of approximately nine thousand. Among all the experiences that the students could be engaged in, were many involving Astronomy. These experiences were prepared by professors, together with graduate and undergraduate students, who are pursuing their degrees in all branches of science, including Astronomy. Although there is this incredible team of faculties and graduate and undergraduate students working together; the target is the students of the high and elementary schools. We certainly focus on the engaging and encouraging of students to experience scientific work at first hand.

This flourishing program is continuing to grow and to become strong. It has matured in the sense that now our professors have prepared an excellent didactic material that can, together with the hour/class teaching, prepare high school and elementary school students for a better understanding of science; particularly, helping in this way for a better education in Astronomy.

The main event of the Encounters lasts five days in the School of Science of ULA, but subsidiary events are spread all over the year and around the country through trips that our faculty members undertake with our students. Thereby, they reach places where students perhaps cannot attend the main event during its celebration. As a successful program, it can be interesting to see if other countries can adopt this method to recruit or to trigger the interest of students to pursue their studies in the Sciences.

Keywords. Spread of astronomy teaching, target elementary and high school education, impact in the country, astronomy in Venezuela

---

1. Introduction

The new changing time that is forth see in countries and universities, have converge in the need that our *Alma Mater* must be actively incorporated in the developing process that has affected the region and our country as a whole. In this sense, as faculty members, compromised to support, develop, and foment education, in the year 2000 we presented

† Present address: Universidad de Los Andes, Facultad de Ciencias, Núcleo La Hechicera, Mérida, Venezuela

a proposal to the Physics Department of the School of Science of the Universidad de Los Andes (ULA), that very well fits with the changing process that is incubated in the educational environment. At that moment, its name was "Encounter with Physics" (Rosenzweig & Escalona 2001, and Rosenzweig *et al.* 2001). It is important to point out that in this Department, most of the students pursuit their undergraduate and graduate theses, including astronomy and astrophysics topics. The idea of implementing this event was taken onto consideration not only by ULA, but by other private and public institutions devoted to the development of education and science in the region. In the overall, the main goal is to generate a proposal addressed to the quality of the process teaching/learning of basic Science, including the different fields, as astronomy as well.

A growing interest from the faculty members and our students, together with the interest of high and elementary school students, produced a successful attendance during the next consecutive years. As a matter of fact, the Encounters with Physics were celebrated until 2003, but something new was happening; that is, the participation of other faculty members from the Departments of Biology, Chemistry, and Mathematics. The enthusiasm was the same, the invited students could observe a growing amount of experiences making possible that we could trigger their interest to see not only how science is every where, but also the reason for which the phenomena can occur, and to allow the interaction with the speaker and to give their own opinion. This way, we could transform the high and elementary school student from a guest to a performer. On the other hand, our faculty members, as well as our students, had to improve the way to explain the phenomena, the laws that are involved in order of not misleading the young minds of our invited. Due to the inclusion of the other basic sciences, starting the year 2004, in its fifth edition (Rosenzweig 2004), the Encounter with Physics, started to be called Encounter with Physics, Chemistry, Mathematics, and Biology (hereinafter "Encounters with Science"), with experiences involving astronomy as well.

In section 2 we will explain the motivation, section 3 will show the strategy required for the creation and sequence of the event, section 4 will show some impressive statistics, section 5 a chain reaction from the schools, other universities and regions, as a consequence of the encounters, and section 6, the future.

2. Motivation

The very first motivation to pursuit an event of this magnitude is related with a common problem that is facing the science in the whole world. That is, the deterioration of its teaching and the so pronounced lack of interest from the students point of view, and the lack of conscience about its importance in our lives, as well. It is a world wide need, especially in Venezuela, to promote not only the innovation, but the development of technology in all areas (Rosenzweig & Escalona 2001, and Rosenzweig *et al.* 2001); thereby, the so badly called hard sciences cannot be immune to this need. For this mainly reason, we dont want to spear any effort in order to support all type of events that can contribute to the spreading and better performance in the teaching of these disciplines.

It is sad, but certainly truth, that sometimes the lack of motivation starts from the teacher itself, perhaps due to the fat that he/she cannot find a support from the universities or from the educational system. Thereby, we are eager to promote the encouragement not only to the high and elementary school students, but also to their teachers in order to have a better performance in the process of teaching/learning of science. The motivation to organize these events goes beyond of just having an exposition of experiences. As time goes by, we want to converge in the elaboration of specific projects that involve the conclusion of a modern didactic focused in the good learning of this important field of knowledge.

In order to seal our motivation, important facts have been present since the beginning, that is we have committed ourselves so these events can survive time, and to have a systematic close look on schools and high schools through guidance, to give advises, corrections, and proposed researches in their educational projects. Our Professors and undergraduate students, as well as graduate students, have elaborate books (Memorias del I Encuentro con la Física 2001 and Memorias del IV Encuentro con la Física 2004), so that we can distribute to the schools as a guide for them to elaborate theirs own experiments. These memoirs include many experiments in the biology, mathematics, chemistry, physics and astronomy.

3. Strategy involved for the creation and sequence of the events

In what follows, a brief explanation of the main steps that should be taken into account, in order to crystallize the motivation, of a better learning/teaching of science.

(a) To encourage and to motivate the faculty members that do believe in the lack of interest and deterioration for the learning of science. A better preparation of the students before their entry in the University, will easier their work and give them the certainty of a successful teaching at this level.

(b) To encourage the university students as well, in order to be a good helping hand for the faculty members. Making them to understand that they will create a "dynamical duo" with one target: the primary and high school students.

(c) To contact the educational District responsible of the schools, and make them understand the benefits of a well prepared student population.

(d) To have a well prepared personnel, as a link between the university and all the schools. By well prepared it is understood that this personnel knows the member of the school districts and the statistics in each school. By the former, we can assure the green light for schools to participate. By the latter, we can know how many students of each school can be invited from the relative point of view (a necessary limited amount of students is imposed but, the ones that will attend can serve as multipliers for the others).

(e) To have an appealing material to promote the event. Every body, especially youngsters are easier to enrol when they see a nice and well prepared material.

(f) To search for financial support from inner and outer organizations. The former being the university, and the latter, both private and public organisms must be involved. Usually, the events require a registration fee, but in many cases schools dont have high budgets. For this, the Encounters dont have any fee for registration.

(g) To do an effort in order to give to the invited students a certificate of attendance. Some students might discard it but others can keep them as a reminder of an important event and, why not, the decision to be enrolled in this branch of knowledge as is the basic sciences.

(h) To make an extra effort in order to give to the invited students a demonstration of care by giving them some refreshments. Some times, bakery shops are among the institutions eager to help. It is important to notice that many students come from very far away and they must make a very early trip with no chance to have a decent meal.

(i) To have in mind, above all, that we must show to the invited students that the School of Science is not a hard place and that there are living professors and students, eager to open their helping arms in order to make understand that science is just a way to understand nature.

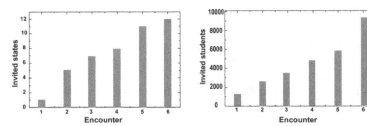

Figure 1. Statistic that show how the event has been growing as part of the national interest for this initiative

4. Statistics

In the year 2000, we were dreamers thinking that science could fulfil the interest of young people, perhaps with potential to do so but not knowing so. It seemed an impossible mission to trigger their intellectual creativity and to vanish their fear for something that was really unknown for them. Nevertheless, in its sixth edition, the School of Science of ULA has proven that this task called Encounters with Science, a complete pedagogic program, is worthwhile and that it was a certainty to have this dream. The seed has spread and is continuing to grow and to give fruits all over the country. Now a day, we are positively convinced that the plantation of science must be pursuit since the early stages of the educational stair, and we are now obliged to continue to organize this event as long as it takes. We are optimist in this assignment since, year after year, there are more professors and students enthusiastic to design and implement demonstrative and didactic experiences, several of them truly original, and all of them having a common factor: to show in a very illustrative, and simple way, the beauty of the phenomena that involve science. The integration of all the disciplines has shown that indeed it has triumphed.

Not only the integration of all the disciplines and the enthusiasm of the professors, and students, have shown that the Encounters are here to stay, but the incredible interest of so many schools and high schools to attend. Our guests, the high and elementary students, want to come every year. In the year 2000, only 1200 students from the State of Mérida came to the School of Science of ULA; last year, in its sixth edition, approximately 9000 students from 12 states of Venezuela came. The statistics are shown in the histograms that follow (figure 1). Also, in figure 2, the map of Venezuela shows how the event has expanded all over the country. We must point out that, because of financial problems, the students find it hard to come every year. We are trying to elaborate agreements, in order for other institutions to join this work.

5. A chain reaction

Besides the unquestionable success of this event during five days on the grounds of the School of science, not only the students from our careers are the actors but also elementary, high schools and even very young children are selected to present their work in front of the contemporaneous invited. The selection of the young guests is performed because now, the elementary and high school students do their own Encounters with Sciences because we have triggered this interest. In their event, we go and select the best experiments due to their creativity and, of course, we take into account the preparation of the students. It is truly a beautiful spectacle to see all the schools in a Science festivity, friends cheering up the others as a demonstration of support and, above all, the pride to

Figure 2. Map of Venezuela where shaded states indicate the ones that participated in the event and from where our professors and students came

go to the renowned School of Science of ULA, and be able to be in an equal position of the students of the wrongly called the hard sciences.

As far as we know, already five schools are doing their own events for approximately four years, and the interest is growing every day. We have the knowledge that in other states, farther than the surroundings of Mérida, there are more schools devoted to extend the main event taking place at ULA. With these schools, we are articulating the Web of Encounters in order to create a big research platform composed by students and teachers of public and private schools and the School of Science; thereby, to conduct a hand by hand work with the final aim to give to our professional help in the teaching of the sciences, that will allow to have criteria that will be incident in its better learning processes.

But this "festivity" does not even end at this point; as a matter of fact we have been spreading out workshops for the capitation and actualization of teachers, in several states of Venezuela, bringing interesting oral presentations and, in a smaller scale, some experiments that we present in the main event. This task is performed with an alliance with other universities and institutions. Per example, we have been presenting this work in the state of Nueva Esparta (Porlamar) with the Universidad de Oriente; in Barinas with the Universidad de Los Llanos Ezequiel Zamora; in Trujillo with the extension of ULA in this state; in Zulia (Maracaibo and Santa Bárbara) with La Universidad del Zulia and Universidad del Sur del Lago, respectively; in Caracas with the Museum of Science; in Táchira with local schools and School Districts; and, in Falcón (Dabajuro) with the Universidad Francisco de Miranda. In this last place, a curious and gratification issue arose when very humble people from this small village, requested that we present the event in a church! In figure 2 we show the places were the events have taken place.

Also, we have prepared a didactic material (books, manuals, promotional kits, didactic toys, etc.) that resumes all the scientific experiments that we show in the several activities that we have mentioned, and we give it to all the schools with no cost.

Well, no wonder our Quijote de La Mancha dream called Encounters with Science has earned, in 2004, the first edition of the Prize for the Scientific Spreading, Mention: Best Initiative. All the state of Mérida, and specially ULA, was very happy and proud for this honor because it is a truly event of and for the people and for the next generation of scientist (hopefully). As a matter of fact, in its second edition a special mention was given to some students of the School of Science of ULA for being implementing, with dedication and professionalism, the experiments during the Encounters with Science.

6. The future

It is obvious that the Encounters with Science have now a national status and a clear success due to several reasons: (a) the ever growing interest of our faculty members and our students, (b) the interest from primary and high school students, and (c) the financial help from several organisms in ULA (School of Science, several organisms under the tutoring of ULA's President, Academic and Administrative Vice-Presidents), Fundacite-Mérida, FONACIT, OPSU, Laboratorios Valmorca, Fundación Polar, other private and official organisms, and individuals in general. For the latter, it seems a long list and, perhaps, we can think that we get a lot of aid; the reality is not so, every year we have a lack in the financial sense that makes us think that we will not be able to cope with the main event and all the other activities already described. But, we do a large effort to not let down such an important mission for education. We do this effort with the clear conviction that doing the incredible task of designing, creating, explaining in a simple way some experiments that trigger the imagination of young minds; we will leave a legacy that will last in time. A future would be certain in obtaining funds from even international organizations and to be able to insert this event in the official organisms that have the heavy duty to make Science as one of the attractive options for a career.

We hope that this opportunity that we have been given to spread the news over this event in and internationally way, can help us to find sensible people that can adopt part of the financial problem that we face.

Acknowledgements

PR is thankful for the invitation of the IAU GA Special Session 5, specially to J.B. Hearnshaw; and, for the financial support of the IAU GA; CDCHT-ULA (under project SE-C-02-06-05); H. Ruiz and M. Bonucci, Academic and Administrative Vice-Presidents of ULA, respectively, and J. Andérez Director of the Inter-institutional Relations of ULA.

References

Memorias del I Encuentro con la Física 2001, in: P. Rosenzweig, O. Escalona, A. Noguera, V. Sagredo, and A. Paniagua (eds.), Facultad de Ciencias, ULA, pp. 184

Memorias del IV Encuentro con la Física 2004, in: V. García (ed.), , Facultad de Ciencias, ULA, pp. 115

Rosenzweig, P. 2004, in: N. Pulido (ed.), *Investigación Revista del Consejo de Desarrollo Científico, Humanístico y Tecnológico*, V edicin de un acto pedagógico de pertinencia social. Sembrando Ciencia (CDCHT - ULA), No. 10, p. 56

Rosenzweig, P. & Escalona, O. 2001, in: N.Pulido (ed.), *Investigación Revista del Consejo de Desarrollo Científico, Humanístico y Tecnológico*, Encuentros con la Física (CDCHT - ULA), No. 5, p. 34

Rosenzweig, P., Escalona, O., Noguera, A., Sagredo, V. & Paniagua, A. 2001, *I Congreso Regional para la Enseñanza de la Física*, Primer Encuentro con la Física (UNET), p. 51

Astronomy for the developing world
IAU Special Session no. 5, 2006
J.B. Hearnshaw and P. Martinez, eds.

© 2007 International Astronomical Union
doi:10.1017/S1743921307006680

Developing astronomy in Cuba

Ramón E. Rodríguez Taboada[1]

[1]Department of Astronomy, Institute of Geophysics and Astronomy,
212 St. No. 2906, La Lisa, C. Habana 11600, Cuba
email: ramone@infomed.sld.cu

Abstract. The development process of astronomy in Cuba is analysed from the point of view of a person actually working in astrophysics. It is concluded that the key word for astronomy development is "engagement" between a mature partner and the new developing group. It can not be forgotten that only astronomers do astronomy and that should be the main goal: to develop people working in astronomy.

Keywords. Sociology of astronomy, astronomy in Cuba

1. Introduction

This talk is devoted to describing the present Cuban situation in astronomy development, from the viewpoint of a person actually working in astrophysics. To make easier to understand the situation, I will begin with a brief historical introduction. Later, the up-to-date situation is presented and the topics relevant to astronomy development are analyzed.

Arising from national needs, astronomical calculations are the only "native-born" branch of astronomy in Cuba. Sun and Moon rise and set, culmination and elongation times of Polaris, local circumstances of eclipses, and other relevant data for the public and economic activities were developed early in the beginning of the 20th century with the foundation of the National Observatory. The National Observatory conducted the calculation of tides and other services too.

With the space age advent and the space race between Russia and USA developing, new needs arose. Located in the western hemisphere, Cuba was an observational platform capable to provide the Soviet Union with the 24 hours solar patrol needed by its Space Agency System to protect men in orbit. This was the beginning of a very fruitful development of solar research in Cuba. Russia installed the instruments, trained the people to operate them, and gave the academic environment needed to develop the scientific work in solar physics, space weather, and related topics. Considering this experience, I will comment on astronomy development in Cuba.

2. Historical review

2.1. Solar research development in Cuba

First was the task we were given. We were faced with a very clear task to perform: namely, a solar radio patrol for proton event diagnosis.

To perform such a task, several institutes of the Soviet Academy of Sciences decided to install radiometers in Cuba. The radiometers were to provide the data to perform the scientific and practical work, but they required a qualified crew to operate and maintain them. The solution to this problem led to the development of solar research in Cuba.

Figure 1. Astronomical Data for Cuba

Figure 2. Havana Radio-astronomical Station, Cuba for SGD reports

In the beginning, the main task was to observe. Highly qualified teams of specialists from Soviet institutes visited the Institute of Astronomy in Cuba, to install equipment and to prepare the people to be in charge in Cuba. So we were qualified to operate and analyze the observations, and to calibrate and maintain the equipment and control their accuracy.

Very soon it was clear that people of the Havana Radio-astronomical Station could provide some kind of collaboration in solar research, and Soviet researchers leading projects involving the Havana Radio-astronomical Station observations included them in their work. For me, this was the turning point in the Cuban development of solar research. The Soviet groups investigating some edge problems in solar physics included Cuban specialists in their research projects, and received them for six-month periods in their institutes to provide the academic environment allowing to raise their qualification and later to prepare doctoral theses in accordance with their experience.

2.2. *Non-solar astronomy research*

The Cuban astro-climate is not good to develop an observational base. For this reason we have only participated in sporadic observations of planetary objects, as in the Soviet project Fobos and during the observational campaign of Halley's comet. These works were mainly astrometrical. Stellar variability was considered as a stellar research line to develop, but it was not successful in the absence of a strong partner to support the activity.

Figure 3. Scintillation telescope MEXART of IG-UNAM

Figure 4. Staff for University for Every Body course of astronomy

3. The present

We are trying to develop stellar astronomy in collaboration with institutions capable to provide both the academic and technical environment. The IAC (Instituto Astrofsico de Canarias) received a Cuban researcher to prepare his doctoral thesis working within a project studying symbiotic systems, and Guanajuato University will receive another one to prepare her doctoral thesis in water masers in young stellar objects. But to continue developing stellar astronomy, we need to influence the public (an official) opinion to convince people they need groups working in astronomy.

How to do that? Publishing. Giving conferences talking about OUR work, not only like spectators of the science. Showing science is culture in modern times. Showing that projects in Astronomy can be cheap. This is very important! Astronomy is not a luxury for rich countries. The Cuban government is promoting a general increase of qualification and knowledge in the population. Cuba has a very complex educational system that includes TV lectures for elementary and secondary (junior and senior) levels in the formal structures of the national courses, but there are college level courses in the Old Adult University and not formal courses by TV programs too. Included in this effort there is a series of TV courses provided by a program named "Universidad para Todos" (University for Everybody). One of the courses passed by the Cuban TV during 2005 is "Elements of Astronomy". It consists of 30 TV programs of one hour duration each, covering a wide spectrum of topics. The course is accompanied by a tabloid with the more important information of the lectures. The volume of the tabloid is of about 100 letter type pages including the figures.

3.1. *Real possibilities*

I consider the Virtual Observatory concept the most appropriate in the near future, but to develop this possibility it is necessary to have an internet connectivity level that is not commonly provided in Cuba. Probably the more viable way to introduce this mode of astronomical research is adding people to the developing astronomical (solar and stellar) projects working in this way on collateral aspects of the main line. But any way, some kind of training is necessary and the need of a mature partner is evident. No matter, there is a lot of open and free software for astronomical use, the solar virtual observatory as it is in fact developing, is based on IDL, a system that is expensive.

Another research direction that we could develop is computational astronomy. But in this case we have not made contact yet with any specialist interested in accepting a Cuban researcher to work on solar system dynamics, galaxy interactions and related aspects.

3.2. *Contributions to astronomy development*

I have noted that in many cases researchers in Latin America look for partners ONLY in developed countries. We have some experience in South-South collaboration. Once Russia was not able to maintain the collaboration at the highest level, and requirements for solar patrol and research diminished. So we searched for common research interests in the area and found that the National Autonomous University of Mexico in its Institute of Geophysics was working in solar research with similar goals. After a few years of collaboration and academic exchanges supported by CONACyT grants, we have developed a strong contact with Mexican solar researchers and Cuban engineers are working in MEXART, a Mexican project including collaboration with India and Japan. Given that the formation of qualified specialists is one of the main aspects of astronomical development, we are working to include a post-graduate level course of Astrophysical Plasmas Basics that we give in the Faculty of Physics of the University of Havana as in the INAOE (Puebla, Mexico) doctoral courses.

4. Concluding remarks

From my experience "engagement" is the key word for astronomy development in developing countries. Efforts to develop astronomy in developing countries succeed only if an astronomer with experience in the field guides the firsts steps of those people that are initiating this work in the fields where they have experience. Medical doctors do medicine, engineers buildings, etc. If you give resources to doctors they will use it to develop medical assistance, if to engineers to build. Only astronomers can do astronomy, so the resources that you want to be dedicated to develop astronomy should be given to astronomical projects including astronomers and/or "astronomers to be" from the developing countries.

Astronomy can not be developed without an appropriate academic environment, and we do not have it at present. Some countries in Latin America have better conditions to collect together small groups of people working in astronomy, but their more mature researchers have not taken conscience of their importance and possible influence in the development of astronomy in our area. It is not "only" about financial resources, it is also about "real collaboration" with mature partners and common research goals. From my point of view, only by working, and making astronomical research, will astronomy be developed. Science is just one, and it seems to me improper to waste any human resources available.

Astronomy for the developing world
IAU Special Session no. 5, 2006
J.B. Hearnshaw and P. Martinez, eds.

© 2007 International Astronomical Union
doi:10.1017/S1743921307006692

Planetario Habana:
a cultural centre for science and
technology in a developing nation

Oscar Álvarez[1]

[1]Ministry of Science, Technology and Environment, Havana, Cuba
email: oscar@citma.cu

Abstract. Astronomical education in Cuba is not widespread in the educational system; nevertheless the public interest in sciences in general but particularly in Astronomy issues is very high, as it has become reflected by the attention paid to educational and scientific program broadcasts in the national television channels. The "Planetario Habana" Cultural Centre for Science and Technology, which is under construction, is aimed at guiding the interest towards basic sciences and astronomical formation of the people, in the most populated and frequented area of the country. A key objective of this project shall be serving as an instructive motivation and entertainment for the casual or habitual visitors to these facilities, offering them the possibility to enjoy vivid representations, play with interactive amusement equipment and listen to instructive presentations on astronomy and related sciences, all guided by qualified specialists.

Another fundamental purpose shall be the establishment of a plan for complementary education in coordination with schools, in order to allow children and young people to participate in activities enabling them to get into the fascinating world of Astronomy, Exploration of Outer Space and Life as a Cosmic Phenomenon.

The setting up of the Planetario Habana Cultural Centre for Science and Technology is under the general administration of the Office of the Historian of the City of Havana, and methodologically is being led by the Ministry of Science, Technology and the Environment, and will show in operation the GOTO Planetarium G Cuba custom, obtained under a Japanese Cultural Grant Aid. It will develop into a an unparalleled centre in the national environment for scientific outreach and education of these sciences.

Surrounded by the attractiveness of the colonial 'ambience', it shall become a centre for dissemination of information about new discoveries and scientific programs developed at national and international level. Here we present a general view of the project, and its present and future development.

Keywords. Astronomy in Cuba, planetarium, public outreach in astronomy

1. Introduction

The Ministry of Science Technology and Environment of the Republic of Cuba in agreement with the Office of the Historian of the City of Havana are working together to facilitate the setting up of the Planetarium and Cultural Centre for Science and Technology at the Old Town of Havana, towards popularization among the students and general public on Astronomy, Cosmology, Astrophysics, Space Sciences, as well as other Sciences and Technology in general.

San Cristobal de La Habana Villa was founded in 1519 and shortly afterward became the capital of the island. For centuries the city was a strategic centre of colonial Spain. Owing to its characteristics as a port city and its unique geographical location at the mouth of the Gulf of Mexico, Old Havana still houses vestiges of its origins, such as the Alameda de Paula, next to the port, the Plaza de Armas and Cathedral Square,

Figure 1. View of the Havana Old Town Central Park and port entry

the Castillo de la Real Fuerza and the Morro and La Cabana castles, the Palacio de los Capitanes Generales, the Palacio del Segundo Cabo, la Plaza Vieja, where will be located de Planetarium, as well as the ruins of the walls that once surrounded it and the brick streets that have defied time in order to reveal all the greatness of the past. In 1982, the old historical area, the urban extension of 19th-century Havana, the port channel with its two developed shorelines, and the fortifications that served as its defense were declared a human heritage jewel by UNESCO, giving international recognition to the policy of safeguarding the city's historical and cultural value. Thus the foreign and national visitors inevitably explores brick streets and alleys; visits castles, forts, the ruins of city walls, buildings, monuments, museums, churches, mansions, elegant houses and squares; enjoys a view of colonial roofs, stained-glass windows, grates, low walls, balustrades, arcades and balconies in an astonishing profusion of styles. Traditions, legends and history have combined to make Old Havana a historical heritage jewel, renowned for its magnificent buildings, squares and churches, all of which date back to colonial times. City streets are still paved with cobblestones. With the legal autonomy granted by the Council of State of the Republic from Cuba to the Historian's Office in 1993, to self-manage the patrimony starting from the state declaration of the Old Havana as Prioritized Area for the Conservation, the action of this Institution, been founded in 1938, began the application of a new administration model. This new administration model's essence is the culture, and therefore the human being like creator and culture payee. He is the center of that process in development.

To this humanist and social and cultural vision it has been indispensable to add an economic vision, with the purpose of making sustainable the process.

In the structure of the Historian's Office, the new Cultural Center of Science and Technology (Planetarium) will be inserted inside the Cultural Patrimony Division of

Figure 2. Panoramic view of the Plaza Vieja. The pink painted building at the centre is the location where the Planetarium is going to be placed

The Office of the Historian of the City of Havana whose corporate purpose is the one of contributing to the knowledge of the history and the culture of the nation, to preserve the symbols, material and spiritual expressions of the Cuban nationality included in the patrimony to its care and to make them useful when propitiating its knowledge and understanding.

The Cultural Patrimony Division of The Office of the Historian of the City of Havana has also the mission to recover and to preserve the cultural historical memory of the city and especially of their Historical Center. To take care of the conceptual and formal validity of the tangible and intangibleness values, objects of their exercise and especially of the institutional work, through the popularization of the same ones. To plan and to execute activities that propitiate the improvement of the emotional and cultural life of the community, especially of the population's more fragile groups as well as to advise in their specialty to the rest of the dependencies of the Institution, looking after the genuineness and coherence of the artistic and cultural life that it is promoted.

At the moment Patrimony administers more than 30 cultural facilities: Museums, Concert Halls, Aquarium, Theaters.

The development of this project will facilitate to the Historical Center to have one of the places more yearned by the population of the capital and of the whole country. This installation and their appropriate operation lean on the sure the bases of the programs that develop the Ministry of the Science Technology and Environment (CITMA) in the fields of science and the technology, the Academy of Sciences of Cuba (ACC), the Institute of Technologies and Applied Sciences (INSTEC), the Havana University (UH) as well as the Office of the Historian of the City (OHCH). This system has human capital with enough potentialities to assume the setting in march of this Center, as well as to guarantee its economic, functional operation, providing competent personnel, wages, technical insurance, electricity, maintenance and constructive maintenance of the property rehabilitated for this cultural service.

The historical centre of the Havana City where de Planetarium will be located has 2.14 square kilometers and 72 000 inhabitants but the floating population reach 80 000 persons due to its historical and cultural interest for the visitors and for the commercial activity of the zone. Astronomical education in Cuba is not widespread in the educational system; nevertheless the public interest in Astronomy and Space related matters is very high as it has become reflected by the attention paid to educational and scientific program broadcasts by the national television channels. The Cultural Centre for Science and Technology and Planetarium project is aim to guide the interest towards elementary concepts on Basic Sciences and basic astronomical formation of the people, in one of the more populated and frequented areas of the country. A key objective of this project shall be serving as an instructive motivation and entertainment for the casual or habitual visitors to these facilities, offering them the possibility to listen to instructive presentations on Astronomy and related sciences, all guided by qualified specialists.

Figure 3. Cultural centre for science and technology: inner view

Figure 4. GOTO planetarium projector planned for Havana

Another fundamental purpose will be the establishment of a plan for educational opportunities in coordination with schools in order to allow children and young people to participate in activities enabling them to get into the fascinating world of Astronomy and Exploration of Outer Space. The setting up of the Cultural Centre for Science and Technology and Planetarium at the Old Town of Havana, provided with the specially designed GOTO G CUBA Planetarium obtained under the Japanese Cultural Grant Aid, will develop into a nonpareil pole in the national scenario for the scientific popularization and education of these sciences, surrounded by the colonial ambience's attractiveness, becoming a centre for the dissemination of information about new discoveries and scientific programs developed at national and international level.

Personal and institutional collaborations in this project are welcome.

Astronomy for the developing world
IAU Special Session no. 5, 2006
J.B. Hearnshaw and P. Martinez, eds.

© 2007 International Astronomical Union
doi:10.1017/S1743921307006709

Astronomy in Colombia

William E. Cepeda-Peña[1]

[1]National Astronomical Observatory, National University of Colombia
email: wecepedap@unal.edu.co

Abstract. Astronomy in Colombia has been done since the beginning of the nineteenth century, when in 1803 one of the oldest (or maybe the oldest) astronomical observatories of America was built. This is a very beautiful, historical and ancient building. A small dome with a small telescope is also inside the university campus. The observatory marks since then the development of astronomy in Colombia as a professional science. At the present time a Master's programme and a Specialization programme are successfully carried out with a good number of smart young students. The observatory has a staff of eleven professors, all with a master's degree in sciences; two of them have a PhD and in a couple of years five staff members will have a PhD in physics. With some international collaboration, they will introduce in a few years a doctoral astronomical program. There are several research lines mainly in the fields of astrometry, Galactic and extragalactic astronomy, cosmology, astro-statistics and astrobiology. Three research groups have got recognition from the governmental institution that supports the research in sciences COL-CIENCIAS. Several papers have been published in national and international journals. Besides the professional line in astronomy, the observatory sponsors several non-professional Colombian astronomical groups that work enthusiastically in the field of astronomy.

Keywords. Galactic astronomy, AGN, teaching of astronomy

1. Historical background

The National Astronomical Observatory of Colombia, NAO, founded in 1803, is the oldest permanent structure in the Americas that was designed for use of a fixed telescope for astronomical observations. The NAO was conceived by Jose Mutis and built with the intention that Caldas be director.

José Mutis is the central figure in early science in Colombia. Mutis was a physician named 'first botanist and astronomer' by King Carlos III for the Botanical Expedition of 1783, which was equipped with a Sisson quadrant, two Adams theodolites, and two Emery chronometers. They were supplied in Spain with a transit instrument, a telescope by Herschel and two achromatic telescopes, but these were lost in transit. At the College of Rosario, Mutis taught mathematics, physics and, against the mainstream Dominican teachings, he taught Copernican astronomy, possibly the first in America to do so. The site chosen was the garden of the Botanical Expedition. It is close to the equator, and in 1803 there was no other observatory at a higher elevation. The 56-foot octagonal tower, diameter 27 feet, has instrument rooms in successive stories. A second 'staircase' tower is 72 feet tall. There were telescopes in the upper rooms of both towers.

On May 24, 1802, the construction of the building began, and it was completed August 20, 1803. Caldas was appointed director in 1805, and in December 1805 began programmes of astronomical observation, meteorological projects, teaching local students, and he helped start a weekly scientific journal. During July of 1810, political unrest in Colombia reached a critical point. Caldas was forced to abandon scientific work, become a military engineer, and engage in the production of maps and armaments.

Francisco José de Caldas, was a lawyer with an interest in astronomy. Circa 1785, he began measuring meridians and calculating latitudes and azimuths. To mark solstices, Caldas measured the amplitude of the ecliptic. He made many meteorological and geographical measurements, including the determination of local longitude using published ephemerides and astronomical phenomena including lunar eclipses and occultations of stars and moons of Jupiter. For the lunar eclipse of 1797, he used an achromat of 30 inches focus to derive longitude.

The arrival of Alexander von Humboldt in Colombia gave a great impetus to science in the area. In 1801, Humboldt tutored Caldas in meteorology and astronomy, giving him star tables and catalogues, and teaching computation. Humboldt had a significant influence on Caldas the astronomer; instructing him on the use of tables of atmospheric refraction and the use of instruments. He provided Caldas with an octant and a quadrant, and star catalogues and ephemeris. Caldas observed the transit of Mercury of 9 November, 1802, timing the final two contact points.

Caldas was an advocate of independence from Spain and allowed activists to meet at the observatory. As a consequence of this, he was arrested by Spanish troops and executed in 1816. Caldas is a national hero to Colombians, probably a unique example of an observatory director elevated to heroic status. Julio Garavito became director in 1891, and despite being equipped with antiquated instruments, was active in both observation and theory. Studies included mathematical optics, lunar motion, Newtonian mechanics, and criticism of relativity. Further civil unrest caused problems at the National University, and engineering classes moved to the observatory. In 1949, Ruiz Wilches, directed the construction of a new observatory in 1952, in the campus of the National University of Colombia, NUC, with a 4-element apochromatic refractor, of 20 cm aperture and 3 metres focal length, previously used at the Observatory of Marseilles, France.

2. Professional astronomers

The NAO has a staff of eleven professors all with a master's degree in sciences; two of them are PhD and in a couple of years five staff members will have a PhD in physics. The Colombian astronomers received their training in the Science Faculty of NUC.

3. Teaching of astronomy

Although astronomy does not appear in the curriculum of the secondary schools, it is being taught in some universities. The teaching is mainly conducted by lectures and courses. The 'General Astronomy' course is an elective for students of the Physics Department of NUC. The NAO leads since then the development of astronomy in Colombia as a professional science. At the present time a Master's programme and a Specialization programme in astronomy are successfully carried out with a good number of smart young students. In the near future, the NAO plans to introduce a PhD programme in astronomy.

4. Scientific interests

Astronomical research in Colombia were chiefly concentrated in theoretical parts rather than the practical requirements of the country. The field of interest are:
- Fundamental astrometry. They are training at NAO.
- Celestial mechanics. The analytical theory of motion of Earth artificial satellites. They are studied at NAO. Some papers were published in this field.

- Ephemerides. The main tasks were compiling the astronomical calendar and nautical almanac, predicting astronomical phenomena (eclipses, comets), and calculating astronomical elements as a service to different governmental institutions of Colombia.
- Active nuclei of galaxies, AGN.
- Black holes and cosmology.
- Galactic astronomy. A lot of papers were published in this field.

Colombian astronomers also wrote books on: Astronomical courses in general and special relativity; dissemination of astronomical knowledge for example: Astronomy for everyone, spherical astronomy and eclipses.

The Galactic Astronomy Group Stellar Clusters is the principal research group in the NAO. Research is being done in different topics by the Galactic Astronomy Group Stellar Clusters. In the field of open clusters, the research has been focused on membership working from proper motions and proper motions and positions; an astro-statistical approach has been used to obtain membership probabilities using mainly Bayesian theory and the EM or expectation maximization algorithm; the required software has been developed.

Another topic is galaxy groups and group lensing. Weak lensing is a powerful tool to test the large-scale structure of the universe. Due to the low signal, this effect is only a statistical one, but with a wide application that goes from the determination of the cosmological parameters of matter density, the cosmological constant or dark energy, normalization of the power spectrum, the Hubble constant, and to constrain shapes and sizes of the dark matter halos. We use weak lensing to study the dark matter distribution in galaxy groups, together with kinematics and dynamics of the galaxies in the group. We are developing new tools to extend the work made in galaxy clusters, using mainly the tangential shear as a signal of the matter distribution, and we compare our results with the dynamics and some tracers, such as inter-galactic matter.

Another line of research is dark energy and dark matter in galactic halos. The distribution of galaxies in the universe is one of the key astronomical observations (together with the cosmic background radiation and supernova Ia light-curves), which is used to identify the cosmological model that rules the dynamics of the universe. Such observations have shown that the universe is in an accelerated phase with zero curvature; these features indicate a dominance of an unknown relativistic fluid, the so called 'dark energy'. The cosmological constant has been claimed to be the best candidate to represent such a component, but still there are other candidates that also lead to the present universe. These are the so called 'Chaplygin gas', 'dynamical dark energy' or 'quintessence models' (inspired by inflationary models). Hence, the main goal of this line of research is to identify which of these candidates fits in the best way the observations of the galaxy distribution.

Finally, some astrometric work is also done in this research group, looking for the best description of the movement of the Sun in the solar neighborhood.

References

Bateman, A. 1953, *El Observatorio Astronómico de Bogotá*, Universidad Nacional de Colombia, Bogotá.

OAN, 1997 *Plan de desarrollo Observatorio: Astronómico Nacional.* Universidad Nacional de Colombia, Bogotá.

Velázquez, M. 1994, *Caldas.* Molinos Velásquez Editores, Bogotá.

Julieta Fierro

Oscar Álvarez

Astronomy for the developing world
IAU Special Session no. 5, 2006
J.B. Hearnshaw and P. Martinez, eds.

© 2007 International Astronomical Union
doi:10.1017/S1743921307006710

New facilities for astronomy education and popularization in Guanajuato, Mexico

Hector Bravo–Alfaro[1]

[1]Departamento de Astronomía, Universidad de Guanajuato. C.P. 36000, Mexico
email: hector@astro.ugto.mx

Abstract. With relatively low funds, the *Departamento de Astronomía de la Universidad de Guanajuato*, in Mexico, has successfully carried out the modernization of a rustic astronomical observatory. At this site, once it is fully refurbished, it is planned to hold not only public observations, but also to have an ambitious programme of science popularization devoted to scholars.

Keywords. Instrumentation, University of Guanajuato, Mexico

1. Introduction

In 1967 it was officially opened the *Observatorio Astronomico* as part of University of Guanajuato, about 400 km NW of Mexico City (see Fig. 1a). Miguel Izaguirre, an Engineer in topography, was in charge of the site, equipped at that time with two refractor telescopes of short apertures. A dome and a hall for some 40-50 people were built on the roof of the University Central Building (Fig. 1b), after what the Observatory became -unofficially- known as "La Azotea" (the roof). The hall was used by Izaguirre, for more than ten years, for teaching astronomy for engineers.

Figure 1. (a) Guanajuato City is located 60 km E of León, in the very centre of Mexico. (b) View of the centre of Guanajuato City; the white building in the centre corresponds to the *Universidad de Guanajuato*, and the dome is visible on top of it (courtesy of K.-P. Schroeder).

In the early 1980s, Miguel Izaguirre and his two sons went for a more ambitious project: a professional observatory to be installed on the hills surrounding Guanajuato City. By a fruitful collaboration between *Universidad de Guanajuato* and *Universidad Nacional Autónoma de México* (UNAM), a 57-cm Ritchey-Chrétien was installed some 25 km W of Guanajuato City (Fig. 2a). Years later, in 1994, these efforts gave rise to the *Departamento de Astronomía* (DA), where nowadays nine professional astronomers do research and teaching at both undergraduate and graduate level (http://www.astro.ugto.mx).

However by the time Izaguirre retired, in the early 1990s, the public observatory "La Azotea" got no more maintenance nor a programme of modernization. By 2001, when the site was formally placed under the DA administration, only a portable 8-inch Schmidt existed as useful equipment. Since then, we applied a programme of free astronomical observations for non-specialized audience, but it remained to make big reparations in the building and dome, and acquire new telescopes and the necessary furniture to take advantage of the hall as conference room (Fig. 2b).

Figure 2. (a) The 0.57-m at La Luz hill. (b) La Azotea Observatory in August 2006, before the reparations.

2. The Project

In October 2005 the author of this work submitted a proposal to the local Council of Science (*Consejo Nacional de Ciencia y Tecnologia del Estado de Guanajuato* (CONA-CyTEG)), to get the necessary funds to acquire new equipment and furniture. The equivalent of some $30,000 U.S. was allocated to "La Azotea" project. In addition, the University collaborated with the equivalent of $20,000 U.S. to fully repair the building. The needed work started in September 2006 and was finished a couple of months later. At the same time, new telescopes, computing and furniture have been acquired with the aim of developing a Centre for Popularization of Science. So far the Observatory is equipped with a Schmidt Celestron 14", two Celestar 8", and a Celestron (104-mm) refractor, most of them with the necessary filters and screens to carry solar observations too. Finally, on November 22nd, the official reopening of the Observatory "La Azotea" took place. Two views of the site after reparations are shown in Fig. 3.

Once the last details are over, in January 2007, we plan to carry the following activities:

(1) A permanent programme of astronomical observations for a wide audience: in the last years, and even in the former conditions, La Azotea hosted a few hundred visitors per year. This number should naturally increase after the current upgrade.

(2) Regular conference series in different science domains: the Astronomy Department of U.G. organizes a series of five public talks on astronomy, twice per year (one in spring and one in autumn) since 1995. These conferences will now be held at La Azotea, starting in April 2007, proposing astronomical observations at the end of each talk.

(3) By a collaboration with the local department of culture we plan to host scholar groups, three times per week, for astronomical and scientific activities adapted for students between ages 6 and 15. Once this programme is applied the estimated number of students visiting the Observatory might rise above 3000 per year!

Figure 3. (a) The Observatory after the main reparations were accomplished. (b) View of the room after works and refurbishment.

(4) Summer schools in sciences for elementary and high-school teachers. In July 2007 we will carry a "pilot" school, for no more than 20 teachers, in domains such as astronomy, mathematics, physics and biochemistry.

(5) Foundation of a society of amateur astronomers.

3. The network

This observatory will constitute the core of the Astronomy Department's multiple facilities for wide-public astronomical observations. Two other sites, also under the DA's administration are currently at work in the Guanajuato environs: First, the professional observatory at La Luz (see Figure 2a), 25 km west of Guanajuato City, with a 0.57-m Ritchey-Chrétien, fully equipped with two CCDs for imaging (STL-11000, 4018 × 2743 pixels and STIO 2184 × 1472 pixels) and low resolution spectroscopy. It was recently acquired a seven-foot diameter radio telescope capable of continuum and spectral line observations in the L-band (1.42 GHz). All these facilities are mainly devoted to training both, undergraduate and graduate students, but wide public is also host there, every week, between September and April every year since 1999. A third site is currently installed on the roof of the Astronomy Department headquarters, 3 km north of Guanajuato centre, where an optical 0.4-m Dobsonian and a 104-mm refractor are already available. At the moment four astronomers, a Ph.D. student and a part-time engineer for support are more concerned on popularization tasks, working several days per week on one of the different sites.

4. Conclusions

The Astronomy Department of U.G. is in charge of the astronomy branch of the undergraduate programme of physics, and of a graduate (M.Sc. and Ph.D.) programme in Astrophysics which started in summer 2004. We are involving students from these programs in popularization activities in order to do part of their observational training with the help of these facilities. Students in advanced levels of the undergraduate programme will participate in the activities of this Centre as part of their social duties. We expect that "La Azotea" became a centre of science popularization with important impact on students from basic levels, not only in Guanajuato (a city with more than 100 000 inhabitants), but also in the environs, where much bigger cities are nearby located.

Acknowledgements

We would like to acknowledge the CONCyTEG for providing most of the funds of this project, including the fees to participate in the XXVI-GA of the IAU.

Antonieta García

The Charles Bridge, Prague

Astronomy for the developing world
IAU Special Session no. 5, 2006
J.B. Hearnshaw and P. Martinez, eds.

© 2007 International Astronomical Union
doi:10.1017/S1743921307006722

Implementing an education and outreach programme for the Gemini Observatory in Chile

M. Antonieta García Ureta[1]

[1]Gemini Observatory, La Serena, Chile
email: agarcia@gemini.edu

Abstract. Beginning in 2001, the Gemini Observatory began the development of an innovative and aggressive education and outreach programme at its southern hemisphere site in northern Chile. A principal focus of this effort is centered on local education and outreach to communities surrounding the observatory and its base facility in La Serena, Chile. Programmes are now established with local schools using two portable StarLab planetaria, an internet-based teacher exchange called StarTeachers and multiple partnerships with local educational institutions. Other elements include a CD-ROM-based virtual tour that allows students, teachers and the public to experience the observatory's sites in Chile and Hawaii. This virtual environment allows interaction using a variety of immersive scenarios such as a simulated observation using real data from Gemini. Pilot projects like "Live from Gemini" are currently being developed which use internet video-conferencing technologies to bring the observatory's facilities into classrooms at universities and remote institutions. Lessons learned from the implementation of these and other programmes will be introduced and the challenges of developing educational programming in a developing country will be shared.

Keywords. Gemini Observatory, astronomy outreach and education, Chile

1. Introduction

The Gemini Observatory is an international partnership of seven countries including the United States, United Kingdom, Canada, Chile, Australia, Brazil and Argentina. Gemini consists of twin 8-metre optical/infrared telescopes located at two of the best locations on our planet for astronomical research: Hawai'i and Chile. Together these telescopes can access the entire sky.

The Gemini South telescope is located at an elevation of over 2 700 metres on a mountain in the Chilean Andes called Cerro Pachn. This location is about 20 kilometres from the long-established Cerro Tololo Inter-American Observatory (CTIO) which is operated by the U.S. National Optical Astronomical Observatory (NOAO). The Frederick C. Gillett Gemini North Telescope is located on Hawaii's Mauna Kea and is part of the international community of observatories that have been established there to take advantage of the superb atmospheric conditions on this long dormant volcano that rises over 4 000 metres into the dry, stable air of the Pacific.

Both of the Gemini telescopes have been designed to take advantage of the latest technology and thermal controls to excel in a wide variety of optical and infrared capabilities. One example of this is the unique Gemini coating chamber that uses "sputtering" technology to apply protected silver coatings on the Gemini mirrors to provide unprecedented infrared performance.

Gemini's progressive instrument programme keeps the observatory at the cutting edge of astronomical research. By incorporating technologies such as laser guide stars, multi-conjugate adaptive optics and multi-object/integral-field spectroscopy, Gemini is well-suited for investigating the most challenging questions in astronomy today.

The Gemini telescopes have been integrated with modern networking technologies to allow remote operations from control rooms at the base facilities in Hilo and La Serena Chile. With the flexibility of "queue scheduling", researchers anywhere in the Gemini partnership will be assured the best possible match between observation, instrument and observing conditions.

2. Gemini outreach programme

The technology of operating the Gemini telescopes is also integral to many of our education/outreach programmes and experiences. Such an ambitious scientific endeavour requires sustainability, and this is why Gemini has implemented a multi-faceted public information and outreach (PIO) programme at both sites. The Gemini PIO effort provides each community with diverse and innovative educational resources and programming designed to encourage the implementation of science within the local school curriculum, motivate students to study astronomy, create awareness (and actions) about light pollution/dark skies and share our exploration of the many unsolved questions about our universe with both our local and international communities.

Gemini has consistently put considerable effort into educational programming for the two host communities (Hawaii and Region IV of Chile). This has resulted in an increased awareness in our local communities of the benefits of having state of the art observatories nearby, and created consciousness about preserving dark skies to allow other scientific projects to take place in the future.

However, having two telescopes located so far apart gives Gemini a unique set of circumstances that need to be considered when implementing any educational programming. These circumstances also provide a powerful combination of resources and synergies that allow for the development of innovative programmes that are not available to other institutions.

3. Outreach at Gemini South in Chile

Of the two Gemini telescopes, I will focus on Gemini South due to its location in Chile. The country of Chile provides a compelling backdrop given the country's geography, economy and political history. So, how does Gemini fit into this backdrop?

Chile is quite well known for many things, including its wine, skiing, and Easter Island, but the scientific community is probably most familiar with the extraordinary conditions that parts of Chile provide for astronomical observations. As an active participant in the local Chilean community it is our responsibility to make sure that local citizens are aware of how extraordinarily fortunate we are to have the natural resources of clear, dark skies and low humidity. In addition Chile's stable economy and supportive political environment combine to make it possible for international organizations like Gemini to build their astronomical research centres in Chile.

How does Chile benefit from hosting an observatory facility like Gemini? What are the jobs that are available to Chileans at an astronomical observatory? What training and education is necessary to participate in this work? What can students do now to prepare for the type of work at an institution like Gemini? What are the other economic and educational benefits that come from our presence? These are just a few of the many issues that need to be shared with our local communities.

Given Chiles long and thin geography it is frequently said that any project in Chile needs to be adapted to the area or region where it is located. This is one of the biggest and most challenging issues that the PIO department has had to face when a new programme is planned and launched.

Two StarLab portable planetariums have been utilized for educational programming at Gemini South for the past five years. When they were purchased, the instructions clearly stated that the equipment would provide the best experience in a carpeted, air-conditioned room. If that was the case, the programmes wouldnt have been able to run more than three times a year. Instead, the Chilean staff decided to leave one of the StarLab in the city area and one exclusively to cover the remote country locations and other regions in Chile. It is not rare to access the schools riding a horse, mule or the city hall special transportation, and if not outside, to be under an improvised netting. The StarLab programme has also run inside a church, the only local building big enough (barely) to hold the portable planetarium.

When visiting the native Mapuches in the south of Chile, the rain and mud made it impossible to setup outside the schools, so they accommodated the StarLab inside their local hospital.

Teachers have always been Geminis best link to educate students and families with new approaches in astronomy. Once a year about 30 to 40 teachers are trained in basic astronomy, and they are assisted by Geminis staff throughout the school year in order to provide the best resources and educational experiences for their students. The Gemini Observatory has profoundly impacted teachers that have participated in Gemini's StarTeachers teachers' exchange programme. This programme has been offered twice to date, and six Chilean teachers have participated in this in-depth exchange with teachers in Hawai'i.

During the StarTeachers exchanges, the participating educators made reciprocal visits to Gemini host communities (in Chile and Hawai'i), and were given the opportunity to experience the execution of Gemini's astronomical research programmes while sharing insights on each other's culture and educational systems. The teachers also presented live, interactive video-conference classes between their home and partner's students using Gemini's Internet2 technologies.

Beginning in 2005 the Astronomical Society of the Pacific's FamilyAstro programme is also being implemented in Chile. In this programme teachers are trained in different modules that later are introduced at events at each school for the students to share with their families. These events have always been a success, not only in the number of interested families that want to attend the workshops, but also because the families take their kits home and continue playing different games while learning and reinforcing many astronomical concepts.

4. Gemini virtual tours and CD-ROM

One of the major attractions that Gemini provides to any interested resident is a tour/visit to our sites. Unfortunately, because of its high altitude and thin air, children in Hawaii and people younger than 16 years of age are unable to visit the Gemini North telescope. In response to this, Gemini created a virtual tour that provides anybody with access to a computer with CD-ROM capabilities the ability to virtually visit the Gemini telescopes.

This CD-ROM has hours of magnificent photos, animations, videos, and games that enable any user to get the most of the interactive experience. Users can even make a simulated observation using real Gemini data and print a custom image. Currently this

CD-ROM is available in English and Chilean Spanish and a French version is currently under development.

5. Talking to astronomers

For students or anyone who ever wonders what it is like to be an astronomer, or what they are doing at 5 a.m., the 'Live from Gemini' programme can provide some insights. As a pilot programme, 'Live from Gemini' is available to educational groups with access to a high-speed IP address or access to videoconference equipment.

Whether in Chile or Hawaii participants will join a Gemini staff host in one of our control rooms and be able to talk to the astronomers on site and answer any of the questions that students might have wondered about. This is a wonderful opportunity to open the control room doors into the world and allow people to share with scientists events that could be of world-wide interest at that time. An example of this is during the Deep Impact collision with comet Temple I in which students from Chile were able to see and talk with astronomers on Mauna Kea at the same moment the impact was taken place in space and visible from Hawai'i.

In all, education can be a challenging role, but we need to make sure that the current generation of young people are growing up with a new mentality in which they can not only identify a few constellations for fun or deep interest, but also understand the role that science plays in their future, their community and why we need to preserve the resources that we have.

For more information, please visit http://www.gemini.edu.

6. Conclusion

The Gemini Observatory provides the astronomical communities in each partner country with state-of-the-art astronomical facilities that allocate observing time in proportion to each country's contribution. in addition to financial support, each country also contributes significant scientific and technical resources.

The national research agencies that form the Gemini partnership include: the US National Science Foundation (NSF), the UK Particle Physics and Astronomy Research Council (PPARC), the Canadian National Research Council (NRC), the Chilean Comisión Nacional de investigación Cientifica y Tecnológica (CONiCYT), the Australian Research Council (ARC), the Argentinean Consejo Nacional de investigaciones Cientficas y Técnicas (CONiCET) and the Brazilian Conselho Nacional de Desenvolvimento Cientfico e Tecnológico (CNPq). The Observatory is managed by the Association of Universities for Research in Astronomy, inc. (AURA) under a cooperative agreement with the NSF. The NSF also serves as the executive agency for the international partnership.

Astronomy for the developing world
IAU Special Session no. 5, 2006
J.B. Hearnshaw and P. Martinez, eds.

© 2007 International Astronomical Union
doi:10.1017/S1743921307006734

The first two years of the Latin-American Journal of Astronomy Education (RELEA)

Paulo Sérgio Bretones[1] Luiz Carlos Jafelice[2] and J.E. Horvath[3]

[1]Instituto Superior de Ciências Aplicadas, Limeira, Brazil and
Instituto de Geociências / UNICAMP Campinas, Brazil
email: bretones@mpc.com.br

[2]Departamento de Física, UFRN
Natal RN, Brazil
email: jafelice@dfte.ufrn.br

[3]Departamento de Astronomia, IAG-USP
R. do Matão 1226, São Paulo SP, Brazil
email: foton@astro.iag.usp.br

Abstract. We present and discuss in this work the motivations, goals and strategies adopted for its creation and launch of the e-journal Latin-American Journal of Astronomy Education (RELEA). We identify and present the difficulties encountered for the achievement of the proposed objectives and operational issues in this period, together with the adopted solutions (refereeing procedure, periodicity, etc.).

Keywords. Astronomy education

---

1. Introduction

In mid-2002 and after several conversations and exchange of ideas with colleagues, the three signing authors decided to establish and launch a publication devoted to scientific education in Astronomy.

In August 2003 the home page of the Latin-American Journal of Astronomy Education (RELEA) was released, containing a first call-for-papers and information about its goals, Editors and Editorial Board and preferred style for the contributions. Several messages and announcements were sent to international publications to call the attention to the new journal (Bretones, Jafelice, & Horvath, 2003a, 2003b, 2003c). We describe below the main features of the RELEA and some of the most important issues arising about its publication.

2. The RELEA "first light": August, 2004

Technically, the RELEA is an electronic journal produced by the technical support team of the *Instituto Superior de Ciências Aplicadas (ISCA)*, based in Limeira city (São Paulo, Brazil). The contents may be printed and copied for distribution for educational purposes without profit, and giving a proper acknowledge to the source. The RELEA is registered in the ISBN and the adopted editorial procedures for publication demand blind refereeing by two specialized colleagues, selected by the Editors upon submission. There is no *a priori* rule about the number of articles per issue or their length. These and other parameters will be in fact selected by the Editors, Editorial Board and referees, which will asymptotically establish how much and what to publish (Fig.1).

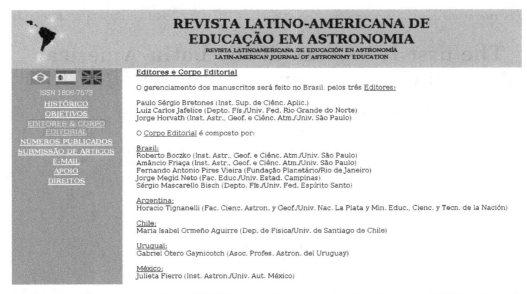

Figure 1. The cover page of the RELEA journal, showing the the Editors and Editorial Board representing their respective Latin-American countries.

The RELEA is composed of the following Sections:
* Culture, History and Society
* Teacher Formation
* Teaching and Learning
* Didactic Resources
* Educational Policies
* News
* Reviews
* Events

Some of the motivations, goals and strategies adopted by the Editors, have been partially covered in the Editorial of No.1, available at the on-line edition

http://www.iscafaculdades.com.br/relea/

and will be discussed in the next section.

3. The motivations and aim for the launching of the RELEA

Our first motivation for the creation of the RELEA was the noteworthy absence of a specific publication in the field in Latin-America. In spite of the existence of some general astronomy journals, their educational contents are marginal at best. This need was in fact pointed out by the President of the Commission 46 of the IAU to one of the editors (PSB) some years ago. Thus, we believe there is a parallelism between the new journal and its established cousins, such as *Gnomon*, *Universe in the Classroom* and *Teaching of Astronomy in Asian Pacific Region*, among others.

One of the consequences of this lack of dedicated journals for the Latin-American region was that, until now, the contributions have been mainly published in more restricted journals or outside Latin-America, mostly in English, and thus being quite atomized and somewhat invisible to potential users. A journal with a representative Editorial Board from the region, with easy access and low production costs (which was one of the main factors defining the adopted electronic format) was deemed important to improve the described situation.

Naturally, the specific reality of the Latin-American education, largely shared by the region countries, is an important factor for the need of dissemination of the contents, preferably in Spanish and Portuguese, but without excluding English at all since the latter became the international *lingua franca* of science today. An important positive feedback is also expected from this launching: in other words, we expect that the RELEA could become an additional instrument to boost the research in Astronomy education, respecting and reinforcing the quality standards accepted by the community, which is the main goal of the blind peer review established as a necessary publishing criterion.

Our own experience and a large amount of evidence suggests a serious lack of reliable and updated astronomical sources which can be employed in the classroom without translation and adaptations. The RELEA attempts to provide methodological and content support for educators, researchers, students and interested readers to fulfil this demand, at least partially.

More generally, an important concern is the advance of scientific illiteracy observed not only in the region, but also in other regions of the world, including developed countries. The paradox of the existence of overabundant information leading to an increasing deficit of basic scientific formation is very complex and will not be discussed here. However, we believe this is especially dangerous for the developing countries, since an impoverished logical thinking complicates the process of decision making. Thus, problems of political, social and economical nature deepen because of this lack of basic education, in particular in sciences.

In spite of these facts, some important features come to help this kind of initiatives. It is quite apparent that teaching of Astronomy has some notorious advantages over other sciences because of its very nature. In first place, Astronomy is a natural science, at least in its historical form. Even without appealing to high-tech developments, a simple and careful observation of the sky planned according to educational purposes (and preferably without unnecessary rigidity) leads to questioning and captures the students and public attention alike. In fact, it is arguable whether an empirical (and less academic) approach is more suitable than the traditional one, given that several researchers put the blame on the latter for the lack of interest in science. The RELEA is expected to be an adequate forum to improve the studies along these lines as well.

Overall, the response of the targeted community has been a bit slow, since the very lack of a well-established forum in the region quenched not only the amount of work dedicated to these matters, but also the practise of publishing the results for general appreciation, criticism and usage. We have experienced a slight increase in the number of submitted papers lately, and hope to be able to publish good quality works increasing the frequency of the issues (presently restricted to one annual edition) in the near future.

4. Conclusions

In summary, the electronic RELEA was launched to stimulate the research in Astronomy education focused in the Latin-American region, and also to act as a resource of useful contents for educational purposes and experiences. It also welcomes contributions from outside the region written in any of the official languages.

Envisioning the long run, we should attempt to constitute a forum to discuss and expose the problems of scientific education, acting as a "meeting point" to help design public policies. This ideal mission is not at odds with the present, more limited role. On the contrary, we feel that a Latin-American initiative may join together independent valuable efforts and experiences towards this common enterprise without loosing the regional focus.

Acknowledgements

We would like to acknowledge the advice and interest of several colleagues that collaborated with this project. LCJ extends his thanks to the UFRN university, in particular to the DFTE, CCET, PPGECNM units. The Director and staff of the ISCA faculties are acknowledged for their support to the new journal and to the person of PSB. We also would like to thank to the IAU for the travel grants that enabled the presentation of this work. JEH is supported by CNPq (Brazil) under a Research Scholarship and Grant which are gratefully acknowledged.

References

Bretones, P.S., Jafelice, L.C. & Horvath, J.E. 2003, *The Astronomy Education Review* 2, 14
Bretones, P.S., Jafelice, L.C. & Horvath, J.E. 2003, *IAU Commission 46 Newsletter* 59, 1
Bretones, P.S., Jafelice, L.C. & Horvath, J.E. 2003, *Gnomon* 23, 2

Paolo Bretones

Astronomy for the developing world
IAU Special Session no. 5, 2006
J.B. Hearnshaw and P. Martinez, eds.

© 2007 International Astronomical Union
doi:10.1017/S1743921307006746

Observing programmes of the Observatorio Astronómico Los Molinos, Uruguay

Gonzalo Tancredi[1,2], S. Roland[1], R. Salvo[1], F. Benitez[1,2], S. Bruzzone[1,2], A. Ceretta[1], E. Acosta[1]

[1]Department of Astronomy, Universidad de la República
Iguá 4225, 11400 Montevideo, Uruguay
email: gonzalo@fisica.edu.uy

[2]Observatorio Astronómico Los Molinos
Cno. de Los Molinos 5769 y Cno. Uruguay, 12400 Montevideo, Uruguay
email: oalm@fisica.edu.uy

Abstract. The Observatorio Astronómico Los Molinos (OALM) is the only professional observatory in Uruguay. Its research activity is dedicated to the study of Small Bodies of the Solar System, i.e. comets and asteroids. It is one of the few observatories in the southern hemisphere dedicated to this topic. The observational programmes that we are conducting at present are described, putting especial emphasis in the results obtained last year.

Keywords. Astronomy in Uruguay, Observatorio Los Molinos, comets and asteroids

1. Introduction

The Observatorio Astronómico Los Molinos (OALM) was inaugurated in 1993 and belongs to the Innovation, Sciences and Technology Office (DICYT) of the Culture and Education Ministry (MEC). It is supported by the Department of Astronomy of Universidad de la República. It is the only professional observatory in Uruguay. It is located in the northern border of Montevideo, the state capital. Therefore, the light pollution issues are worrisome. The latter fact lead us to build another observing facility placed in the country side, in the Province of Maldonado (250 km from Montevideo).

The scientific activity of OALM is focused on Solar System Minor Bodies (i.e. comets and asteroids), being one of the few observatories dedicated to this topic in the southern hemisphere (see Fig. 1 with the distribution of the observatories that have contributed more than 100 astrometric reports to the Minor Planet Center in 2005). The main instruments placed in the OALM' facilities are CCD-equipped 35 and 46cm telescopes used for astrometrical and photometrical programmes. The acquisition of the latter instrument, *Centurion 18 (Astroworks)* give birth to the research programme called *Búsqueda Uruguaya y Seguimiento de Cometas y Asteroides - BUSCA*. Also in the OALM facilities are placed other observing instruments with their respective domes that belong to amateur astronomers. A detail information about our facilities is found in our institutional website: *http://oalm.astronomia.edu.uy*

In the framework of the BUSCA programme, the following activities are developed in the observatory:

- Follow-up and confirmation of Near-Earth Objects (NEOs).
- Search for NEOs in the direction of their orbits radiant.
- Photometric follow-up of comets.
- Photometric profiles of asteroids in cometary orbits (ACOs).

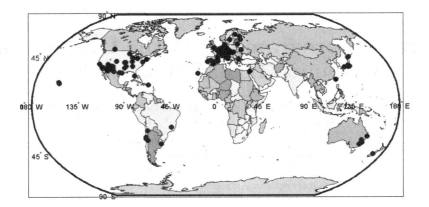

Figure 1. Distribution of observatories that have contributed more than 100 astrometric reports to the Minor Planet Center in 2005. There are 138 observatories in the northern hemisphere, and 15 in the south. OALM is ranked #7 in asteroids reports and #4 in comets reports in the southern hemisphere.

In the following sections we describe in detail the previous observing programmes and present some results.

2. Follow-up and confirmation of Near-Earth Objects (NEOs)

Through the astrometric reports done in OALM we contribute to the improvement in the determination of the orbital elements of NEOs and the estimate of their impact risk. We observe recently discovered asteroids (i.e. the confirmation task), or already discovered ones that need further observations (i.e. the astrometric follow-up). The confirmation process briefly consists in checking the object list in the NEO Confirmation Page (NEOCP) of the Minor Planet Center (MPC) every night. The object is then observed in the predicted ephemeris and the astrometry is done and reported back to the MPC. An inspection of the physical appearance is done in order to find possible cometary comae. In such a case, the object should be catalogued as a comet. An example of this event was the confirmation of comet C/2006 A1 (Pojmanski) accomplished in January 2006 in OALM (IAUC 8653). (see Fig. 2).

In Fig. 3 the evolution of astrometric reports in the period 1997 - 2005 is presented (*a*) and a distribution among the months for year 2005 (*b*). We notice the increment of reports during year 2005 as a result of the massive usage of the 18 inch instrument (Centurion) for this job. From the monthly distribution of reports, we conclude that the winter season is the less productive epoch in the entire year due to bad atmospheric conditions. Unfortunately this is the epoch in which the ecliptic reaches the highest altitude for our sky.

In Fig. 4 the declination (*a*) and the observed magnitude (*b*) distributions are shown. There is a concentration of reports in the southern declinations because of our selection criteria applied on the observed objects. This criteria is based on the observatory shortage in the southern hemisphere which causes the southern celestial hemisphere to be poorly observed. The most frequent magnitude range is $R \sim 17 - 18$, but seldom reaching magnitudes close to 19.

Figure 2. Image of C/2006 A1(Pojmanski) taken at OALM.

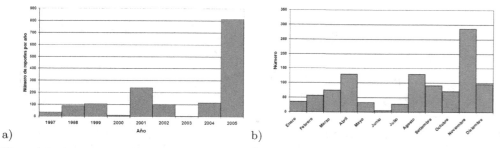

a)   b)

Figure 3. Astrometric Reports submitted to the MPC in the period 1997 - 2005 (*a*) and by month during year 2005 (*b*).

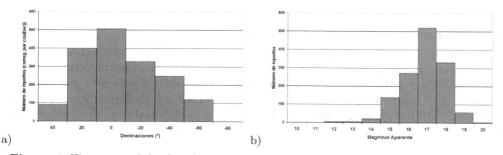

a)   b)

Figure 4. Histogram of the distribution of astrometric reports by declination (*a*) and by apparent magnitude (*b*).

3. Search for NEOs in the direction of their orbits (radiant)

From the study of the direction of approach of Near-Earth Asteroids (NEAs) to the Earth, we found well-delimited zones in a celestial map where these objects could be observed when they are in direction of collision with the planet. We refer to these positions as the radiants of the orbits, in correspondence with the terminology used in meteor science. If we plot the radiants in a sky map in ecliptic coordinates referred to Sun, the radiants are concentrated in two zones: one close to the anti-solar point and another one centered in the Sun. The anti-solar region is commonly explored by the large survey programmes (i.e: Linear). The regions near the Sun, at low ecliptic latitudes, are filled with objects with elongations lower than 30°, making the observations impossible. Elongations suitable for night observations are obtained only for radiants with high ecliptic latitudes.

Since beginning 2006 our survey programme has been concentrated in this region with ecliptic longitudes close to the Sun and high ecliptic latitudes. Due to the annual motion of the Sun, this zone will change their equatorial coordinates giving different configurations during the year. In some months, the densest zones would be in the evening or morning twilight. We generate alt-azimuthal maps of the visible sky at the moment of twilight for each month, centered on the zenith and viewed from inside the celestial sphere. In Fig. 5 evening and morning diagrams are shown for the beginning of each season. The densest and highest region are then selected to point the telescope for NEA searching.

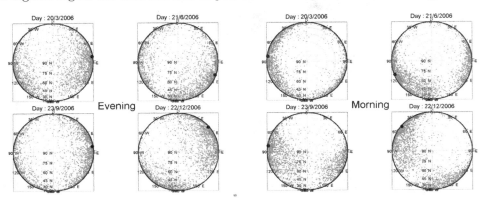

Figure 5. The distribution of NEAs' radiant projected on horizontal coordinates for different epochs of the year, half an hour after (before) sunset (sunrise). The region to look is the highest concentrations of small dots opposite to the large black dot (the anti-Sun).

4. Photometric follow-up of comets

The number of astrometric reports of comets between 1997 and 2005 followed the same increase trend as shown in Fig. 3a for NEOs. Concerning comets, astrometric reports are very important for the continuous improvements of the orbital elements and the determination of non-gravitational forces. The non-gravitational parameter can be directly associated with the comet gaseous production, which is obtained from the perihelia light curves (see below), to estimate the mass of the comet(Rickman *et al.* 1987).

4.1. *Photometry - Perihelion light curve*

We are observing every comet passing through perihelion whenever the magnitude is brighter than $R \sim 18$. In addition to study the action of non-gravitational forces, the light curves are useful to monitor any abrupt magnitude change like outbursts, in order to understand the underlying physical processes like fragmentation (splittings) or the exothermic phase transition of crystalline to amorphous ice.

The photometry of an extended object like the cometary comae is done in the following way: first the astrometric report is done with the software *Astrometrica* †. Second, the photometry is done with the software *Focas* ‡. The latter uses the multi-box photometry technique. The photometry is done at several boxes of fixed angular size centered on

† http://www.astrometrica.at/
‡ http://www.astrosurf.com/cometas-obs/_Articulos/Focas/Focas.htm

the comet coordinates. The sizes of these boxes are 10"x10", 20"x20", 30"x30", 40"x40" (Fig. 6).

Figure 6. Fixed angular size boxes are shown centered on the comet. Starting from the left, the box sizes are: 10"x10", 20"x20", 30"x30", 40"x40".

In Fig. 7 the perihelion light curve of comet *37P/Forbes* is shown. It was obtained using the OALM's telescopes during year 2005.

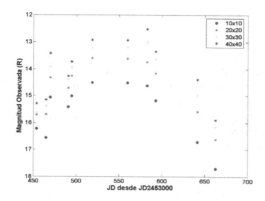

Figure 7. Perihelion light curve for comet *37P/Forbes*

Light curves obtained with our small telescopes will allow us to anticipate when the comet will be inactive. We then plan to use larger telescopes to observe and estimate the nuclear magnitude of the comet, as a continuity of the observational programme and the compilation work of cometary nuclear magnitudes that we have been conducting (Licandro *et al.* 2000, Tancredi *et al.* 2006).

5. Photometric profiles of asteroids in cometary orbits (ACOs)

Physical and dynamical evolution models of comets suggest the possible total deactivation of the gas and dust emission. This occurs by the formation of a dust crust that quenches the gas sublimation. There are some comets with very low levels of activity (Tancredi *et al.* 2006), a fact that support the existence of dormant comets. Analyzing the distribution of orbital elements for both comets and asteroids, a dynamical criteria was designed in order to classify objects between these two populations. Roughly hundred asteroids were found in cometary orbits and they require a detailed photometric study in order to determine signs of very low activity or a dormant phase.

The photometric follow-up of asteroids in cometary orbits consists in making a profile analysis of the asteroid and compare it to stellar profile in the field, in order to look for possible gaseous activity. This is viewed in the CCD image as an enhancement of the cometary profile over the stellar one, due to the coma.

The night of 01/30/2006, the asteroid 2001 ME1 in a cometary orbit was observed at an heliocentric distance of 1.2 AU. The preliminary comparison of the asteroid and the stellar profiles reveals no evident cometary activity. (Fig. 8).

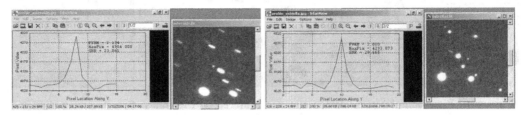

Figure 8. Profile comparison of asteroid 2001 ME1 (left) and a field star (right).

In addition to the photometric studies of comets and asteroids in cometary orbits, we perform the astrometric measurements and report them to the MPC.

6. Conclusions

We have presented the research activities that we are conducting at the OALM with relatively small telescopes (46 cm and 35 cm). We have shown that using these small telescopes for common-day follow-up programmes allows us to make relevant contributions in some scientific niches that are less explored. These small telescopes with easy access are useful for back-up programmes for other more ambitious projects.

We have shown that there are regions in the sky not explored by the big surveys that could be searched for with other small telescopes.

Acknowledgements

OALM is supported by the contributions of the Education and Culture Ministry, and the Sciences Faculty of Universidad de la República and the Programme for the Development of Basic Sciences (PEDECIBA).

References

Licandro, J.; Tancredi, G.; Lindgren, M.; Rickman, H.; Gil Hutton, R. 2000, *Icarus* 147, 161
Rickman, H.; Kamel, L.; Festou, M. C.; Froeschle, Cl. 1987, in *ESA, Proc. of the Int. Symposium on the Diversity and Similarity of Comets*, p 471
Tancredi, G.; Fern'andez, J.A.; Rickman, H.; Licandro, J. 2006, *Icarus* 182, 527

Astronomy for the developing world
IAU Special Session no. 5, 2006
J.B. Hearnshaw and P. Martinez, eds.

© 2007 International Astronomical Union
doi:10.1017/S1743921307006758

A new astronomical facility for Peru: transforming a telecommunication's 32-metre antenna into a radio-telescope

José Ishitsuka[1,2], M. Ishitsuka[**], N. Kaifu[*], M. Inoue[2], M. Tsuboi[3], M. Ohishi[2], T. Kondo[4], Y. Koyama[4], T. Kasuga[6], K. Fujisawa[5], K. Miyazawa[*], M. Miyoshi[2], T. Umemoto[3], T. Bushimata[2], S. Horiuchi[7] and E. Vidal[1]

[1]Instituto Geofísico del Perú,
Calle Badajoz 169, Urb. Mayorazgo IV Etapa, Ate-Vitarte, Lima 3, Perú
email: pepe@geo.igp.gob.pe

[2]National Astronomical Observatory of Japan, 2-21-1 Osawa, Mitaka, Tokyo 181-8588, Japan
[3]Nobeyama Radio Observatory, Minamimaki, Minamisaku, Nagano 384- 1305, Japan
[4]National Institute of Information and Communications Technology 893-1 Hirai, Kashima, Ibaraki 314, Japan
[5]Yamaguchi University, Yoshida 1677-1, Yamaguchi 753-8512, Japan
[6]Hosei University, Kajinocho 3-7-2, Koganei, Tokyo 184-8584, Japan
[7]Swinburne University of Technology, Faculty of Information and Communication Technologies, PO Box 218, Hawthorn VIC 3122, Australia
[*]Retired from NAOJ, [**]Retired from IGP

Abstract. In 1984 an INTELSAT antenna of 32 m of diameter was constructed at 3 370 metres above the sea level on the Peruvian Andes. At the time Entel Perú the Peruvian telecommunications company managed the antenna station, of almost 12 hectares in extension. In 1993 the government transferred the station to the private telecommunications company Telefónica del Perú. Since transoceanic fiber optics replaced radically satellite communications in 2002, a beautiful 32- metre parabolic antenna was finished its tele-communications mission and become available for other use. So in cooperation with the National Astronomical Observatory of Japan we began coordination to transform the antenna into a radio-telescope.

Researches on interstellar medium around Young Stellar Objects (YSO) are possible using methanol maser that emits at 6.7 GHz, so initially we will monitor and survey maser sources at the southern sky. An ambient temperature receiver with $T_{rx} = 60$ K was developed at Nobeyama Radio Observatory and is ready to be installed. The antenna will be controlled by the Field System FS9 software installed in a PC within a Linux environment. An interface between antenna and PC was developed at Kashima Space Research Center in Japan by Mr E. Vidal.

In the near future S-band (2 GHz), X-band (8GHz), 12 GHz and 22 GHz observations are planned.

The peculiar position and altitude of the Peruvian Radio Observatory will be useful for VLBI observations with the VLBA for astronomical observation and geodetic measurements. For Peru where few or almost non astronomical observational instruments are available for research, implementation of the first radio observatory is a big step to foster sciences at graduate and postgraduate levels of universities.

Worldwide tele-communications antennas recently tend to finish their role as tele-communications antennas. Several of them are transformed into useful observational instruments.

Keywords. Maser, Peru, radio-astronomy, YSO

1. Introduction

To perform astronomical observations in developing countries, where astronomy faculties at universities are not yet established, such a 32 m diameter radio-telescope would be an excellent means to develop astronomy and basic sciences. Around the world some successful cases are working well, one in Ceduna of the University of Tasmania, Australia and another in Japan, a 32-m radio-telescope at Yamaguchi City managed by Yamaguchi University.

In Peru Nippon Electric Company (NEC) constructed in 1984 a 32-m parabolic antenna for satellite communication managed by the past telephone company Entel Peru, later in 1993 government transferred the antenna station to the private company Telefónica del Perú. Then in 2000 the Peruvian INTELSAT communication station stopped operations. Two years later, we began conversations with telephone company in order to transfer the antenna to Instituto Geofísico del Perú and to transform into a radio-telescope. Transfer of the antenna ownership from the Telefónica del Perú to the Instituto Geofísico del Perú is under negotiation.

2. Science and education

Since methanol masers around high-mass star forming regions are good tracers to investigate early evolutionary phase of YSO, methanol masers appear to be closely associated with newly formed massive stars (Menten 2002; Minier *et al.* 2003 and references therein); initially we are planning to monitor and survey methanol maser sources.

There are 65 universities in Peru, 28 are public and 37 are private, 14 of the 65 universities have faculty of physics. The new radio-telescope facility will become an important instrument to educate undergraduate students, also will allow perform researches of astrophysics for graduate students. Running a radio-telescope involves antenna technologies, receiver system, down converters, computers and software development, students of other fields would be also educated and trained. We can settle radio-astronomy in Peru putting in operation the 32-m antenna.

3. Collaborations

From the beginnings of our project, National Astronomical Observatory of Japan and National Institute of Information and Communications Technology strongly encouraged us to develop instruments and train our staff. Collaborations made possible to develop and construct equipments to be installed on the radio-telescope. That also made possible to train and educate students, scientists, and engineers that will use in the future the radio-telescope.

4. 6.7-GHz receiver

Peruvian antenna used to transmit at 6 GHz and receive at 4 GHz for tele-communications, so connecting a 6.7 GHz receiver to the transmission feed will be the most efficient way to transform the antenna into a radio-telescope. We can receive 6.7 GHz methanol maser line spectrum signals, with 70 MHz of band width adding continuum observations.

A 6.7-GHz ambient temperature receiver have been prepared at Nobeyama Radio Observatory (Figure 1) and successfully tested in laboratory, from HOT-COLD load receiver temperature T_{rx} was 60 K, then we tested the receiver installing on a 20-metre radio-telescope of VERA Project (Sasao 1996; Honma *et al.* 2000) and could get W3(OH) methanol spectrum (Figure 2).

Figure 1. Ambient temperature 6.7 GHz receiver, $T_{rx} = 60$ K. Designed and built in Nobeyama Radio Observatory of NAOJ

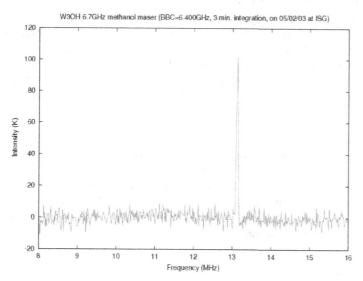

Figure 2. Methanol maser line spectrum of W3(OH)

5. Antenna control system

Being controlled by the Field System FS9 developed by NASA, the telescope is capable of not only single dish observations but also VLBI observations with other telescopes located in different places.

An interface was developed in Kashima Space Communication Center, see Figure 3 it will allow control the radio-telescope from a PC equipped with Linux Debian. Field System FS9 is the software that commands the Antenna Control Unit (ACU) to move the radio-telescope appropriately and perform astronomical observations.

Figure 3. Interface between PC(FS9) and ACU of antenna.

6. Future frequencies and VLBI

Once the Peruvian radio-telescope is tuned to receive 6.7 GHz signals, in the future we are planning to implement other frequencies receivers, such as 2 GHz and 8 GHz to perform Geodetic VLBI observations, and 12.2 GHz and 22 GHz to observe in VLBI mode, observations with other international VLBI arrays are in the scope, include Peruvian radio-telescope, as a VLBA station will improve observations for VLBA, especially for low elevation sources (Horiuchi *et al.*, 2005).

7. Location

The Telecommunications Antenna Station is located over a hill in a beautiful valley called Valle del Mantaro, at 3 370 m above the see level and have a comfortable dry weather. Station used to be called Estacion Terrena de Sicaya.

The coordinates are: Latitude: -12°02'15", Longitude: -75°17'39"

The location is peculiar and interesting to perform geodesy observations, precise measurements of the antenna location will be useful to measure plate tectonics on the region.

8. Conclusions

Re-use of tele-communication antennas as a radio-telescope is one of the way to develop and settle astronomy in countries where lack of observational instruments is a serious problem.

Implementing the first radio observatory in Peru is a big step and foster sciences at graduate and postgraduate levels of universities.

Collaboration of developed countries to develop instruments and educate people are vital. Equipment or financial assistance for developing countries are important to start this kind of project. In the mean time scholarships to train and educate people must be implemented.

International collaborations have to be improved, participation of people of Peruvian radio-telescope in scientific events will level up skills and knowledge in radio astronomy, related sciences and technologies.

Acknowledgements

We acknowledge to each of the staff at Kashima Space Communication Research Center, thanks to them now we have an antenna control system ready to be installed. We also acknowledge to people in Japan that contribute with donations and made possible to have an emergency found to support running costs of our radio-telescope.

References

Horiuchi S., Murphy D. W., Ishitsuka J.K. & Ishitsuka M. 2005, *ASOP* p572
Honma M., Kawaguchi N. & Sasao T. 2000, *SPIE Vol. 4015, Radio Telescope* p624-p631
Menten K. M. 2002, *IAU Symp. 206, From Protostars to Black Holes* p125
Minier V., Ellingsen S. P., Norris R. P. & Booth R. S. 2003, *A&A, Vol. 403* p205
Sasao T. 1996, *4th APT Workshop* p94-p104

José Ishitsuka

left to right: John Hearnshaw, Siramas Komonjinda and Ben Stappers

The Astronomical Clock, Old Town City Hall, Prague

Astronomy for the developing world
IAU Special Session no. 5, 2006
J.B. Hearnshaw and P. Martinez, eds.

© 2007 International Astronomical Union
doi:10.1017/S174392130700676X

Application of the Field System-FS9 and a PC to the antenna control unit interface in radio astronomy in Peru

Erick Vidal[1], José Ishitsuka[1,2] and Yasuhiro Koyama[3]

[1]Instituto Geofísico del Perú,
Calle Badajoz 169, Urb. Mayorazgo IV Etapa, Ate-Vitarte, Lima 3, Perú
email: evidal@axil.igp.gob.pe

[2]National Astronomical Observatory of Japan,
2-21-1 Osawa, Mitaka, Tokyo 181-8588, Japan
email: pepe@hotaka.mtk.nao.ac.jp

[3]National Institute of Information and Communications Technology
893-1 Hirai, Kashima, Ibaraki, Japan
email: koyama@nict.go.jp

Abstract. We are in the process to transform a 32m antenna in Peru, used for telecommunications, into a Radio Telescope to perform Radio Astronomy in Peru. The 32m antenna of Peru constructed by NEC was used for telecommunications with communications satellites at 6 GHz for transmission, and 4 GHz for reception.

In collaboration of National Institute of Information and Communications Technology (NICT) Japan, and National Observatory of Japan we developed an Antenna Control System for the 32m antenna in Peru. It is based on the Field System FS9, software released by NASA for VLBI station, and an interface to link PC within FS9 software (PC-FS9) and Antenna Control Unit (ACU) of the 32 meters antenna.

The PC-FS9 controls the antenna, commands are translated by interface into control signals compatibles with the ACU using: an I/O digital card with two 20bits ports to read azimuth and elevation angles, one 16bits port for reading status of ACU, one 24bits port to send pulses to start or stop operations of antenna, two channels are analogue outputs to drive the azimuth and elevation motors of the antenna, a LCD display to show the status of interface and error messages, and one serial port for communications with PC-FS9.

The first experiment of the control system was made with 11m parabolic antenna of Kashima Space Research Center-NICT, where we tested the right working of the routines implemented for de FS9 software, and simulations was made with looped data between output and input of the interface, both test were done successfully.

With this scientific instrument we will be able to contribute with researching of astrophysics. We expect to into a near future to work at 6.7GHz to study Methanol masers, and higher frequencies with some improvements of the surface of the dish.

Keywords. Masers, radio-astronomy, astronomy in Peru

---

1. Introduction

Based on a 32m antenna used for satellite communications will be converted into a radio telescope just implementing few functions to the tracking system. This proceeding explains briefly about development of an interface to make the antenna to track celestial bodies that will be useful for radio astronomy.

Figure 1. Laptop Compaq nx9005 running Field System FS9

2. Field system FS9

Field System FS9 released by NASA for VLBI stations is a free license software that runs on GNU Linux Debian. Field System FS9 was made to administrate VLBI equipments, using: GPBI, and SERIAL interfaces, is also flexible to use other input output ports as LAN data exchange, only implementing necessary codes.

Now a days Field System FS9 is being used in many radio observatories, that allow us join them for coordinated observations.

2.1. *Installation of the field system FS9.*

Installing FS9 system is not difficult and it would be easily made with the following steps:
-   (*a*) Implement a Desktop PC
-   (*b*) Install Linux Debian OS
-   (*c*) Configure environment for Field System FS9 with FSADAPT
-   (*d*) Install Field System FS9
-   (*e*) Configure hardware settings
-   (*f*) Configure procedure files
-   (*g*) Write program for local station.

See Figure 1.

2.2. *Linking PC - FS9 and 32-m antenna*

According to technical information of Nippon Electric Company, the antenna maker, it is not possible to exchange control signals to the ACU trough serial RS-232 or parallel connection with any other device, it is because the antenna was built for geostationary satellites. However mechanics of antenna allows to be used for continuous tracking movements, as we need for astronomical observations.

3. Interface

3.1. *Interface overview*

Based on the requirements of Field System FS9, and the mode of operations of 32m antenna of Peru, it was necessary to build an interface to translate instructions of Field

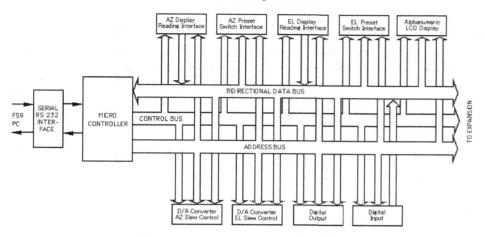

Figure 2. Block diagram of interface.

System FS9 control system to the ACU. The interface device must be able to translate serial text format commands into: digital pulses, analog signal and Binary Coded Decimal (BCD) data. Also read analog, digital and BCD information from the control panel of antenna and translate to the equivalent text format information compatible with Field System FS9.

3.2. *Design of the hardware*

Based on the capabilities of the antenna control system and operations mode of 32m antenna of Peru, the interface must be implemented with a microprocessor. The Interface includes the following ports to interact between PC-FS9 and ACU.

- Processor:
  - RISC micro-controller: Controller of interface.
- Input/Output Ports:
  - Serial RS-232 Interface: PC- FS9 communication port
  - 1 Digital input: State reading of ACU
  - 1 Digital Output: Command send to ACU
  - 2 Channel analog output: Slew control
  - 2 Channel parallel output: New angle data for antenna
  - 2 Channel parallel input: Read current angle of antenna
  - 1 Alphanumeric LCD: Display interface state
  - Expansion port: Future use.

See the Figure 2.

4. Interface software

4.1. *Implementing the software*

Interface device will control the ACU receiving information trough serial RS-232 port and changing the text instructions into analog and digital signals to slew azimuth and elevation motors, activate, deactivate functions of antenna, get status information from the panel of the ACU and encode the read status to text commands compatible with FS9 control system.

5. Conclusions

Lately many large parabolic antennas that was used for satellite communications are discontinued, and no longer used. Most of antennas were designed just for geostationary satellites, and they are not implemented with tracking software neither to remote control the antenna, even though their mechanical systems are capable to do it. Present document shows that it is possible to do, attaching the appropriate hardware and making a communications antenna working for radio astronomy.

Appendix A. Technical information of the 32-m antenna

- Type :Cassegrain
- Diameter :32m
- Max speed : 0.3deg/s
- Frequency :4GHz Tx, 6GHz Rx
- Antenna Mount :Wheel & Track
- Main Dish accuracy
  - Without wind :1.0mm rms
  - Winds 13 to 20m/s :1.26mm rms

Appendix B. Technical Information of the Interface

- Main Processor :PIC16F877
- I/O Ports:
  - Digital
    — two 20bit (Az, El read)
    — one 16bit (Status read)
    — one 24bit (control to ACU)
  - Analog
    — two 16bit D/A converter (drive Az/El)

Appendix C. Control computer

- Hardware
  - Compatible IBM Personal Computer
  - GPIB Card
  - Serial Interface
- Hardware
  - Operative System :Linux Debian
  - Control Software :FS9 Field System

Acknowledgements

For making possible developing radio astronomy in Perú, I would like to thank to: Consejo Nacional de Ciencia y Tecnología Concytec - Perú, for financing my air ticket to Prague, to Instituto Geofísico del Perú - IGP, for allowing me to participate at IAU-GA-2006, National Astronomical Observatory of Japan NAOJ, for financing partially my stay in Prague. People of Radio Astronomy Application Group of National Institute of Information and Communications Technology NICT - Kashima, Japan for their encourage and kindness during my stay in Japan for constructing Peruvian Antenna's interface.

Section 3:

Astronomy in Africa

Astronomy for the developing world
IAU Special Session no. 5, 2006
J.B. Hearnshaw and P. Martinez, eds.

© 2007 International Astronomical Union
doi:10.1017/S1743921307006783

Building capacity for astronomy research and education in Africa

Peter Martinez[1]

[1]South African Astronomical Observatory, P.O. Box 9, Observatory 7935, South Africa
email: peter@saao.ac.za

Abstract. About 1.5% of the world's professional astronomers are based Africa, yet in terms of research output, African astronomers produce less than 1% of the world's astronomical research. The advent of new large-scale facilities such as SALT and HESS provides African astronomers with tools to pursue their research on the continent. Such facilities also provide unprecedented training opportunities for the next generation of African astronomers. This paper discusses recent efforts to develop astronomy education and research capacity on the continent. Various capacity-building initiatives are discussed, as well as the lessons learnt from those initiatives.

Keywords. Astronomy, Africa, education, research

1. Introduction

In his survey of astronomy around the world, Heck (2000) presented geographical distributions relating to various aspects of astronomy world-wide. The striking feature in all Heck's maps is what he described as "the desperate emptiness of most of the African continent." Of the 52 countries in Africa, only nine (*viz:* Algeria, Egypt, Ethiopia, Kenya, Libya, Mauritius, Morocco, Nigeria and South Africa) were listed by Heck as having some astronomy-related organizations. No other region of the world has such a dearth of activity in astronomy. A closer look at the nature of the facilities in these countries reveals that only Egypt, Namibia, Mauritius and South Africa and possess operational large-scale astronomical research facilities.

Professional astronomy in Africa is dominated by Egypt in the north and South Africa in the south. South Africa invests more in astronomy annually than all other African countries combined. This is reflected in the scientific output, which is greater than that of all other African countries combined – by a wide margin. The historical development of astronomy in South Africa, up until 1994, has been reviewed by Feast (2002). Whitelock (2004) reviewed developments in post-apartheid South Africa, from 1994 to 2004.

In this paper we will take a closer view of the status of professional astronomy in the rest of Africa and we will review some of the lessons learnt from various capacity-building initiatives that have taken place in recent years. This study encompasses the 46 countries in continental Africa and the independent island states Cape Verde, Comoros, Madagascar, Mauritius, Sao Tome and Principe and Seychelles – a total of 52 countries.

The focus of this study is on the development of professional astronomy at universities and research establishments. We do not include astronomy activities conducted purely in the amateur sphere, important though they are. Our reason for maintaining this distinction is that the pursuit of astronomy as a scientific discipline in a country requires a much higher level of professional and political commitment than the popularization of astronomy. Indeed, virtually all of the institutions involved in astronomy teaching or research in Africa are also highly active in the popularization of astronomy, but that is not their *raison d'être*.

Table 1. Table 1: IAU membership statistics for Africa as of September 2006. Source: IAU website *www.iau.org*

National Members	Members			% Members National		% Members All IAU	
	Male	Female	Total	Male	Female	Male	Female
Egypt	51	6	57	89.47	10.53	0.52	0.06
Morocco[1]	7	0	7	100.00	0.00	0.07	0.00
Nigeria	9	1	10	90.00	10.00	0.09	0.01
South Africa	58	6	64	90.63	9.38	0.59	0.06
Individual Members							
Algeria	3	0	3	100.00	0.00	0.03	0.00
Ethiopia	1	0	1	100.00	0.00	0.01	0.00
Mauritius	1	0	1	100.00	0.00	0.01	0.00
Total Members							
All Africa	130	13	143	90.91	9.09		
All IAU	8451	1322	9773			86.47	13.53

Notes:
[1] Morocco has interim membership status.

2. Organized astronomy communities in Africa

A good way to depict the presence of organized astronomy in Africa is to consider membership of the International Astronomical Union (IAU). Table 1 depicts the IAU membership statistics for Africa as of September 2006. The total African membership of the IAU amounts to 143 persons in 7 countries, corresponding to 1.5% of the total IAU membership. The African membership is dominated by South Africa (45%) and Egypt (40%), followed by Nigeria (7%) and Morocco, Algeria, Ethiopia and Mauritius, all with 5% or fewer members. It is also instructive to compare IAU membership changes in recent years. IAU membership has increased by 13% from 127 to 143 during the past triennium. This reflects the growth in astronomical communities in South Africa and Nigeria. The membership numbers in the other countries have remained static during this period.

The dearth of IAU adhering countries in Africa is striking. The only adhering countries in Africa are Egypt, Morocco (interim status), Nigeria and South Africa. Fortunately the IAU's rules are flexible enough to permit individual PhD-qualified astronomers to join the IAU in their personal capacity, even if their countries are not yet ready to join as national members. Through this mechanism, individual astronomers in Algeria, Ethiopia and Mauritius are also members of the IAU. I believe it is very important to maintain this admissions policy as it reduces the isolation of scientists who return to their countries in Africa after obtaining their training in astronomy elsewhere.

One might ask, if it is possible for individual scientists to belong to the IAU, why should it matter that African countries have organized astronomical communities? The answer to this becomes apparent when one considers that the map of IAU member countries reflects the more prosperous African nations in a Gross Domestic Product (GDP) map of the continent. Economic prosperity in the information society is closely linked to the ability to harness science and technology as a "smart user." Virtually all countries are critically reliant on space applications these days. The basic space sciences and their supporting technologies underpin the ability of a country to utilize space applications programmes for development. Countries that belong to the IAU are countries that recognize the importance of astronomy as a national activity to their future well-being.

3. Tracers of research activity in space science

The IAU membership figures, though instructive, do not present the complete picture as far as the distribution of individual scientists in Africa because not all of them are members of the IAU, either through a national adhering organization or in their personal capacity. A better way to gauge the distribution of active individual space scientists is to examine the literature, and also the online usage of the astronomical literature.

Martinez (2006) discussed an analysis of the literature archived in the NASA Astrophysics Data System (ADS) database of abstracts for the period 1973–1996. Out of a total of 181 808 papers in the ADS data base, 1339 (0.74%) had a principal author or at least one co-author affiliated to an institute in an African country. The top five countries identified in this study were South Africa, with 77% of the African papers, followed by Egypt (14%), Nigeria (4%) and 19 other countries with fewer than 1% of the papers. It is not surprising that the top five countries in this list are also the top five countries by GDP in Africa. This sample is very incomplete, and serves only to indicate that researchers producing papers relating to some aspect of astronomy are more widely spread throughout Africa than a simple inspection of IAU membership statistics would suggest. These are the countries that IAU Commission 46 can reach out to in the future.

Recent enhancements to the query forms in the ADS make it possible to query the ADS on-line for affiliations by country. However, comparison of the results of such a search with the publication statistics for South Africa, which are well known to the author, indicate that this type of query returns an incomplete sample at best. Moreover, the results are contaminated with papers that pertain to Earth observation or to non-astronomical topics in physics. Thus, ADS publication counts are at best a noisy data set with which to study astronomical research activity in Africa. Their value lies in revealing the isolated scientists that are active in research as potential local partners for future capacity-building interventions.

A more reliable metric of current astronomical research activity is to examine the country of origin of accesses to the ADS data base. As of this writing, a project is in progress to record these accesses on a month-by-month basis. The early results are encouraging. In October 2003, the Working Group on Space Sciences in Africa published a special issue of *African Skies / Cieux Africains* on the ADS, including remote access by email (Eichhorn *et al.* 2003). Subsequent to the publication of that issue, a number of African countries started to access the database (Fig. 1). As internet connectivity in Africa improves, and as more African astronomers become aware of the ADS, we may expect that the ADS will see greater usage by African astronomers. This is an absolutely critical service for the development of professional astronomy in Africa, as most African institutions do not have astronomical libraries. The ADS is a major contributor towards reducing the isolation of African scientists.

Kurtz *et al.* (2005) analysed the utilization of the ADS data world-wide as a function of GDP and they found a simple relation that ADS use per capita is proportional to GDP per capita squared. Figure 3 of Kurtz *et al.*'s paper is instructive. They plot the number of ADS queries per million inhabitants versus GDP per capita. The only African countries to appear in their plot are Egypt, Morocco, and South Africa, which appear to conform to the relation

$$(ADS\ use) \propto (GDP\ per\ capita)^2.$$

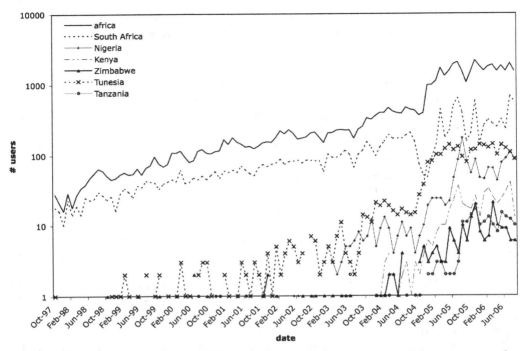

Figure 1. Usage of the NASA ADS by scientists in African countries. Figure courtesy of Guenther Eichhorn, ADS.

It is not surprising that these are also three of the four IAU Member countries in Africa.

4. New large-scale facilities

Although the prospects for astronomy in the continent as a whole are not encouraging, there are some notable improvements compared to the situation as it was ten years ago. The improvements are most noticeable in the southern African region, where political stability, government support for science, and excellent astronomical conditions have attracted a number of large-scale international astronomical facilities. Namibia hosts the international High Energy Stereoscopic System (HESS), an array of four imaging atmospheric Cherenkov telescopes for the investigation of cosmic gamma rays in the 100 GeV energy range (Hinton, 2004). HESS is the premier facility of its kind in the world. The Southern African Large Telescope (SALT) is a 10-m optical telescope inaugurated in November 2005 by South Africa and 10 international partners. SALT is currently the largest single optical telescope in the southern hemisphere. South Africa is also currently developing the Karoo Array Telescope (KAT), a 1% technology demonstrator of the Square Kilometre Array (SKA), which South Africa is bidding to host. In addition to these large-scale facilities a number of smaller robotic telescope facilities are coming on-line, which may be accessed remotely over the internet by scientists in Africa (see for example Martinez *et al.*, 2002). For the first time, these facilities provide an opportunity for African scientists to perform cutting-edge research on the continent and in the context of their own national environments. The challenge is to create the right conditions to attract and retain bright young scientists in their own national environments after they complete their studies.

5. Capacity building initiatives – lessons learnt

Over the years, a number of institutions and international organizations have conducted capacity building activities to develop astronomy in Africa. The IAU supports capacity building in Africa principally through the activities of its Commission 46, Astronomy Education and Development. Activities within this Commission are organized into nine Programme Groups (Isobe 2003). The capacity-building activities of COSPAR are conducted through the COSPAR Panel on Capacity Building. This Panel has conducted a series of workshops in developing countries aimed at increasing the utilization of space archive data (Arnaud & Willmore, 2006 - this volume). The first such workshop in Africa was held in Durban, South Africa, in July 2004 on the topic of X-ray astronomy. For many years, the United Nations Office for Outer Space Affairs has been arranging the UN/ESA Workshops on Basic Space Science (Haubold & Wamsteker 2004) which have been attended by a considerable number of African scientists.

So, what have all these activities amounted to? Why is there still a dearth of astronomy research and training activities in Africa? There are no simple answers to these questions. However, it is possible to distill out of my experiences of capacity building in a variety of African countries over the past decade a series of lessons learnt and some basic conditions which must be fulfilled before astronomy can thrive in a country.

Seek out fertile ground in which to plant a seed

By "fertile ground" I mean several things. Firstly, an adequate infrastructure is a *sine qua non* of sustainable capacity building for research in astronomy. This includes things such as computers, internet access, library facilities, and so on.† Secondly, this must be matched by an institutional environment with a supportive hierarchy. Without committed institutional support, there is no hope of a capacity-building initiative becoming sustainable in the long run as it will collapse the moment the external support ends or the key person(s) withdraw for any reason. Thirdly, there should be cohesion and a common unity of purpose within the developing scientific community for capacity building initiatives to take root. Fourthly, there should be a supportive (or at least non-obstructive) national environment. For example, in a supportive environment, there are means to remove or minimize administrative and fiscal impediments such as import duties on donations of scientific equipment or books.

Capacity building is about people, not about equipment

Don't begin a capacity building programme with the installation of an observatory before the following are in place: (i) a clear scientific plan for how to utilize the telescope, (ii) a clear technical plan to support the telescope, (iii) adequately trained people in place. A capacity-building initiative that is based around equipment, rather than around people is doomed to fail. When contemplating the installation of new facilities as part of a capacity-building programme, it is important to ensure that the necessary human capital is developed through training in the operation, purposeful use and maintenance of equipment, before such equipment is put in place. Given the limited resources available for capacity building, an approach that works well is to train the trainers. We have used this approach to introduce astronomy into the undergraduate physics curricula at various African universities, and we are starting to see more African students enrolling for postgraduate degrees in astronomy on the continent and elsewhere.

---

† Note that I did not include a telescope in this list. This is not because I believe that telescopes are unimportant, but here we are talking about a *basic* infrastructure for research. Without these elements in place, a telescope will not be scientifically productive.

Invest in young people

Capacity-building opportunities should target young people who are not burdened with administrative or other duties and have more time to drive developments from the bottom up. The same lesson applies also to the scientists doing the capacity building. Many young professionals are keen to share their expertise with colleagues from developing countries and they generally have the mobility and time to do so. The challenge to the older scientists in developed and developing nations is how to engage most effectively with the capacity-building process in such a manner as to allow their younger colleagues to achieve the desired sustainable results.

Capacity building is a process, not an event

Capacity building activities must be part of a long-term programme. Activities such as capacity building workshops will not lead to sustained research activity if they are not part of a long-term programme. Hence the importance of strategic partnerships among the scientific unions, the scientific community in the developing country and the development aid sector. It normally takes several years of engagement to build up some level of sustainable activity. This requires firm commitment from all partners in the face of occasional setbacks and failures, even in the best supported scenarios.

Succession planning is important

The developing world is replete with examples of small groups that flourished under the guidance of an inspired individual for a period, and then faded out of existence when that person left, retired or passed away. When a new capacity-building initiative starts, it is not unusual for it to be championed by the senior academics or scientists, who have the managerial experience and academic credentials to obtain support from their university authorities or government. However, these same scientists have a career investment in an area outside of astronomy and are never likely to have the time to become independently competitive astronomers themselves. But they will inspire their students to study astronomy at postgraduate level, and it is those students who will one day return to establish a solid, independent astronomy research group. It is thus very important to develop a succession plan as part of the long-term capacity building strategy. This gives the institution some guarantee of continuity, but at the same time it also entails institutional commitment to the young scientists, who know that there will be a position to return home to. This is a fundamental aspect of the commitment to astronomy that we discussed in Section 2.

Match new facilities to education and research needs

Often capacity building in astronomy focuses around the acquisition of a small telescope. Small (< 0.5m) telescopes have an important role to play in undergraduate teaching and student training. Regular access to such telescopes by students and the public can do much to promote astronomy in a developing country. However, above about 0.5m aperture I believe one needs to consider whether investing in a telescope is the best way to promote internationally relevant astronomical research in a particular environment. For many African countries, I would argue that a good internet connection and access to large-scale facilities elsewhere is far more likely to result in productive research than an ill-equipped telescope at a poor site.

Astronomers should not pretend to be engineers

This may seem an obvious statement, but I have seen many instances where astronomers try to perform functions that engineers are trained for. The result is that they spend all their time fussing over equipment that never quite performs to specification, and they have no time to do any science. Moreover, the students who struggle with such equipment often lose their motivation to study astronomy, or, worse, wind up learning very little astronomy. A small observatory with very limited resources should consider whether its priority is to do astronomy, or to do engineering. If astronomy is the priority, it might be better to acquire a professionally built research-grade telescope and/or instruments, rather than trying to develop this in-house. If this is not a possibility, then it would be better to obtain service observations from telescopes at good sites or to use data from space missions.

Use information technology as much as possible

Information technology is a powerful enabler of research. Access to email and online literature, software and data reduces isolation of scientists and makes them much more productive. For the developing world Open Source is a particularly enabling technology in the sense that, in addition to the cost savings associated with keeping software current, the open nature of the software leads to greater innovation and allows users to adapt it to local demands and steer their own IT infrastructure. Moreover, because the astronomy community is a heavy user of Open Source software, the skills acquired by participants in capacity-building activities can be of benefit not only to themselves, but also to their home institutions.

In the astrophysics domain, most of the literature is available on-line through the NASA Astrophysics Data System (Eichhorn, 2004), and large quantities of astrophysical data are available on-line or on request from a variety of data centres. This provides excellent opportunities to promote research in developing nations without needing to develop costly infrastructure in conditions that are sub-optimal for ground-based astronomy. Moreover, the same IT infrastructure that is required to do cutting-edge astrophysical research can also be utilized for other purposes, so the benefits of the investment are enjoyed by a much wider community of users than just the astronomers.

Capacity building activities should also take into account the general level of preparation of participants. For example, "hands-on" capacity building activities assume a certain degree of computer literacy on the part of the participants. However, one cannot assume familiarity with the computing environments used in astrophysics. Often a carefully planned training activity is compromised when time is lost bringing people up to speed with very basic computer skills. Better screening of prospective participants and/or precursor computer training would enhance the efficacy of such activities in future. Attention also needs to be given the the participants' home computing environments. It does no good to train people to use software tools that cannot run on computers at their home institutions. In this case, perhaps the provision of computers needs to be part of the capacity-building plan. This is where development aid agencies come into the picture.

Promote and nurture regional initiatives and facilities

Regional networks represent the scientific community's determination to organize itself and its activities. Close cooperation between the scientific unions and these regional structures can be mutually beneficial and support sustainable capacity building. Regional networks, such as the Working Group on Space Sciences in Africa, can support capacity-building initiatives of the scientific unions by organising or promoting awareness of key events, such as workshops, and by supporting the follow-up phase afterwards. Regional networks can also assist scientists with accessing facilities in the region and with accessing training and career opportunities on the continent. Africa has two regional centres affiliated to the United Nations for training in space science and technology, one in Nigeria for anglophone Africa and one in Morocco for francophone Africa. These centres form an important part of the constellation of facilities available on the continent for human resource development. Though neither centre has a component of astronomical research at the moment, such a component could be developed in partnership with astronomy institutes elsewhere in the region.

Focus on solving real research problems

Capacity building initiatives (visits, workshops) often focus on equipping the participants with the skills and tools to conduct research in a given field, yet few participants go on to initiate research projects on returning to their home institutions. I believe the reason for this is that these scientists work in isolation, with no idea of what are the relevant problems to tackle. One way to address this is to structure the capacity building activity around some area of research that the group of participants can continue to work on as a network after they return to their home institutions. Establishing networked collaborations is a good way to build up a critical mass of scientists who can support each other and produce science that is relevant and of international quality. This will require ongoing support and encouragement from colleagues in developed countries, but will yield a piece of publishable research and take the group of participants through the whole research cycle.

Form partnerships for capacity building

Working in partnerships allows an organization to leverage its resources with those of other organizations and to coordinate collective efforts for maximum impact. Working in partnership also introduces complications. The different organizations will have their own objectives, programmes and financial cycles, but they are more likely to support initiatives that are supported by other partners as well. The types of interactions that are most likely to lead to sustainable capacity building are those that also attract the support of the development sector and/or government. A good example of this is the COSPAR/IAU Regional Workshop on Data Processing from the Chandra and XMM-Newton Space Missions, held in Durban, South Africa in July 2004. This workshop arose from cooperation between the capacity building programmes of the IAU and COSPAR discussed earlier in this paper, with additional funding from the National Research Foundation of South Africa, the UN Office for Outer Space Affairs, the Abdus Salam Centre for Theoretical Physics and the European Space Agency.

6. Closing remarks

With the advent of new large-scale facilities for ground-based astronomy in Africa, such as the Southern African Large Telescope (SALT) and the High Energy Stereoscopic System (HESS), and a regional climate of enhanced scientific cooperation, African astronomers are now able to access to some of the world's premier astronomical facilities without having to leave the continent. Increasing internet penetration and access to the literature through on-line services such as the NASA ADS is reducing the scientific isolation of African astronomers. Together, these factors should lead to a steady growth of astronomy on the continent. Indeed, IAU membership over the past triennium in Africa has increased by 13%. This growth is encouraging, but it will have to be sustained for many years to build a significant and vibrant astronomical community on the continent.

The development of new facilities in Africa is largely the result of investment by the international community in Africa. In the southern hemisphere the continent has some of the world's best astronomical sites, free of light pollution and radio interference. African countries are beginning to recognize the geographical advantages of the region and are seeking to develop the region as a southern hemisphere hub of astronomy. South Africa is in the process of enacting very progressive legislation to protect present and future astronomical sites from activities that might degrade the astronomical conditions across the electromagnetic spectrum. The region as a whole is backing the South African-led bid to host the SKA on the continent. These are encouraging indications that astronomy is flourishing in at least some parts of the continent.

Developing astronomical research and education capacity in Africa is a long-term activity. The IAU, COSPAR and the UN have played a very significant role through their activities on the continent over the past few decades. The actions of these role-players have all had some long-term effect on individuals, and the cumulative impact of these interventions will take years to manifest themselves. The participants in these activities have often returned home to inspire their students, who in turn aspire to astronomical careers and pursue their studies at major astronomical centres. The capacity-building efforts of organizations such as the UN, IAU, COSPAR and the Working Group on Space Sciences in Africa are thus starting to yield fruit in terms of producing a new generation of African astronomers. Working in partnerships greatly enhances the impact of capacity-building activities. In order to be sustainable, capacity-building activities should form part of a comprehensive long-term programme. Such a programme should address pipeline issues to ensure that interventions by the different role-players are mutually supportive and appropriately phased. In order to facilitate coordination of activities by the different organizations, consideration should be given to the establishment of a capacity-building forum. Such a forum could facilitate improved dialogue between the scientific unions and scientific institutes, the development sector (e.g. UNESCO) and the developing countries to link the players with the technical means (the scientific community) to the communities with the needs (the developing countries) through provision of support for development of infrastructure and operation of projects by the development sector. An organization like the Working Group on Space Sciences in Africa would be well placed to initiate such a development. The challenges are immense, but so, too, are the rewards.

Acknowledgements

I acknowledge with gratitude the generous support received from the following organizations for the various capacity-building initiatives described in this paper: Observatoire Midi Pyrénées, South African National Research Foundation, South African Astronomical Observatory, IAU, COSPAR, United Nations Office for Outer Space Affairs and UNESCO. I also acknowledge the contributions by many colleagues who have participated in the capacity building activities funded by these organizations, as well as the insights I have gained from working with them. This research has made use of NASA's Astrophysics Data System.

References

Arnaud, K., Willmore, A.P., in: J. Hearnshaw & P. Martinez (eds.), *Astronomy for the Developing World*, Cambridge University Press, ... This volume, pXX.
Eichhorn, G., Accomazzi, A., Grant, C.S., Kurtz, M.J., Murray, S.S. 2003, *African Skies* 8, 7
Eichhorn, G. 2004, *Astron. and Geophys.* 45:3, 7

Feast, M.W. 2002, in: A. Heck (ed.), *Organizations and strategies in astronomy III,* Astrophysics
    and Space Science Library, Vol. 280. Dordrecht: Kluwer Academic Publishers, ISBN 1-4020-
    0812-0, p. 153

Haubold, H.J., Wamsteker, W. 2004, in: W. Wamsteker, R. Albrecht & H. Haubold, (eds.),
    Developing Basic Space Science World-Wide: A decade of UN/ESA Workshops, Kluwer
    Academic Publishers, Dordrecht, p. 3

Heck, A. 2000, *Ap&SS* 274, 733

Hinton, J.A. 2004, *New Astron. Revs* 48, 331. See also http://www.mpi-hd.mpg.de/hfm/HESS/

Isobe, S. 2003, in: A. Heck (ed.), *Organizations and strategies in astronomy IV*, Kluwer Academic
    Publishers, ISBN 1-4020-1526-7, p. 189

Kurtz, M.J., Eichhorn, G., Accomazzi, A., Grant, C., Demleitner, M., Murray, S.S. 2005,
    J. American Soc. for Information Science and Technology 56:1, 36. Also available at
    http://cfa-www.harvard.edu/ kurtz/jasist1-abstract.html

Martinez, P., Kilkenny, D.M., Cox, G., *et al.* 2002, *MNRAS* 61, 102,

Martinez, P. 2006, in: A. Heck (ed.), *Organizations and Strategies in Astronomy VI,* Kluwer
    Academic Publishers, Dordrecht, p.39

Whitelock, P. 2004, in: A. Heck (ed.), *Organizations and Strategies in Astronomy V,* Kluwer
    Academic Publishers, Dordrecht, p.39

Peter Martinez

Astronomy for the developing world
IAU Special Session no. 5, 2006
J.B. Hearnshaw and P. Martinez, eds.

© 2007 International Astronomical Union
doi:10.1017/S1743921307006795

Astronomy in the cultural heritage of African societies

Paul Baki

Department of Physics, University of Nairobi,P.O Box 30197, 00100, Nairobi, Kenya
email: pbaki@uonbi.ac.ke

Abstract. The African perspectives of astronomy are explored from the point of view of using indigenous knowledge of the night sky for purposes of addressing local challenges such as food insecurity and periodic natural weather phenomena such as droughts and floods.The local ethnic groups use stellar positions, and plant and animal behaviour changes for purposes of forecasting the weather and climate for the coming seasons.These traditional indicators give rise to an inter-disciplinary discourse that could benefit the community in environmental protection measures and boost the tourism industry in some countries in Africa.

Keywords. Traditional indicators, eco-system, eco-tourism, cultural value of astronomy

1. Introduction

Most African societies have developed indigenous astronomical knowledge largely for understanding and predicting seasonal weather changes. These communities depend mostly on rain-fed agriculture for subsistence farming, so they use their knowledge of the day and night sky to forecast rainfall and to predict periodic phenomena such as floods and droughts. In this paper we discuss a few traditional tools that are used by some ethnic communities in East Africa to interpret astronomical phenomena for solving their local problems. These traditional methods rely on the interaction of plants and animals with the terrestrial environment. Their scientific value needs to established and recognized.

2. Traditional biological and astronomical Indicators

African communities combine their knowledge of plant and animal behavioural changes together with their night sky knowledge to predict the weather and climate for the coming season. These communities recognize that some plants and animals are more sensitive to changes in the atmospheric conditions than others. Traditional forecasting complements modern meteorological forecasting and is still the major source of weather and climate information for farm management in the rural areas. In this discussion we focus on the traditional forecasting methods used by the Luo community who live around Lake Victoria in Kenya and Tanzania. In this part of Kenya and Tanzania, there are two distinct wet seasons. Short rains occur from October to December and long rains from March to May.

Plant indicators

Certain types of plants are known to shed their leaves to signal the onset of dry conditions, or they flower before a wet season begins. The shedding of leaves is an indication of water stress conditions associated with dry conditions. The trees shed their leaves to reduce evapo-transpiration and grow leaves when the rains approach. These behavioural changes have been used to predict the weather and climate for the coming season. Among the plants with these observed properties are:

(i) Those plants that shed leaves to indicate an impending dry season are: *Terminii browni, Ficus sur & Kigelia africana* – trees that shed leaves twice a year to mark distinct dry conditions around Lake Victoria region. The plants grow leaves when a wet season is approaching.

(ii) Those that flower to indicate an impending change of season are:

(*a*) *Zephranthus* - a field flower that appears a week or two before the onset of rains. The flower appears white during rainy season and pinkish during dry periods.

(*b*) **Blue Lotus (or Water Lily)** - this plant grows in water but will never blossom during the dry season. Its flowering is normally an indication that a wet season is approaching and that the rains will be adequate. If the coming rains will be poorly distributed this plant does not flower at all.

Animal indicators

The behaviour of certain animals is believed to indicate changes in the weather:

(i) The bird **Robin Chat** disappears for several months and only reappears when a rainy season begins. The swallows **Hirundo Abyssinia** and **Hirundo Smithic** exhibit circular movements in the sky when the rain is forming. Certain seasonal cries of birds are also believed to communicate changes in the weather.

(ii) The absence of **frogs** and **toads** indicates a dry season. When frogs stop croaking during the rainy season, even when it is still raining, it is an indication of the onset of a dry spell.

(iii) Movements of **ants** indicate that a wet or rainy season is approaching.

(iv The appearance of **snakes,** and other **reptiles** and wild animals around houses is an indication of the prevalence and continuity of a dry spell.

Astronomical indicators

The movement of stars has also been related by the Luos to the weather and change of seasons. The constellation, **Orion**, is classified by the Luos as the "male constellation" and the **Pleiades** as the "female constellation". Their appearance in the sky is linked with an impending change of season.

(i) The appearance of the female constellation indicates the cultivation season, while the appearance of the male constellation signals a decline in rains showing the start of dry season or harvesting.

(ii) It has been noted that the appearance and positioning of the **Milky Way** (called *Rip-* in Luo), especially in April, is normally an indication of the impending dry season. These traditional indicators are still the most widely used methods for farm management and food production. An understanding of link between these traditional indicators of weather/climatic changes and astronomy could change peoples' view of astronomy to see it as a practical discipline that can help put food on the table rather than an esoteric science that it is perceived be.

3. Cultural and economic value of astronomy to society

In the African perspective it seems that the best way to spread knowledge in astronomy is to begin by appreciating its cultural value. A possible path to follow is to:

(a) establish the scientific value of these traditional indicators in an interdisciplinary project involving both astronomers and biologists, and then to incorporate them into the standard astronomy curriculum;

(b) emphasize the traditional role that astronomical knowledge has played and continues to play in agrarian societies in Africa.

These kinds of initiatives will:

(i) Encourage the protection of the specific flora and fauna used as traditional indicators and this will be a good strategy for sustainable environmental protection. By linking indigenous astronomical knowledge to indigenous biological knowledge, the importance of astronomy will be recognized and astronomy will be seen as an intrinsic part of African culture, rather than as an esoteric 'foreign' pursuit.

(ii) Promote greater awareness of African perspectives of the cosmos, which will be of interest not only to the local inhabitants, but also to international visitors. Moreover, the generally lower levels of light pollution in African countries creates opportunities for people to re-aquaint themselves with the night sky in a way that is not possible for city-dwellers in the northern hemisphere.

(iii) Translate to a boost in revenue collection, especially for some African countries, like Kenya, that rely substantially on eco-tourism for their foreign exchange earnings. Thus most African governments might find the need to support programmes in astronomy if it is to bring that much needed capital for development.

4. Conclusion

Indigenous African astronomical beliefs and uses of astronomical knowledge need to explored and recognized. Most, if not all, African ethnic groups have their own usage of sky knowledge and it would interesting to find out scientific basis of the traditional indicators that the these groups (such as the Luos of Kenya and Tanzania mentioned in this paper) use to predict the weather, climate and periodic natural phenomena such as droughts and floods.

For purposes of spreading astronomy education in Africa, it would be necessary tap into these traditional values of astronomy and incorporate them into the standard astronomy curriculum, so that the role of astronomy in tackling local challenges continues to be recognised. This would in turn lead to the conservation of the environment and a boost in the tourism sector. Once the value of indigenous astronomical knowledge is recognised, and it is seen as an intrinsic part of African culture, it might be possible to use this to secure funding for the development of astronomy from the various African governments.

Acknowledgements

I would like to thank the organizers of this Special Session, John Hearnshaw and Peter Martinez, for inviting me to give a presentation and the IAU in general for sponsoring me to attend the 26th General Assembly of the IAU.

References

Cerdonio, M. & Noble, R.W. 1986, *Introductory Biophysics,* World Scientific Publishing Co, 40

Paul Baki

l to r in centre: Julieta Fierro, Barrie Jones and Patricia Rosenzweig

Astronomy for the developing world
IAU Special Session no. 5, 2006
J.B. Hearnshaw and P. Martinez, eds.

© 2007 International Astronomical Union
doi:10.1017/S1743921307006801

Astronomy education in Morocco–a new project for implementing astronomy in high schools

Hassane Darhmaoui and K. Loudiyi

School of Science & Engineering, Al Akhawayn University in Ifrane, Morocco
email: H.Darhmaoui@aui.ma, K.Loudiyi@aui.ma

Abstract. Astronomy education in Morocco, as in many developing countries, is not well developed and lacks the very basics in terms of resources, facilities and research. In 2004, the International Astronomical Union (IAU) signed an agreement of collaboration with Al Akhawayn University in Ifrane (AUI) to support the continued, long-term development of astronomy and astrophysics in Morocco. This is within the IAU programme Teaching for Astronomy Development (TAD). The initial focus of the programme concentrated exclusively on the University's Bachelor of Science degree programme. Within this programme, and during two years, we were successful in providing adequate astronomy training to our physics faculty and few of our engineering students. We also offered our students and community general astronomy background through courses, invited talks and extra-curricular activities. The project is now evolving towards a wider scope and seeks to promote astronomy education at the high school level. It is based on modules from the Hands on Universe (HOU) interactive astronomy programme. Moroccan students will engage in doing observational astronomy using their personal computers. They will have access to a worldwide network of telescopes and will interact with their peers abroad. Through implementing astronomy education at this lower age, we foresee an increasing interest among our youth not only in astronomy but also in physics, mathematics, and technology. The limited astronomy resources, the lack of teacher experience in the field and the language barrier are amongst the difficulties that we will be facing in achieving the objectives of this new programme.

Keywords. Astronomy education, interactive astronomy

1. Introduction

Morocco is a developing country where a little less than half of its population is illiterate and a little over 50% are more than 15 years old. The unemployment rates are particularly high for young adults between the ages of 15 and 34 years. The recent national transition to a more open economy incites considerable improvement of the Moroccan educational system. There is an increasing need for a skilled work force to support the country's economic growth. In order to achieve that, a stronger scientific education for the youth is essential. During the past decade, it was a national challenge to make education accessible to all, but the lack of resources made it very difficult for the Ministry of National Education (MNE) to equip most middle and high schools with science laboratories. The MNE was also unable to supply science teachers with all the necessary demonstration kits and pedagogical tools. This reflected negatively on the overall number of engineers, scientists, medical doctors and skilled technicians graduating each year. These numbers are below the international average. It is found that the science level and background of a large proportion of Moroccan youth either in middle or high schools is not up to the expectations set by the MNE. In 2003, the US National Center for Education Statistics published a worldwide comparative study which ranks countries in terms

of education in the field of sciences and mathematics (Gonzales *et al.* 2004). This showed
that Moroccan eighth-graders perform far below the international average. Moroccans
ranked 41st among 46 other participating countries. A closer look at this international
study shows that Moroccan students are mainly lacking observational and basic data
analysis skills. Astronomy education, so far ignored in the Moroccan science curricula, is
proposed here as a key learning activity, that would enhance the scientific skills of the
students. It could also play a major role in increasing youth interest in physics, mathematics and technology, hence producing the desired skilled work force needed by the
emerging market. The project will also contribute to the popularization and development
of astronomy in Morocco.

2. Astronomy education in Morocco

Astronomy education in Morocco is currently limited to masters and doctoral programmes. There are four universities that offer graduate programmes in the space sciences
(Casablanca, Marrakech, Rabat and Oujda). Research interests focus mainly on helio-
seismology, astroclimatology, astroparticles, cosmic radiation, near Earth objects and
search for supernovae, and site testing for the ELT. There are two small observatories
in the country. The Oukaimeden Observatory, near Marrakech, was founded in 1988 in
the High Atlas Mountains and has 60-cm and 35-cm telescopes. The Rabat Observatory
was founded in 1999 and has a 51-cm telescope. In the science curricula of middle and
high schools, astronomy is almost absent. It appears only in the middle school physical
sciences course (8th grade), where students have their first acquaintance with astronomy
through a single short chapter titled "Elementary notions about astronomy" (Radi *et al.*
2004). In this chapter, students get a few glimpses of the scope of astronomy and its
fundamentals, such as the difference between a star and a planet, the characteristics of
the different planets in the solar system (diameter, distance from the sun, moons, period
and temperature), and one paragraph about famous Muslim astronomers.

3. A new project for implementing astronomy in high schools

We propose a new project to introduce interactive astronomy in the Moroccan high
school curriculum in order to improve students' observational and data analysis skills,
and to increase their interest in science and technology. We believe that introducing
astronomy at this level will boost the development of the space sciences in the country.
The project centres on the use of internet astronomy. Students will have access to an
automated telescope from their personal computers and will be able to investigate the
universe without being at any observatory. They will be able to request observations,
download images and analyze them. They will engage in doing science by using their
own data, which is quite an exciting learning activity for most students. This educational
programme will be based on *Hands-On Universe* (HOU), a programme developed at
the University of California (Ferlet & Pennypacker, 2006). HOU enables students to
investigate the universe while applying tools and concepts from science, mathematics,
and technology.

The introduction of this programme into the Moroccan high school curriculum will
not be easy. Astronomy is viewed by most science teachers as a difficult subject that
needs advanced technology and resources. A lot of effort must be put into changing this
negative perception to help teachers appreciate the richness of this subject in educational
examples, activities and tools that would enhance Moroccan pupils' observational and
data analysis skills. Currently, the AUI is proposing to be a site for the University of

California Lawrence Hall of Science Real Astronomy Experience telescope and also for a one-meter Las Cumbres Observatory Telescope. We expect that once either of these telescopes is installed, growing interest in introducing astronomy in the Moroccan science curriculum for middle and high schools will emerge. This will coincide with the Ministry of National Education project GENIE that aims at introducing computer laboratories and internet access in 3000 middle and high schools by the year 2009. Internet access to the automated telescopes at AUI, or any HOU telescope, would then not be a problem for these schools. The newly established Soft Center for Citizen Empowerment at AUI will contribute to the development of the necessary Arabic IT support material in astronomy education and will train teachers on the use of the HOU telescopes.

4. Conclusion

Interactive astronomy is an inexpensive science activity that suits Moroccan youth. High school students will engage in doing observational astronomy and developing scientific reasoning and methodology from their personal computers. They will be part of a worldwide network of students using the same technology and working on similar scientific activities. These common activities will engage them discussions with peers in other countries, thus building international ties while sharing the excitement of scientific pursuits. They will also develop their communication skills in foreign languages. We foresee that the project will substantially contribute to the popularization and development of the space sciences in Morocco.

Acknowledgements

We gratefully acknowledge financial support from IAU for H. Darhmaoui to attend the 26th General Assembly in Prague. We also thank AUI for backing and supporting this project. The authors would also like to thank Khalil Chamcham, James White, Michele Gerbaldi and Carl Pennypacker for their continuous assistance in promoting astronomy in Morocco through various activities at AUI.

References

Gonzales, P., Guzmán, J.C., Partelow, L., Pahlke, E., Jocelyn, L., Kastberg, D., & Williams, T. 2004, *Highlights From the Trends in International Mathematics and Science Study (TIMSS)2003(NCES 2005005)*, U.S. Department of Education, National Center for Education Statistics. Washington, DC: U.S. Government Printing Office. Also available at http://nces.ed.gov/timss/results03.asp

Ferlet, R., & Pennypacker, C. R. 2006, In: Heck, A. (ed), *Organizations and Strategies in Astronomy,* Volume 6, Strasbourg Astronomical Observatory, France. Astrophysics and Space Science Library Volume 335. ISBN 1-4020-4055-5. Published by Springer, 2006., p.275-286

Radi, M., Marzouki, M., Essaghir, A., Abdelkarim, B., Jizar, A., 2004, *Fi Rihab Al Ulum Al Fiziyaiya,* Publisher: Es-ssalam Al Jadid - Addar Al Aalamiya lilkitab, Morocco.

Peter Martinez

Visit to Ondrejov Observatory near Prague during the IAU General Assembly

Astronomy for the developing world
IAU Special Session no. 5, 2006
J.B. Hearnshaw and P. Martinez, eds.

© 2007 International Astronomical Union
doi:10.1017/S1743921307006813

Astronomical extinction over the ELT Moroccan sites from aerosol satellite data

El Arbi Siher[1,2]†, Zouhair Benkhaldoun[2] and Aziza Bounhir[2,3]

[1]Faculté des Sciences et Techniques, Département de Physique, BP: 523, Béni Mellal, Maroc.
email: siher@ucam.ac.ma

[2]Faculté des Sciences Semlalia, Département de Physique, LPHEA, Marrakech, Maroc.
email: zouhair@ucam.ac.ma

[3]Faculté des Sciences et Techniques, Département de Physique Appliquée, Marrakech, Maroc.
email: bounhiraz@yahoo.fr

Abstract. Two Moroccan sites have been selected to be characterized for the ELT telescopes. Those sites are in the Atlas, between Oukaïmeden (where the national observatory is situated) and the Canary Islands. For this preliminary study, we have used the Nimbus-7 TOMS aerosol index to derive the astronomical extinction. This work builds on the findings of an earlier study which established the link between these two parameters over the Canary Islands.

Keywords. Site testing, atmospheric effects, telescopes

1. Introduction

In 1988, the astrophysics group of the University of Marrakech was funded for the construction of a national astronomical observatory. The first site chosen for this project was Oukaïmeden, located close to Marrakech in the High Atlas mountains at an altitude of 2700 m. Several studies were made for the qualification of this site. Those studies allowed our group to develop experience in the field of site selection and site testing.

The geographical situation of the Atlas mountains makes them ideal candidate sites for the future Extremely Large Telescope (ELT). We therefore mobilized the members of our group to undertake qualitative studies of the selected sites. Several studies have started, namely, the measurement of the astronomical seeing, the observation of clouds and the influence of dust, coming from the south, on the state of the atmosphere. Our research focusses on two issues: (i) the characterization of the site from the point of view of the passage of the aerosols, and (ii) the determination the physical nature of these aerosols, an issue still under discussion (Siher, *et al.* (2002); Varela, *et al.* (2004)).

In an earlier paper (**?**), we showed how tropospheric aerosols can influence the astronomical extinction. We used the Nimbus-7 TOMS aerosol data and the observed astronomical extinction at the Observatorio del Roque de los Muchachos (ORM) on La Palma in the Canary Islands to establish this relationship.

In this work, we use the correlation found by Siher *et al.* (2002) to extract the astronomical extinction from the aerosol index over the period 1979–1992. Fortunately, all ELT Moroccan candidates sites are in the same TOMS/Nimbus7 pixel area. This method, not yet proved for sites other than ORM, gives a preliminary indication of the quality of these sites.

† Present address: Faculty of Sciences and Techniques, Department of Physics, BP: 523, Beni Mellal, Morocco.

2. Presentation of the ELT candidate sites

A long list of a potential candidate sites for the ELT telescope was shortened in the framework of the ELT design study and the site working group proposed to focus on the study of four sites:

- Moroccan Anti-Atlas mountains: Adrar-n-Aklim (2531m, 30.1°N, 8.2°W) and Jebel Lekst (2359m, 29.7°N, 9°W);
- ORM, La Palma, Canary Islands;
- north of Paranal in Chile;
- The north western part of Argentina.

The Moroccan anti-Atlas mountains are close to the Atlantic Ocean and are influenced by the wind coming from the west.

3. Preliminary determination of astronomical extinction from aerosol index over Moroccan ELT sites

The correlation between aerosol index (AI) and astronomical extinction (AE) as found by Siher *et al.* (2002) is given by:

$$AE = AI * 0.16 + 0.14 \qquad (3.1)$$

This equation obtained for the ORM observatory is still under verification for other sites.

Figure 1 shows the evolution of the astronomical extinction for the proposed ELT sites, derived from the aerosol index using equation 3.1 for the period 1979–1992. The Nimbus-7 TOMS aerosol indices (AI) are greater than about 0.7 (absorbing aerosol), which corresponds to a threshold of about 0.2 mag/airmass for the derived AE values.

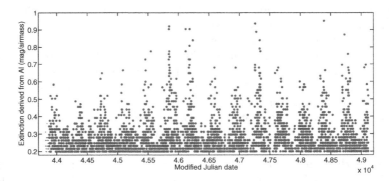

Figure 1. Evolution of the derived extinction from the aerosol index over the ELT sites

Figure 3 is a histogram showing the distribution of extinction values derived from the aerosol index. In this figure, we note the presence of a one-year period showing the seasonal variation in extinction. The high values occur in April-May-June and the minima occur in the winter. By contrast, in the High-Atlas mountains (Oukaïmeden), the high values occur in the summer (June-July and August) (Siher, *et al.* (2002). Figure 2 clearly shows this seasonal effect by presenting the monthly variation, modulo one year, of the astronomical extinction derived from the aerosol index over the ELT sites. This figure shows that the predominant intervals are less than 0.25 mag/airmass; therefore, we can say that this location is not strongly influenced by the Saharan dust.

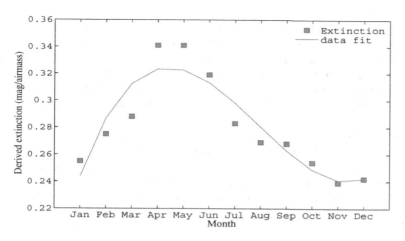

Figure 2. Monthly variation, modulo one year, of the derived extinction at the ELT sites. The continuous line is a fit showing a clear seasonal variation.

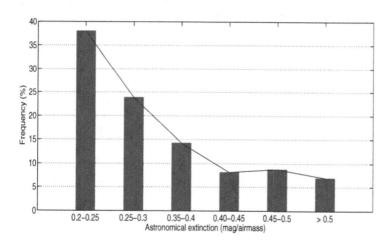

Figure 3. Percent of predominating intervals of extinction

Table 1. Astronomical extinction over Moroccan ELT sites and ORM. Values are derived from aerosol index.

Year	79	80	81	82	83	84	85	86	87	88	89	90	91	92
ORM	.38	.39	.40	.36	.37	.39	.41	.39	.36	.43	.39	.38	.35	.38
Moroccan ELT sites	.27	.26	.27	.27	.27	.31	.30	.27	.29	.31	.29	.28	.28	.28

To compare our sites to the ORM observatory, we present in Table 1 the yearly means of the extinction derived from aerosol index over both ELT sites and the ORM site. As a caveat, we caution that ORM values represent only summer period means, but, for the Moroccan ELT sites, values are calculated over the entire year. This difference results in a higher derived annual mean extinction over the ORM observatory compared to the Moroccan sites.

4. Conclusions

This study uses the correlation between aerosol index and astronomical extinction established over the ORM observatory to derive the extinction over the Moroccan ELT sites. We find that this derived extinction exhibits a seasonal variation and that the extinction is predominantly less than 0.25 mag/airmass. We further conclude that the Moroccan ELT sites are not strongly influenced by the Saharan dust. This study confirms that these Moroccan sites are good potential sites for the ELT.

Acknowledgements

We thank the European Southern Observatory, especially Marc Sarazin, for the financing our participation in the General Assembly of the IAU in Prague, the IAU Secretary General for paying the registration fee and the Cadi Ayyad University for the financing of the airfare. Our thanks also to NASA for the use of the Nimbus-7 TOMS data.

References

Siher, E.A., Ortolani, S., Sarazin, M., Benkhaldoun, Z. 2002, *SPIE* 5489, 138
Siher, E.A., Benkhaldoun, Z., Fossat, E. 2002, *Experimental Astronomy* 13, 159
Varela, Antonia M., Fuensalida, Jesús J., Muñoz-Tuñón, C., Rodríguez, E., José, M., García-Lorenzo, B., Cuevas, E. 2004, *SPIE* 5489, 105

Astronomy for the developing world
IAU Special Session no. 5, 2006
J.B. Hearnshaw and P. Martinez, eds.

© 2007 International Astronomical Union
doi:10.1017/S1743921307006825

Effect of altitude on aerosol optical properties

Aziza Bounhir[1,2]†, Zouhair Benkhaldoun[2], El Arbi Siher[2,3] and L. Masmoudi[4]

[1]Faculté des Sciences et Techniques, Département de Physique Appliquée, Marrakech, Maroc
email: bounhiraz@yahoo.fr

[2]Faculté des Sciences Semlalia, Département de Physique, LPHEA, Marrakech, Maroc
email: zouhair@ucam.ac.ma

[3]Faculté des Sciences et Techniques, Département de Physique, BP: 523, Béni Mellal, Maroc
email: siher@ucam.ac.ma

[4]LETS, Faculté des Sciences, avenue Ibn Battouta, BP: 1014 Rabat, Maroc

Abstract. The ELT project is currently under way in Europe and North America. Astronomical sites depend critically on sky transparency and on aerosol loadings. A quantitative survey of aerosol optical properties at candidate ELT sites is an essential part of the site selection process. There are basically two methods to characterize aerosol properties: ground based measurements and satellite measurements. In this paper we will establish a full climatology of two sites very close to each other, but at a difference of 2300m in altitude: Izaña and Santa Cruz located in the Canary Islands. Both have sun photometers from the AERONET network. We also use the aerosol index determined from TOMS satellite data to determine how aerosol optical properties vary with altitude. We establish a correlation between the TOMS index and the aerosol optical thickness in both sites. Aerosol optical properties show very good correlation between Izaña and Santa Cruz. As a result we establish a set of relationships helpful to characterize sites at elevated altitude from data of neighbouring sites at low altitude.

Keywords. Site testing, atmospheric effects, aerosols, sun photometer

1. Introduction

It is important for astronomers to identify high-quality observatory sites. An exhaustive approach towards prospecting for good sites is to make measurements at all potential candidate sites. However this is very difficult and time consuming. The ideal is, if possible, to make measurements at accessible low-altitude places near the candidate sites and then to extrapolate the results to more elevated places. An important study is to see how geophysical data affecting astronomical observations vary with altitude. In this paper, we focus on aerosol optical properties at two sites located in the Canary Islands; Izaña at an altitude of 2367 m and Santa Cruz at an altitude of 52 m and separated from each other by some dozen kilometres.

2. Data and method

The instruments used for ground measurements are the CIMEL sun photometers from the AERONET Network. These radiometers make measurements of direct sun and diffuse sky radiance within the spectral range of 340–1020 nm (Holben, *et al.* (1998)). The direct

† Present address: Faculty of Science and Technology, Department of Applied Physics, Marrakech, Morocco

A. Bounhir *et al.*

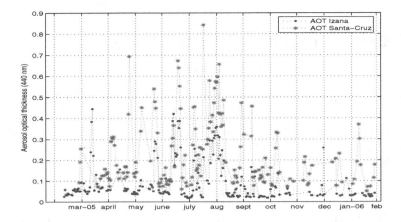

Figure 1. Daily values of the aerosol optical thickness (AOT) at Izaña and Santa Cruz.

sun measurements are acquired in eight spectral channels 340, 380, 440, 500, 670, 870, 940, 1020 nm. The 940 nm band is used to estimate total precipitable water vapour content (WVC). The bandwidths of the interference filters vary from 2–10 nm. The aerosol optical thickness (AOT) is computed from the Bouguer-Beer-Lambert law. The Angstrom parameter (ANG) is derived from a multispectral log-linear fit to the classical equation: $AOT \propto \lambda^{-ANG}$.

The sky radiance almucantar measurements are acquired at 440, 670, 870 and 1020 nm. A flexible algorithm for retrieval of aerosol physical properties developed by Dubovik & King, (2000) was used for retrieving aerosol size distribution over a range of sizes from 0.05–15 μm, together with spectrally dependent complex refractive index and SSA (Single Scattering Albedo) from spectral and sky radiance data. An automated and computerized cloud screening algorithm (Smirnov, *et al.* (2000)) was applied to direct sun measurements.

The use of satellite observations is the most efficient way to determine aerosol physical properties on large temporal and spatial scales. Among these instruments, the Total Ozone Mapping Spectrometer (TOMS) has the capability to sense aerosols (Hermann, *et al.* (1997)) and derive their optical properties (Torres, *et al.* (1998), (2002)) over both land and ocean, through the aerosol index (AI).

3. Results

The period of study extended from March 2005 until February 2006. We will characterize atmospheric optical conditions by the aerosol optical thickness at 440 nm (AOT), the water vapour content (WVC) and the Angstrom parameter (ANG) (870-440 nm). Daily AOT values in Fig. 1 depict very low values for Izaña compared to Santa Cruz, except during summer time, where dust events occur. The annual mean for Izaña is 0.08, which is 2.8 times less than the annual mean of Santa Cruz (0.23). From this Figure we can deduce that the aerosol layer is below Izaña's altitude most of the time, except in summer when dust events occur. During July and August the aerosol layer is higher than 2400 m as reported by (Hsu, *et al.* (1999)).

The annual mean of the Angstrom parameter is 1.2 for Izaña and 0.6 for Santa Cruz. Small-particle aerosols dominate Izaña's atmosphere. The Angstrom parameter at Izaña

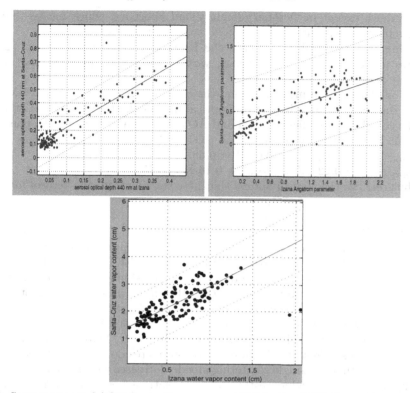

Figure 2. Scattergrams of AOT Santa Cruz versus AOT Izaña, ANG Santa Cruz versus ANG Izaña and WVC Santa Cruz versus WVC Izaña.

is around 1.5 most of the time except in summer during dust events. The annual mean water vapour content is 0.5 cm for Izaña and 2.2 cm for Santa Cruz.

The aerosol optical thickness (AOT) frequency distribution is very narrow at Izaña with a modal value at 0.05 and broader at Santa Cruz with a modal value around 0.1. About 70% of Izaña's AOT occur below 0.05, which denotes good observation conditions. About 70% of Santa Cruz's AOT occurs below 0.25.

The Angstrom parameter histogram is bimodal at both sites. The modes at Izaña are 0.3 and 1.6. The second one is dominant. At Santa Cruz the modes are 0.3 and 0.9.

The water vapor content histogram modes are 0.3 cm at Izaña and 2.5 cm at Santa Cruz. About 95% of the occurrence is below 1.2 cm for Izaña and higher than 1.2 cm for Santa Cruz.

Figure 2 shows scattergrams of the relation between Santa Cruz and Izaña concerning aerosol optical depth, Angstrom parameter and water vapour content. The resulting relationships show notable correlations (correlation coefficients of 0.92, 0.78 and 0.66, respectively).

$$\text{AOT}_{SantaCruz} = 1{,}5(\pm 0.15)*\text{AOT}_{Iza\tilde{n}a} + 0.07(\pm 0{,}02) \; R{=}0.92$$
$$\text{ANG}_{SantaCruz} = 0.35(\pm 0.098)*\text{ANG}_{Iza\tilde{n}a} + 0.25(\pm 0{,}11) \; R{=}0.66$$
$$\text{WVC}_{SantaCruz} = 1{,}6(\pm 0.31)*\text{WVC}_{Iza\tilde{n}a} + 1{,}33(\pm 0{,}2) \; R{=}0.78$$

No specific correlation was found for monthly mean values of the single scattering albedo (SSA). Monthly SSA values at 440 nm vary from 0.95 to 0.6 at Santa Cruz and from 0.9 to 0.7 at Izaña. The values at Santa Cruz are most of the time higher than the Izaña values, except in October, January and February, which means that Izaña's aerosols are more absorbent.

Concerning the relationships between the Izaña and Santa Cruz aerosol size distributions, we found very high correlations, varying from 0.97 for June and decreasing progressively to 0.6 for December.

Concerning the satellite measurements, the relationships between the TOMS aerosol index (AI) and the corresponding aerosol optical thicknesses (AOT) give satisfying correlations: the correlation coefficient of AI and AOT for Izaña is 0.68 and the one for Santa Cruz is 0.75. The relationships concerning AOT and AI are:

$$AOT_{Izaña} = 0.12*AI_{Izaña} + 0.05 \ R = 0.68.$$
$$AOT_{SantaCruz} = 0.22*AI_{SantaCruz} + 0.16 \ R = 0.75.$$

We can thus retrieve the aerosol optical thickness starting from the aerosol index signal.

4. Conclusion

In this work we have characterized the aerosol optical properties of Izaña and Santa Cruz during a year. For that purpose we used the AERONET data. We have established the climatology of both sites. We find linear relationships between Izaña and Santa Cruz aerosol optical properties; aerosol optical thickness, water vapour content and the Angstrom parameter give good correlations (92%, 78,5% and 66,4%, respectively). Size distribution correlates well (R varying from 98% to 60%) for June, July, August, September, October, November and December. January, February, March, April and May give no correlation. The single scattering albedos of both sites do not seem to correlate. One surprising thing is that the single scattering albedo decreases with increasing wavelength, even during dust events. We have established correlations between TOMS Aerosol Index (AI) and the aerosol optical thickness (AOT) (R at Izaña is 68,5% and at Santa Cruz 75,5%). Now the question is, can we use these relationships in other locations close to the Canary Islands, like the Atlas mountains of Morocco, for example?

Acknowledgements

We thank ESO, especially Marc Sarazin, IAU and the Cadi Ayyad University for financial suport. Our thanks to TOMS and AERONET for the use of their data.

References

Dubovik, O., King, J.M.D. 2000, *J. of Geophys. Res.* 105, 673

Hermann, J.R., Barthia, P.K., Torres, O., Hsu, C., Sefter, C. 1997, *J. of Geophys. Res.* 102, 16911

Holben, B.N., Eck, T.F., Slutsker, D., Tanr, D., Buis, J.P., Setzer, A., Vermote, E., Reagan, J.A., Nakajima, T., Lavenu, F., Jankowiak, I., Smirnov, A. 1998, *Remote Sensing of Environment* 66, 1

Hsu, N.C., Herman, J.R., Torres, O., Holben, B.N., Tanr, D., Eck, T.F., Smirnov, A., Chatenet, B., Lavenu, F. 1999, *J. of Geophys. Res.* 104, 6269

Smirnov, A., Holben, B.N., Eck, T.F., Dubovic, O., Slutsker, I. 2000, *Remote Sensing of Environment* 73, 337

Torres, O., Barthia, P.K., Hermann, J.R., Ahmad, Z., Gleason, J. 1998, *J. of Geophys. Res.* 103, 99

Torres, O., Herman, J.R., Barthia, P.K., Sinyuk, A. 2002, *Advances in Space Research* 29, 1771

Section 4:

Astronomy in Eastern Asia and the Pacific

Astronomy for the developing world
IAU Special Session no. 5, 2006
J.B. Hearnshaw and P. Martinez, eds.

© 2007 International Astronomical Union
doi:10.1017/S1743921307006849

Astronomy in Asia

Boonrucksar Soonthornthum

National Astronomical Research Institute of Thailand (NARIT), Ministry of Science and Technology, Thailand
and
Department of Physics, Faculty of Science, Chiang Mai University, Chiang Mai 50200, Thailand
email: boonrucksar@narit.or.th

Abstract. Astronomy in Asia has continuously developed. Local wisdom in many Asian countries reflects their interest in astronomy since the historical period. However, the astronomical development in each country is different which depends on their cultures, politics and economics. Astronomy in some Asian developing countries such as China and India are well-developed while some other countries especially in south-east Asia, with some supports such as telescopes, training, experts etc. from some developed countries, are trying to promote relevant research in astronomy as well as use astronomy as a tool to promote scientific awareness and understanding for the public. Recently, a new national research institute in astronomy, called the National Astronomical Research Institute of Thailand (NARIT), with a 2.4-metre reflecting telescope has been established in Thailand. One of the major objectives of this research-emphasis institute would aim at a collaborative network among South-East Asian countries so as to be able to contribute new knowledge and research to the astronomical community.

Keywords. Asian developing countries, SEAAN, NARIT

1. Introduction

It has been aware that the social and economic growth of the countries is dependent on an essential emphasis on education, science and technology. Recently, there has been a great surge in an international effort to foster the development of science and technology in various countries all over the world. International and regional co-operation in basic science research, including astronomy, is promoted in order to ensure high scientific standards. International collaboration has had a very profound impact on research and education in astronomy which, in turn, leads to new discoveries of theories and knowledge that are responsible for our understanding of complex mechanisms of the universe, the behavior of all celestial objects and the evolution of the universe to its present state. Thanks to IAU which strives to promote vital worldwide network and regional co-operation in astronomy to ensure that astronomy becomes a truly shared asset.

As the title suggests, my discussion will be wrapped only on Asia. Asia is one of the world's biggest continents made up of, so to say, extremely developed, least developed, and developing countries. There are significant social and economic differences among these countries. The difference in the scientific and technological infrastructure and in the popularization of science and technology in the two groups of countries are the most important causes of differential social and economical levels. Lack of appreciation of the importance of science as an essential ingredient of economic and social development, and weak educational and scientific infrastructure are the basic problems of the developing countries. The same situation applies to astronomy in Asia, both on basic and advance levels.

Developed countries have long recognized the necessity and importance of infrastructure and critical size of human resources in the area. In these countries funding scientific enterprise is widely accepted as a vital and long-term investment to a knowledge-based society. The fact is reflected in the amount of investment of the countries the field. Whereas, in developing countries, astronomy has been treated as a 'marginal activity' and perceived as an 'ornament'. Thus, astronomy in developing countries has principal shortcomings in its funding and supporting facilities. Inadequate infrastructure is a critical factor which creates solid barriers to advancement of astronomy in Asian developing countries.

2. Basic problems encountered in promoting advancement of astronomy in Asian developing countries

I shall try to summarize some common major factors which have stood up as the obstacles for promoting astronomy in Asian developing countries. The details, of course, are different according to social, cultural, historical, and political elements of each country. They are as follows:

1. Lack of national policy on long-term investment in the field.

2. Lack of planning, management, and support of activities from the government. This, obviously, is a consequence of 1.

3. Lack of concrete strategies and directions for research work which, in turn, leads to lack of commitment to retain high-level astronomers and researchers to maintain their productivity

4. Lack of vision to encourage collaboration with well-established astronomical institutes or of-like organization in other countries.

5. Lack of infrastructure that is comparable to international standard.

6. Inadequacy to reach the critical size of human resources and the amount of investment in the area.

7. Limited access of relevant information through journals, textbooks, especially internet and electronic communication which now play an integral part of all scientific endeavours.

3. Strategic plans for development of astronomy in Asian developing countries

It may be of advantage if we try to pinpoint the strategic plans for development of astronomy in Asian developing countries.

1. Bridging the gap between astronomers and the government.

2. New attitude of astronomers for the development of astronomy in the country.

3. Promote the national awareness and the investment for the development in astronomy in the country.

4. Roles of universities: teaching, research and produce good quality graduates.

5. Providing of the critical mass of researchers and supporting staff in astronomy in the country.

6. The development of standard astronomical infrastructure in the country.

7. Astronomy network in the region mentoring with well-established astronomical institutes.

Table 1. The number of researchers in south-east Asian developing countries in 2005

Country	IMF Quota (millions of SDR)	Population (million)	Wealth	Astronomers/Population[1]
Indonesia	2079.3	222.8	9.33	0.08(17)
Malaysia	1486.6	25.4	58.52	0.28(7)
Philippines	879.9	83.1	10.59	0.04(3)
Thailand	1081.9	64.2	16.85	0.05(3)
Singapore	862.5	4.3	200.58	0.70(3)
Vietnam	329.1	84.2	3.91	0.04(3)
Myanmar	258.4	50.5	5.12	-
Laos	52.9	5.9	8.97	-
Cambodia	87.5	14.1	6.21	-
Brunei	215.2	0.4	538.00	-

[1] http//www.iau.org/MEMBERSHIP.10.0.html

4. New attitudes for developing countries

Today, in the context of an increasingly globalize trend towards the era of knowledge-based society, developing countries have to embrace a new attitude. Instead of 'asking for donation and support' from developed countries, the concept has gradually switched to the two C's, namely, 'collaboration' and 'co-operation' between countries. Both developed and developing countries have to develop symbiotic relationships to prosper. This means that developing countries have to maneuver the standard of science and technology of their countries to meet the international standard so that collaboration may work out properly. Astronomy is one of the sciences that developed countries have endeavoured to promote and implement in Asian developing countries, focusing on sustained aid to improve the quality and the potential to ensure international standards in research work and academic activities. This has been consolidated in the framework of IAU which is one of the main agencies in the field.

5. Roles of universities

The essential prerequisite to any progress is an early recognition of necessity of good education in the area. Therefore the roles of universities in development of astronomy cannot be underestimated. The critical size of human resources and the amount of invest-ment in the area of astronomy illustrates how astronomy is of neglected importance in south-east Asian developing countries. Universities in these countries face shortage of re-searchers. So, universities need to provide more academic staff and support more training to the existed academic staff. Moreover, universities need to initiate more national and international research collaboration. Table 1 shows number of researchers in south-east Asian developing countries in 2005. Wealth of the country is defined by the ratio of IMF Quota and the population and Astronomers/Population is the number of astronomers, as the individual member of the IAU (in the parenthesis), divided by the population per million of each country, see Hearnshaw (2001). The table shows that the number of astronomers in developing countries is very less and some countries have no astronomer at all. The lack of professional astronomers is one of a critical problems encountered in promoting advancement of astronomy in Asian developing countries.

6. Situation in Thailand: the establishment of the National Astronomical Research Institute of Thailand

Here, I would like to give you some information about the situation in my country. In relevance to this, a new astronomical facility has been established in Thailand recently. On July 20th, 2004, the Thai cabinet approved in principle a proposal to establish a National Astronomical Research Institute of Thailand (NARIT) to commemorate King Rama IV's 200th birthday and also to celebrate the 80th birthday of the present King of the Kingdom of Thailand, King Bhumibhol. The main facility of NARIT is the National Observatory which is located on the very top of the highest mountain at the altitude of 2550 metres above the sea level, named Doi Intanon, in Chiang Mai province of northern Thailand, which is also renowned for the superb climate and tourist attractions. The National Observatory will operate a 2.4-metre reflecting telescope with alt-azimuth driving system. This system has been synchronized with automatic motion of the dome structure, so that observation can be carried out efficiently. The National Observatory and the 2.4-metre reflecting telescope are under construction and expect to be operated in the beginning of the year 2009. Figure 1 shows the conceptual design of the National Observatory of Thailand and Figure 2 shows the drawing of the 2.4-metre reflecting telescope which is under construction by EOS Space Systems Pty. Ltd.

Apart from the main observatory and research buildings, there is one more building associated with NARIT, namely, the Information Technology and Training Center for Astronomy (see the conceptual design in Figure 3) which is located at the Office of the Headquarter of Doi Intanon National Park. The centre will provide information and a learning experience in astronomy for the publics, the facilities for remote operations of the National Observatory, a telescope maintenance workshop and accommodation for astronomers.

The office of the Headquarter of NARIT will be located in one of the Chiang Mai University campus which is about 5 km from the centre of Chiang Mai city. The mission of this research-emphasis institute is to provide the collaborative network for developing and strengthening knowledge in astronomy so as to meet the international standards and will also provide education and learning culture in astronomy for the public and thus encourage them to seek further involvement with science and technology.

7. The South-East Asian Astronomical Network

Most of the developing countries in Asia are now aware of the importance of science and technology, including astronomy, and are now attempting to work on sustainable development of basic science. But the awareness does not necessarily make it easy to develop and popularize astronomy, especially in south-east Asian developing countries. As can be seen, the rate of growth and the rate of productivity in promoting astronomy in Asian developing countries vary significantly from region to region. Moreover, south-east Asian developing countries have significantly slower rate than other Asian developing countries.

One major barrier for developing countries in south-east Asia to tackle development issues has been the absence of a defined venue for getting involved. There are few resources outlining the key steps towards cooperation or initiating advancement efforts.

The astronomical network then, is a mindset sorely needed for the development of astronomy challenges facing south-east Asian developing countries and an ever-increasingly globalized world.

Figure 1. the conceptual design of the National Observatory of Thailand at the summit of Doi Intanon mountain, Chiang Mai, Thailand

Figure 2. The 2.4-metre reflecting telescope which will be installed at the National Observatory of Thailand

In view of that, I would like to take this opportunity to present the proposal to establish effective mechanisms for nurturing shared development among south-east Asian countries – 'The South-East Asian Astronomical Network (SEAAN)' – with well-established astronomy institutes in some developed countries as mentors. The network will act as a regional platform to create a fruitful advancement in the field of astronomy for member countries. In this way existing resources may be shared and we can work together towards international cooperation. No doubt the IAU will play a central role in pushing this project towards its goal.

Figure 3. The conceptual design of the Information Technology and Training Center for Astronomy which is located at the Office of the Headquarters of Doi Intanon National Park, Chiang Mai, Thailand

The network, if ever put into action, will make possible for member countries to share information, responsibility in advancement of astronomy in south-east Asia region. Moreover the network will help increasing research capacity and building on achievements.

References

Hearnshaw, J.B. 2001, in: Alan H. Batten (ed.), *Astronomy for Developing Countries*, IAU Special Session at the 24th General Assembly, p. 23–27

Astronomy for the developing world
IAU Special Session no. 5, 2006
J.B. Hearnshaw and P. Martinez, eds.

© 2007 International Astronomical Union
doi:10.1017/S1743921307006850

Astronomy in Thailand

Busaba Kramer

National Astronomical Research Institute of Thailand,
Physics Building, Chiang Mai University, Chiang Mai 50200 Thailand
email: busaba@narit.or.th

Abstract. During the last few years, Thailand has seen a significant change in the way astronomical research and education is pursued in the country. The government has approved the establishment of the National Astronomical Research Institute of Thailand (NARIT) which aims to develop not only astronomical research but also astronomy education at all levels, both in formal and informal education. A framework of national key projects exists which includes national facilities, national collaborative research networks, teacher training and public outreach programmes. Examples of these programmes will be presented in this paper.

Keywords. Astronomy education, Thailand, astronomy development, developing country, NARIT

1. From the past to present

Astronomy has been one of the interests of the Kings of Siam (former name of Thailand) since the 1600s. The first astronomical observatory was built in 1685 during King Narai's Era. In the 1800s, King Rama IV studied western astronomical text books and precisely calculated the location of the total solar eclipse in 1868. Figure 1 shows the observatory and location of the total solar eclipse in Thailand in 1868 and the portrait of King Rama IV and his equipment. Apart from astronomy, he also brought modern scientific education to the kingdom, therefore he is considered as the "Father of Thai Science".

Modern theoretical research in astronomy only started in Thailand in the 1930s at Chulalongkorn University. One of a few observatories which has been fully utilized since the early days is Sirindhorn Observatory at Chiang Mai University, established in 1977 and equipped with 0.4-m Cassegrain telescope. Since 1996, the observatory also has a 0.5-m Cassegrain telescope, however the light pollution in the nearby Chiang Mai City has been too high to observe for research purposes. Today, observatories at several other universities have been established with 0.4-m class telescopes but not all of them are fully utilized due to the lack of personnel trained in astronomy. Indeed, there are less than 30 professional astronomers nationwide, most of them being university lecturers. Not all of them are still active in research. Thailand now has 26 well established universities with a faculty of Science and 41 newly established universities (transformed from teacher colleges in 2004).

In school education astronomy was mostly neglected until 2001 when astronomy was added to the National Science Curriculum with strong support from the astronomy community. However, as a result Thailand now faces a serious shortage in teachers being capable of teaching astronomy, so that teacher training has been an important aspect in the plans to further develop astronomy in Thailand.

In 2004, the government approved the establishment of the National Astronomical Research Institute of Thailand (NARIT), under the Ministry of Science and Technology. NARIT is a research- emphasis institute providing a collaborative network for developing and strengthening knowledge in astronomy, so as to meet international standards. The

Figure 1. Centre – Image of King Rama IV; left – equipment and western astronomy text-books which he used in calculating the eclipse path; right – group photo with the King at the observatory and map of the eclipse path in 1868.

Figure 2. Thailand and locations of Doi Inthanon and National Observatory site.

institute also promotes education and learning culture in astronomy for the public and thus encourage them to seek further involvement with science and technology.

The main facility of NARIT is the National Observatory which is located on the very top of the highest mountain at the altitude of 2 550 metres above the sea level, named Doi Intanon, in Chiang Mai province of northern Thailand, which is also renowned for the superb climate and tourist attractions. The National Observatory will operate a 2.4-metre reflecting telescope with an alt-azimuth drive system. This system has been synchronized with automatic motion of the dome structure, so that observation can be carried out efficiently. The National Observatory and the 2.4-metre reflecting telescope are under construction and are expected to be operated in the beginning of the year 2009. Figure 2 shows the conceptual design of the National Observatory and its location on the map of Thailand.

2. Roles of NARIT

NARIT plays an important role in the development of astronomy in Thailand. Its aims are not only to establish the national facility but also to develop astronomical research and education in astronomy in Thailand on the national level, from school to advanced research.

2.1. *Astronomical research*

2.1.1. *National Policy*

Being an organization directly under the Ministry of Science and Technology, NARIT is directly involved in the national science policy for the development of astronomical research in Thailand. The institute exercises a 4-year rolling strategic plan (*e.g.* 2004–2008) which supports the National Science and Technology Policy.

2.1.2. *Infrastructure*

Infrastructure is crucial for the development of astronomical research. NARIT aims to establish the integrated infrastructure which will support the need of both research and technology transfer. These are the National Observatory, a research centre, data and IT centre, a Learning Centre for Astronomy, and Technical Support and Incubation Centre for Telescope Technologies.

2.1.3. *Human resources*

Human resources are one of the most important key factors for the development of Thailand as a country. One of the National Key Success Factors is therefore to increase the ratio of the number of scientific researchers in Thailand to the number of population to 8:10 000 within 10 years (i.e. by 2016). Therefore, with a rough estimate for the population of Thailand of ∼60 million people, the target for the number of scientific researches in Thailand by 2016 is about 48 000. Estimating that about 2% of all researchers should work in the fields of astronomy, this implies a total of about 960 astronomers and astrophysicists. This is a very ambitious goal considering that this number should be reached in 10 years from a starting number of less than 30 astronomers. However, this number may indeed be required as a critical mass to self-sustain astronomical research and, in particular, to train a sufficient number of teachers for more than 40 000 schools all over Thailand.

In order to reach this goal, NARIT is working on planning strategies in developing human resources for astronomical research in Thailand. The strategies are as the following:

(1) Allocating PhD Scholarships to study abroad. By working in world class astronomical institutes, students will gain the most advanced research knowledge and related technologies. Furthermore, after graduation and returning to Thailand, the young researchers will continue their link with the international organizations, therefore enabling them to pursue high quality research and high productivity while working back home.

(2) Creating Postdoctoral Research Fellowships and Senior Research Fellowships to recruit young and senior researchers from the international market. These human resources will help to improve the capability of producing PhD and MSc students in Thailand. Moreover, they will also create networks and links to the international astronomical community as well.

(3) Promoting MSc and PhD graduate courses among Thai universities. The PhD graduates will be fed in to the research community while the MSc graduates will be fed in to both the research community as research assistants and highly qualified school teachers.

(40) Promoting a number of lectureship positions in astronomy at the universities which currently do not have astronomy courses, especially in the recently established (41) universities.

(5) Promoting the number of the nation's IAU individual members and the exchange of astronomers through a stop-over astronomers' programme and hosting astronomy meetings and workshops.

2.1.4. *Networking and collaborations*

Networks and collaborations are crucial to the development of astronomical research, both nationally and internationally. Within Thailand, universities will be encouraged to join as networks to improve their research and education capacity, while NARIT plays an important role in providing national facilities and facilitating the networks. One of the most needed outputs from the networks will be the PhD and MSc collaborative programmes.

NARIT also aims to host national and international meetings and workshops in astronomy. In March 2007, NARIT will host the annual astronomy conference, Thai National Astronomy Meeting 2007 (TNAM2007) in Chiang Mai. Since the number of astronomical research community in Thailand are not so high, this meeting will be held as part of the annual national physics conference. On the international level, the institute is now planning to host the first South-East Asia Astronomical Network Workshop (SEAAN2007). More information about the network can be found in Soonthornthum's paper in these proceedings. The institute is also planning to host the International Pulsar Workshop *(http://www.nari.or.th/pulsar2007)* in Krabi, during March 29 – April 7, 2007 and the 8th Pacific Rim Conference on Stellar Astrophysics in Phuket in 2008.

2.2. *Education in astronomy*

The role of NARIT is not only in the development of astronomical research but also in education in astronomy in Thailand on the national Level The scope can be extended from all-age school education to the public understanding of science.

2.2.1. *National policy*

Similarly to the national policy for astronomical research, NARIT is also involved in the policy making for astronomical education by exercising a similar 4-year rolling strategic plan.

2.2.2. *NARIT outreach programme*

Despite being a newly established institute, NARIT has already carried out several outreach programmes since its first year. Example of the current outreach programmes are the following:

(1) Astronomy exhibitions – organized during the National Science Weeks in Bangkok which are hosted by the Ministry of Science and Technology annually. In these 12-day exhibitions, more than 300 000 students attend from schools nationwide.

(2) Mobile exhibition units – containing posters, astronomical models, telescopes and equipment and learning material which can be transported to remote schools in other provinces.

(3) Online media and e-learning – the institute is working closely with several partners to develop online materials for school and public, see http://www.astroschool.in.th and http://www.stkc.go.th for example.

(4) Media and press – the institute is running regular radio shows and newspapers articles.

(5) Family Astro-Sunday – under the government policy to encourage the family to spend time together during the weekend, the institute organizes star parties at the Sirindhorn Observatory on Sunday twice a month.

2.2.3. *Formal learning*

Astronomy has been included in National Science Curriculum since 2001. However, there is still a lack of personnel, knowledge and learning materials in astronomy. NARIT is working closely with the Institute for the Promotion of Teaching Science and Technology (IPST), Ministry of Education in the development of the above factors. The key projects include curriculum development, distance learning in astronomy, certified teacher training programmes, school and teacher networks, and promoting gifted and talented students in astronomy. During Nov 30 – Dec 9, 2007, NARIT together with the Promotion of Academic Olympiads and Development of Science Education Foundation (POSN) organizes the 1st International Olympiad on Astronomy & Astrophysics (IOAA) in Chiang Mai to promote interest and education in astronomy and astrophysics of high-school students.

2.2.4. *Informal learning*

Learning science outside school is one of the best ways to discover and to be inspired by science. NARIT aims to promote astronomy in informal learning in collaboration with the National Science Museum (NSM), an organization under Ministry of Science and Technology, and gives academic advice in astronomy to other science museums and planetarium nationwide. The institute also realizes the important role of amateur astronomers and astronomy clubs and societies in astronomy development in Thailand.

3. Conclusions

From the background of astronomy in Thailand from past to present, it can be said that now is a Golden Era for the development of astronomy in Thailand after NARIT has been established with strong supports from the government. While astronomy helps to promote scientific awareness of the community, industrial links and technology transfer are important also. Given the current need for of additional teachers, lecturers and researchers in astronomy, there are good opportunities for pursuing careers in astronomy in Thailand. Collaborations and networking are always important for the development.

Boonrucksar Soonthornthum

Busaba Kramer

Proceedings Title IAU Special Session 5
IAU Special Session no. 5, 2006
J.B. Hearnshaw and P. Martinez, eds.

© 2007 International Astronomical Union
doi:10.1017/S1743921307006862

Astronomy in Vietnam

Nguyen Quynh Lan[1] and Nguyen Dinh Huan[2]

[1]Department of Physics, Hanoi National University of Education, Hanoi, Vietnam
email: nquynhlan@hnue.edu.vn
[2]Department of Physics, University of Vinh, Vietnam
email: ndhuan@hotmail.com

Abstract. In this paper we summarize the history and current state of astronomy in Vietnam. Future plans for the development of astronomy education and research in Vietnam are summarized and some avenues suggested. These include the construction of a planetarium in Hanoi, and the development of a curriculum for an undergraduate major and a Master's degree in astrophysics at some Universities, as well as the development and operation of an observatory in Hanoi.

Keywords. Astronomy, developing countries, education, Vietnam

1. Introduction

Vietnam has a long history of teaching humanities. It is a country with high respect for a culture that is deeply rooted in its traditions. At the time of the Vietnamese feudal dynasty, there were some offices for research in astronomy and the maintenance of a calendar. The first observatory was not built in Vietnam until the 20th century. It was located at Phu Lien, Hai Phong province about 120 km from Hanoi on a beautiful site near the sea. In addition to astronomy, geophysical and meteorological observations were made there. Before 1950, astronomy was officially taught in the last class of secondary school. In war time, however, in order to shorten the education programme, astronomy was no longer a compulsory course for secondary schools. That situation remained unchanged until two years ago.

Since 1975, astronomy has developed in many regions in Vietnam, but it is still rather unconnected. Most Vietnamese astronomers have little opportunity to improve their knowledge and keep up with the field and some have moved on to other fields. After the first visit to Vietnam was made in 1991 by Professor Yoshihide Kozai, the former President of the IAU and a former director of NAO of Japan, the Vietnamese Astronomical Society was created. Since then Vietnamese astronomers have begun to communicate with each other to improve on astronomical research and education. Also since that time, and with the help of support from the IAU, there are now many activities underway to develop astronomy in Vietnam.

2. Support from the IAU and the international community

After visits to Vietnam were made by many foreigner professors of the IAU, the Vietnamese Astronomical Society received a number of facilities from several countries. Vinh University received a 21-cm refractor telescope from the Royal Astronomical Society of Canada with the help of Prof. A. Batten, and three telescopes from Nishi-Harima Astronomical Observatory arrived with the help of Prof. T. Kogure. A planetarium with a capacity of 100 seats and a projector were received as a Cultural Grant-in-Aid from

Japan through the support of Prof. Kitamura. With the help of Prof. Y. Kozai, the Hanoi National University of Education also received a 40-cm Meade Schmidt Cassegrain telescope (computer-controlled with a CCD ST-7) and a 10-cm Takahashi reflector with an H-alpha filter from the Sumimoto Science Foundation of Japan. A small solar radio interferometer was also donated by Prof. Nguyen Quang Rieu. The University of Education in Ho Chi Minh City has also received a 20-cm telescope and a CCD through the support of Prof. Y. Kozai and the Gunma Observatory. In addition, a PC, printer, and slide projector have been received and a number of books and journals have been sent to many universities in Viet Nam.

The main teaching task for astronomy faculty members in the various Vietnamese universities of Education has been to train the teachers for the on-going compulsory astronomy courses. Teachers' workshops have been organized which gradually introduce up-to-date astronomical topics and review the associated physics. Through the TAD programme of IAU Commission 46, some summer schools and workshop from 1997 to 2003 have been organized in Hanoi, Vinh and Thai Nguyen for faculty members and physics students. Annual workshops on the 'Teaching of Astronomy' have been organized. These are intended for astronomy teachers at universities, for graduate students who expect to start a career in astronomy, and for secondary school teachers with experience in teaching astronomy. Besides the support of the IAU, there have been help and collaboration from the University Paris 6, Observatory de Paris, University of Notre Dame and the conference series 'Rencontres du Vietnam' to organize some courses on astrophysics for school teachers and students. These were held at the Hanoi National University and the Hanoi National University of Education in 2004 and 2006. These activities were very useful for both the university lecturers and school teachers. They have provided a chance for these teachers to update their knowledge on astrophysics and to communicate with foreign faculty. Through these astrophysics courses a few students have the chance to study astronomy at a more advanced level. Especially important is that through these activities Vietnamese astronomers have been able to select the best students and send them abroad into MSc and PhD programmes in astronomy. At the present time there are five students who have completed their PhD in astrophysics, but some of them are still doing postdoctoral research in the USA, France and Taiwan. There are also other students who are still doing their PhD work in the USA and France.

The Vietnamese Astronomical Society has highly valued the assistance of the IAU for the development of astronomy in Viet Nam via TAD and other programmes. Thanks to these programs, some astronomers and teachers have had the opportunity to study and to improve their knowledge. Universities have also received telescopes along with many books and journals. Through this generosity the quality of astronomy teaching at Vietnamese universities has significantly improved and some of our best students are being inspired to careers in this fascinating subject.

3. Some activities for astronomy development

3.1. *Education and outreach*

After more than ten years of offering this opportunity to the Ministry of Education, secondary school students now have a chance to know and understand astrophysics – a field that is still very new for everyone in Vietnam. Vietnamese astronomers have written an astronomy textbook appropriate for the course on the astrophysics to be required in the natural sciences branch of secondary schools. A 10th grade course on Kepler's laws and a 12th grade course on the introduction to the Solar System, Galaxies and Cosmology has been approved for the natural science branch of secondary schools. Also, there are three specialized subjects for the 10th, 11th, and 12th grade on the celestial sphere, telescopes, and astrophysics.

Almost all of the Universities of Education in Vietnam have a curriculum of astrophysics courses for undergraduate students in physics who plan to start a career in astronomy and become teachers. There is a course on fundamental astrophysics with 60 class-hours on theory and 15 class-hours for observation with small telescopes for the 3rdh year students. Especially in the Hanoi National University of Education which has the 40-cm telescope, there are two courses on astrophysics, one is on fundamental astronomy for the 3rd year students with 60 class-hours and the other is astrophysics for the 4th year students with 60 class-hours plus a lab for observations. Every year, each of the universities selects about 5 to 10 senior students to do research and write their theses in astrophysics. Although these students are very interested in astrophysics, they cannot study for an advanced MSc or PhD degree because Vietnam still has no graduate degree programme in astrophysics. Since the 1970s to 1980s there was only one graduate student with a major in astrophysics and another twenty PhDs who were trained in Russia and Eastern Europe. By now, however, they have moved on to different fields and different regions in Vietnam. In the following decade, no one could get a PhD in astrophysics or astronomy. The number of PhDs since 2000 is very small. Therefore, no university can attract enough students and faculty to create a graduate degree programme in astrophysics.

The task of developing astronomy research and education in Vietnam is still very difficult. We really very much need more support from the IAU and other countries to train professional astronomers. At present, Vietnamese universities train future school astronomy teachers and provide small telescopes for observations, carefully selected books, journals for their libraries, and effective materials for audio-visual and hands-on presentations. Universities also provide access the internet, so as to optimize this facility for aiding Vietnamese astronomy. Many universities have a website in astronomy, Astronomical dictionary and a popular Astronomy Bulletin have also been published. The activities of the Vinh Planetarium has been much improved by the addition of two shows which were donated by the Davis Planetarium (Baltimore, USA) as well as many DVD videos. A new text book 'Astrophysics', was printed in 2000 and re-published 2002 and 2003, in Vietnamese and English on facing pages. That, together with books, journals, and educational software are the main resources for the teachers.

A national professional workshop on 'Astronomy and Culture' was held in Hanoi in 2003 in order to disseminate astronomical knowledge and dissipate mysticism among the Vietnamese people. The Vietnamese Astronomical Society collaborated with VTV3 of Vietnamese Television to carry out TV programmes on astronomy in order bring astronomy to everyone and everywhere in Vietnam. The telescopes in Hanoi and Ho Chi Minh City universities provide an opportunity to impart knowledge to science students on the nature of science, the role of measurement, and the need for judgement regarding the quality of data and reliability of their interpretation. When there is any event such as an eclipse, or a transit of the Sun by Venus or Mercury, the observatory at the Hanoi National University of Education always organizes observations for students and anyone who has a passion for astronomy. The 40-cm Meade Schmidt Cassegrain telescope is large enough that students can experience the excitement of scientific inquiry by participating in observational research projects. But the observatory is located in the centre of Hanoi city and a CCD is the image detector, so it is still limited as a resource to observe and analyze data. On the other hand TAD sent one staff member to the Gunma Observatory, Japan for a few months. That was very instructive, but it was still not enough time for him to obtain advanced knowledge on observing techniques.

Nowadays the Vietnamese young people are very interested in astronomy and astrophysics. So there are some amateur Clubs of Astronomy in Vietnam. The members of these organizations are comprised of not only physics and astronomy students. They are from almost all universities and high schools in Vietnam. They organize annual activities such as seminars, and field trips for outdoor observations with small telescopes. They have created a website with fundamental knowledge on astronomy for college and high school students.

The project of the building of a planetarium in Hanoi was started a few years ago with the support of the French government, but is not yet completed. Hanoi is the capital of Vietnam with three million people, so the planetarium will be very important and as a means to bring astronomical knowledge and excitement to the public.

3.2. *Some activities on research*

A National Committee on Space Technology Research and Applications will be established in Viet Nam in the near furture in order to organize, supply concrete guidance, and carry out the strategy of Space Research and Applications until 2020. Together with this National Committee, a National Institute of Space Technology will be established. Training courses for professionals in Space Technology, including telecommunications, and satellite applications, is a main objective of the laboratories on Space Technology etc. The building of an infrastructure for Space Technology involves a center for receiving and analyzing satellite images; a satellite positioning station system, launching and using of the VINASAT geo-static satellite in 2008; the receiving of technology transfer on small satellite technology; completing the design, manufacturing, and launching of a small Earth observing satellite and corresponding control centers on Earth. The centre will also cooperate on research and training with countries having advanced Space Technology; and work on the manufacturing of some hardware and software. The objective is that by 2020, Viet Nam can build advanced Earth stations; have the capacity for small satellite manufacturing technology; can design and manufacture small satellites to observe the Earth; have its own rocket technology, and have a high quality staff. Remote sensing technology had been introduced into Vietnam since the 1980s and has progressed significantly. Application fields at various line agencies are mostly in land use and land cover mapping, environment monitoring, management and assessment, topographic mapping, disaster prevention, and geology, among others. Some projects are being set up at the Ministry of Science and Environment to apply remote sensing for land use, land cover, and mapping of the whole country by 2005. In addition, several research projects at the National Centre of Science and Technology of Vietnam address natural resources management and environment impact assessment.

Recent astronomical research in Vietnam has concentrated on preparing ephemerides, conducting astrometry, studying the motion of artificial satellites, and assisting surveys of the Vietnamese territory. Astronomers have published primarily in Chinese, Russian, and Eastern European journals. Some have written books on astronomy and astrophysics for use at the university and high-school levels; for the dissemination of astronomical knowledge; and for use in practical problems in astronomy. The analytical theory of the motion of Earth's artificial satellites and stellar astrometry are the main themes of research in astrometry in low latitude. They are studied at Vinh University. Some research has been published in Acta Astronomica Sinica (P.R. China) and the Russian Astronomical Journal.

With the facilities at the Hanoi National University of Education we have done some research such as:

- UBV photometry: Using the 16-inch LX 2000 filter with UBV photometry and CCD ST-7, CFW - 8 and RGB to do some research on stars, clusters, galaxies, and proto-planets.
- Solar observation is another possibility. In Vietnam we have places where there are more than 200 days of sunshine yearly and a large number of days when the Sun's latitude at noon is higher than 60 degrees. With the new period of solar maximum starting from 2007, we can use telescopes with an H-alpha filter and a 16-inch LX 2000 solar filter to photograph, analyze and study solar activity.

The theoretical astrophysics group in Hanoi National University of Education has support and collaboration with the Center for Astrophysics and the Joint Institute for Nuclear Astrophysics (JINA) at the University of Notre Dame, USA.

One research topic is on some candidates for self interacting dark matter (SIDM) in unified theories beyond the standard model of particle physics. To test this we are developing calculations of smoothed-particle hydrodynamic simulations of galaxy and large-scale structure formation in the universe. In another project this work also shows promise toward an accurate formulation of the effects of local inhomogeneities on the expanding universe. We also have done research on the possibility that a cosmic bulk viscosity can be produced from dark matter decay which could solve the important dark energy problem of the universe and we have researched possible particle-physics candidates for this decay. Another project involved studies of the possible evidence for white dwarf stars with strange matter interiors (strange dwarfs). Some results on these research topics have been published in Europhys. Lett. and the J. Phys. G; and Astrophysics and Space Science.

The cosmic ray group at the Institute of Nuclear Physics is a part of the international Pierre Auger Project. A lot of papers in this field have been contributed by this group. At the annual International conferences on Astrophysics and Cosmology from 2000 to 2006 of the 'Rencontres du Vietnam', there have been good opportunities for Vietnamese scientists to obtain up-to-date information and to communicate with other scientists around the world.

4. Conclusion

After more than fifteen years of effort to develop astronomy in Vietnam with the support of IAU. Astronomy in Vietnam has step by step made some progress in education and research on Astronomy. But it is still very far behind the developed world. Astrophysics involves fields which are very important for research, study, knowledge and culture. It is a part of modern physics so Vietnam sincerely wishes to have a graduate degree programme in astrophysics at least at the master's level, but we still do not have enough professional people in this field. To develop astronomy in Vietnam we will need more help and donations from the IAU and other organizations to complete the building of the planetarium in Hanoi to help gain popular acceptance an knowledge of astronomy. Also, the further training of professional astrophysicists both here and abroad is very important for us to facilitate the development of research in astrophysics in Vietnam.

Acknowledgements

We would like to thank and acknowledge the IAU support for travel to attend the SPS5 workshop on Astronomy for the Developing Word in the General Assembly of the International Astronomical Union in Prague, Czech Republic 2006.

Astronomy for the developing world
IAU Special Session no. 5, 2006
J.B. Hearnshaw and P. Martinez, eds.

© 2007 International Astronomical Union
doi:10.1017/S1743921307006874

Astronomy development in Thailand: the role of NARIT

Boonrucksar Soonthornthum, Busaba Kramer* and Saran Poshyachinda

National Astronomical Research Institute of Thailand,
Physics Building, Chiang Mai University, Chang Mai 50200 Thailand
*email: busaba@narit.or.th

Abstract. Astronomy development in Thailand has improved significantly during the last few years. The government has approved the establishment of the National Astronomical Research Institute of Thailand (NARIT). Roles of NARIT in the development astronomical research and astronomy education in Thailand includes a national framework, national facilities, collaborative research networks, teacher training and public outreach programmes. The new 2.4-metre reflecting telescope will serve not only astronomy community in Thailand but also in Southeast Asia.

Keywords. Astronomy education, Thailand, astronomy development, developing country, National Observatory

1. Background

In 2004, the Royal Thai government approved the establishment of the National Astronomical Research Institute of Thailand (NARIT), under the Ministry of Science and Technology. NARIT is a research-emphasis institute providing a collaborative network for developing and strengthening knowledge in astronomy, so as to meet international standards. The institute also promotes education and learning culture in astronomy for the public and thus encourage them to seek further involvement with science and technology. The review of astronomy in Thailand from past to present and the roles of NARIT in both astronomical research and education in astronomy are presented in Kramer's paper (in these proceedings). In this paper we report the recent development of the Thai National Observatory.

2. Thai National Observatory

The main facility of NARIT is the National Observatory which is located on the very top of the highest mountain in Thailand at the altitude of 2 550 metres above mean sea level. It is named Doi Intanon, in Chiang Mai province of northern Thailand, which is also renowned for the superb climate and tourist attractions. Seeing tests and weather monitoring at the site have been carried out since March 2006. The average seeing is 0.7 arc sec (Poshyachinda, private communication).

The National Observatory will operate a 2.4-metre reflecting telescope with an alt-azimuth drive system (see Figure 1). This system has been synchronized with automatic motion of the dome structure, so that observations can be carried out efficiently. The National Observatory and the 2.4-metre reflecting telescope are currently under construction and are expected to be operational in the beginning of the year 2009. The time-line of the development and list of astronomical instruments are presented in Table 1.

Table 1. Time-line and list of astronomical instruments of the Thai National Observatory

Year	Activities	Astronomical instruments
March 2006 -	Seeing tests	
July 2006 -	Construction of enclosure and telescope	
October 2008	Installation of telescope	
January 2009	First light	4k × 4k CCD camera
2010	Full operation	High resolution échelle spectrograph

Figure 1. The 2.4-metre telescope which will be installed at the National Observatory

3. Summary

NARIT plays an important role in the development of astronomy in Thailand. Its aims are not only to establish the national facility but also to develop astronomical research and education in astronomy in Thailand on the national level, from schools to advanced research. The national facility will be equipped with modern astronomical instruments and will serve not only the astronomy community in Thailand but also in Southeast Asia.

Astronomy for the developing world
IAU Special Session no. 5, 2006
J.B. Hearnshaw and P. Martinez, eds.

© 2007 International Astronomical Union
doi:10.1017/S1743921307006886

Collaboration and development of radio-astronomy in Australasia and the South-Pacific region: New Zealand perspectives

Sergei A. Gulyaev and Tim J. Natusch

Centre for Radiophysics and Space Research, Auckland University of Technology, Private Bag 92006, Auckland 1142, New Zealand
email: Sergei.Gulyaev@aut.ac.nz; Tim.Natusch@aut.ac.nz

Abstract. As a result of collective efforts of an Australian–New Zealand VLBI team, the first New Zealand VLBI system was developed, and a series of test observations between New Zealand and Australia conducted. The equipment and techniques used to conduct New Zealand's first VLBI observations are discussed and results of work in Australia and New Zealand to obtain fringes and the image of the source (PKS1921-231) are presented. The road map for New Zealand radio-astronomy as well as New Zealand involvement in the SKA is discussed.

Keywords. Radio-astronomy, VLBI, SKA

1. Radio-astronomy and VLBI in New Zealand

Radio-astronomy in New Zealand has links stretching back to the work of Elizabeth Alexander on solar emissions in the 1940s, and John Bolton and Gordon Stanleys mobile cliff interferometer. This latter instrument was used in New Zealand in 1947 to obtain rising and setting records of various sources. It was the first direct collaboration between New Zealand and Australia in this field. Measurements made near Sydney and Auckland determined the positions of several strong radio sources for the first time. This allowed their optical counterparts to be identified as supernova remnants and external galaxies and not "radio-stars" as previously had been assumed. A number of New Zealand pioneers, Gordon Stanley and Bruce Slee and more recently, New Zealanders Dick Manchester (ATNF) and Peter Napier (VLA) are among world famous radio-astronomers.

Current work in radio-astronomy is centred on the newly formed Centre for Radiophysics and Space Research (CRSR) located at Auckland University of Technology (AUT). AUT has undertaken the task of developing VLBI facilities in New Zealand. This work is being conducted in collaboration with Swinburne University of Technology, Melbourne and with the support of the Australia National Telescope Facility and the University of Tasmania.

A grant of $NZ300 000 was obtained from the New Zealand Ministry of Economic Development (MED) to support this work. The MED sees VLBI activity as a potential user of the Advanced Network, which in turn forms a vital component of the Governments Digital Strategy. A grant application for $5M for establishment of New Zealand Radio Telescope National Facility (RTNF) was recently submitted to the New Zealand Government. It is proposed to install an integrated radio-astronomical and geodetic VLBI station in the North Island and upgrade an existing old 11-m Earth station in Southland into a small radio-telescope. The main investment required is one modern 12–16 metre radio telescope, and two hydrogen atomic clocks – the standard of time and frequency

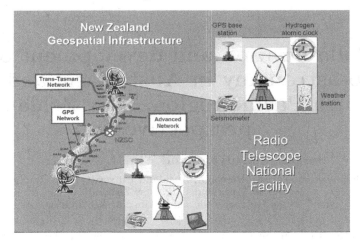

Figure 1. The concept of the Radio Telescope National Facility as a part of New Zealand geospatial and ICT infrastructure.

essential for VLBI. The proposed RTNF, along with the existing geodetic (GPS) network, Advanced Research Network (broadband) and New Zealand Supercomputer will comprise a world class national geospatial infrastructure (figure 1).

2. Regional collaboration and development

Radio telescopes in the Asia-Pacific region form a natural network for VLBI observations, similar to the very successful networks in North America (Network Users Group) and Europe (European VLBI Network). New Zealands VLBI facility, which we are developing since 2005, has the potential to strengthen the Asian-Pacific VLBI network and its role in astronomy, geodesy and geoscience (figure 2). It will positively influence regional and international activities in geoscience and geodesy that advance New Zealand's national interests.

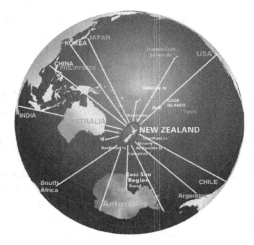

Figure 2. New Zealand's unique geographic location allowing VLBI with six continents and a number of countries in the South-Pacific region.

Figure 3. 6-m radio telescope (Auckland) used in the first Trans-Tasman (New Zealand–Australia) VLBI observations in 2005.

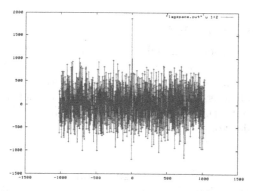

Figure 4. The first VLBI fringe obtained from New Zealand–Australia test in 2005.

A self-contained radio-astronomy system for VLBI, including a 1.658 GHz (centre frequency), 16 MHz bandwidth RF system (feed and downconversion system locked to a Rubidium maser and GPS clock), an 8-bit sampler with digitization system, and a disk-based recording system built around a commodity PC was developed in New Zealand Centre for Radiophysics and Space Research (Gulyaev, Natusch *et al.* 2005). This was designed as a portable system for use on different radio telescopes, since the one thing that New Zealand lacks at the moment is ready access to a large collecting area, fully steerable antenna.

A number of Trans-Tasman tests has been conducted in 2005–2006 between the CRSR system installed on a 6-m dish located in Auckland (figure 3) and the Australia Telescope Compact Array in Narrabri, Australia. This work has been successful, with fringes located from the recorded data (figure 4) and a high resolution image of the quasar PKS1921-231 synthesized (figure 5). New Zealand has demonstrated the capacity to contribute to modern radio-astronomical and VLBI research (Gulyaev, Natusch *et al.* 2006).

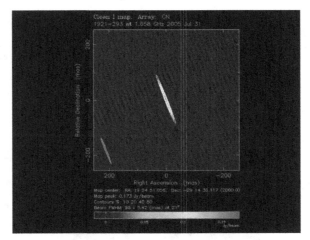

Figure 5. An image of PSK1921-231 synthesized from data obtained from New Zealand–Australia VLBI test in 2005.

Experiments were recently conducted with Kashima Radio Observatory (Japan); new tests are planned with Korea and Fiji. Plans have been made to build a new 16-m antenna in New Zealands North Island and to upgrade an 11-m dish in the South Island (see figure 1).

3. New Zealand and the SKA

New Zealand's geographic location makes it a natural place for possible extension of the Australian SKA from 3000 km to almost 6000 km, significantly increasing the maximum spatial resolution of the SKA telescope. Four trial sites with excellent radio environments, favourable natural conditions, and well developed infrastructure and communications have been selected as potential SKA sites. One of them (Warkworth, North Island) is protected as a radio quiet zone.

Acknowledgements

We would like to thank our VLBI colleagues from Swinburne University of Technology, Australia Telescope National Facility, University of Tasmania, and Kashima Space Research Centre.

References

Gulyaev, S., Natusch, T., Addis, B., Tingay, S. & Deller, A. 2005, *Southern Stars* 44, 12.
Gulyaev, S., Natusch, T., Tingay, S., Deller, A., West, C., Ellingsen, S., McCulloch, P., Reid, B., Baynes, E. & Tzioumis, T. 2006, *ATNF News* 58, 8.

Astronomy for the developing world
IAU Special Session no. 5, 2006
J.B. Hearnshaw and P. Martinez, eds.

© 2007 International Astronomical Union
doi:10.1017/S1743921307006898

Mutual collaboration between Institute of Technology Bandung, Indonesia and Gunma Astronomical Observatory, Japan

Osamu Hashimoto[1], Hakim L. Malasan[2], H. Taguchi[1], K. Kinugasa[1], B. Dermawan[2], B. Indradjaja[2], and Y. Kozai[1]

[1]Gunma Astronomical Observatory, Takayama, Agatsuma, Gunma 377-0702, Japan,
email: osamu@astron.pref.gunma.jp

[2]Department of Astronomy, Institute of Technology Bandung, Indonesia
email: hakim@as.itb.ac.id

Abstract. Institute of Technology Bandung (ITB), Indonesia and Gunma Astronomical Observatory (GAO), Japan have been proceeding with several programmes of mutual collaboration in the fields of astronomical research and education since 2002. ITB with Bosscha observatory has a great interest in education of astronomy for public people as well as in the university education and research of their own, and GAO is a public observatory operated by Gunma prefecture local government equipped with a 150-cm reflector and some smaller telescopes, which are capable of scientific research of high grade. We will report some of our cooperative activities including the remote accessing of the telescopes of each observatory by each other, which can provide opportunities for astronomical experiences of the opposite hemisphere for various people of each country. Some scientific collaboration works such as common instruments and data analysis systems developed on both sites are also reported.

Keywords. Instrumentation: miscellaneous, instrumentation: spectrographs, astronomy in Indonesia, remote observing, public education in astronomy

1. Introduction

Gunma Astronomical Observatory (GAO) was established in Takayama village, which is located about 140 km northwest of Tokyo (N 36° 35′ 47″, E 138° 58′ 22″), by Gunma prefecture local government in 1999. Its main telescope is a 150-cm reflector on an azimuth-elevation mounting, which is one of the most advanced telescopes in Japan with some powerful observation instruments, such as an infrared camera and a high resolution spectrograph (Hashimoto *et al.* 2002, 2005). GAO is not operated only for scientific researches by professional researchers but also for public education of general people. An eye piece system for public star gazing is equipped even on the 150-cm telescope on the basis of the latter point of view. There are more telescopes prepared for researchers and public visitors; a 65-cm reflector on an equatorial mounting, six 25–30-cm reflectors, and some 15-cm refractors.

Institute of Technology Bandung (ITB) is the most leading university in Indonesia in the fields of science and technology. It has the largest department of astronomy in Indonesia as well as an outstanding academic astronomical observatory, Bosscha observatory which was established in 1928. Basically ITB and Bosscha observatory are academic institutions for scientific research and university education of astronomy, but ITB is much interested in the public education as well (Wiramihardja 2003).

There are some common points in the characteristics of ITB and GAO. In fact, both are expected to have similar functions in both scientific research and public education of

Signing ceremony on 1 July, 2002 at GAO

MEMORANDUM OF AGREEMENT
Between
GUNMA ASTRONOMICAL OBSERVATORY
Gunma Prefecture, 6860-86 Nakayama, Takayama, Gunma, JAPAN
and
INSTITUT TEKNOLOGI BANDUNG
Jalan Tamansari No.64 Bandung 40116, INDONESIA

1. Due to their geographical locations, expertise in research and education, academic diversity and dedication to public service, Gunma Astronomical Observatory and Institut Teknologi Bandung, within their perspective countries are in an excellent position to cooperate in the fields of Observational Astrophysics and Science Education programs.

2. The undersigned recognize that collaborative efforts will be of mutual benefit and will contribute to an enduring international linkage of aforementioned institutions for both scientific cooperation and assistance.

3. The said parties cooperate in the field of astrophysics and science education. This cooperation will take place on the basis of equal footing and mutual benefit, and under the name *Japan-Indonesia Astronomical Research and Education.*

Gunma, 1 July 2002

Kusmayanto Kadiman
Rector
Institut Teknologi Bandung

Yoshihide Kozai
Director
Gunma Astronomical Observatory

Figure 1. Memorandum of Agreement between ITB and GAO, signed on 1 July 2002.

astronomy. Based on such common circumstances of us, ITB and GAO have kicked off our mutual collaboration programs. Dr Kusmayanto Kadiman, Rector of ITB and Prof. Yoshihide Kozai, Director of GAO have signed the Memorandum of Agreement for the collaboration programs on 1 July 2002 at GAO (Fig. 1).

Since then some concrete programs have been put into practice as our mutual collaboration activities. They are (1) development and operation of a high resolution spectrograph on the GAO 150-cm reflector; (2) development and operation of identical low resolution spectrographs both for Bosscha observatory and GAO telescopes; (3) development and operation of the same reduction and analysis systems both at ITB and GAO for sharing the observational data and the methods with which those data can be handled; (4) mutual collaboration in the studies of astronomy and astrophysics with the use of above mentioned facilities; (5) development and operation of remote telescope system between ITB and GAO, and (6) mutual exchange programmes of staff members of both sides.

2. Facilities for scientific research

As a part of scientific programs in our collaboration, a high resolution spectrograph GAOES (Gunma Astronomical Observatory Echelle Spectrograph) is designed and built for the GAO 150-cm telescope (Hashimoto *et al.* 2005). It provides an optical spectrum of a spectral resolution ($\lambda/\delta\lambda$) up to 100,000 for wavelengths of 360–1 000 nm on a 4096 × 2048 pixel CCD detector, covering a wavelength range of about 190 nm by a single exposure (Fig. 2). Such a performance indicates that GAOES can be regarded as one of the most advanced instrument of this type for a 1–2-m class telescope. With the use of this new instrument we have already started some scientific works. For example, studies of binary systems (Puri Jatmiko *et al.* 2005) and evolved stars in the post-AGB phase (Takeda *et al.* 2005) are actively going in these years.

Two low resolution spectrographs GCS (Gunma Compact Spectrograph) for the GAO 65 -cm reflector and BCS (Bosscha Compact Spectrograph) for the Bosscha 60-cm refractor or 45 cm reflector have been built (Fig. 3). Spectral resolution of those spectrographs

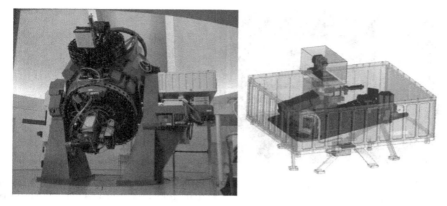

Figure 2. GAOES at Nasmyth focus of the GAO 150-cm telescope (left), and a schematic view of its inside (right).

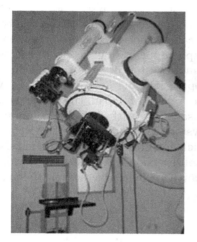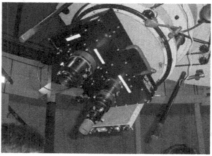

Figure 3. GCS on the GAO 65-cm reflector (left), and BCS on the 60-cm refractor at Bosscha observatory.

is about 500 or 2000 which can be chosen by two different gratings. As those two spectrograph are completely the same, common methods can be developed and applied in the observation and handling of the data at both observatories. We are not intending to use these spectrographs only in various scientific researches, but in some activities for education activities for students and public people.

Also computer systems which are basically the same have been developed both at ITB and GAO for sharing the data of our collaboration studies and for providing the tools for reduction and analysis of those data in the same manner (Kinugasa *et al.* 2005). We are expecting very effective developments in our cooperative studies with those common facilities.

3. Remote telescope system and public education

The location of Bosscha observatory S 6° 49′ 32″, E 107° 36′ 58″ indicates that stars in southern hemisphere that can not be seen in Japan, can be observed at Bosscha observatory, and the north stars which can not be seen in Indonesia vice versa. This point

Figure 4. About 100 people gathering at GAO in Japan for the remote star observation of the southern hemisphere from Indonesia (left), and a real time image of ω Cen projected on the screen at GAO (right), which is sent from a remote control telescope at Bosscha observatory.

is very much useful in our collaborative activities, especially in the public education, as well as in the scientific research. In order to take such a geophysical advantage between different hemispheres, we have developed a remote access system of the telescopes between ITB and GAO (Taguchi *et al.* 2005). It can provide special opportunities for the astronomical experiences of the opposite hemisphere for various people of each country. In June 2006 we had an experiment of the remote system for the public education. It was completely successful with about 100 people gathering at GAO (Fig. 4). While it was rainy at GAO at that time as it was the wet season of Japan, the participants could enjoyed the southern hemisphere of Bosscha observatory with real time communications between us directly made through the remote system. We could realize that such an event is really exciting and impressive for all the participants. In fact, most of them raised their hopes for the next occasion of such a remote observation of the distant world. We are planning another remote observation of the northern hemisphere from Indonesia next December.

Acknowledgements

We would like to acknowledge the continuous encouragement by Prof. B. Hidayat in our cooperative activities.

References

Hashimoto, O., Kingugasa, K., Nishihara, E., Taguchi, H., Malasan, H.L., Kurata, T., Okuda, H., Shimizu, M. & Kozai, Y. 2002, *Proc. The 8th IAU Asian-Pacific Regional Meeting*, Vol.II, p. 7

Hashimoto, O., Malasan, H.L., Taguchi, H., Kurata, T., Yamamuro, T., Takeyama, N., & Shimizu, M. 2005, *Proc. The 9th IAU Asian-Pacific Regional IAU Meeting 2006*, p. 295

Kinugasa, K., Malasan, H.L., Indradjaja, B., Hashimoto, O., Taguchi, H., Kurata, T., Kozai, Y., & Hidayat, B. 2005, *Proc. The 9th IAU Asian-Pacific Regional IAU Meeting 2006*, p. 102

Puri Jatomiko, A.T., Malasan, H.L., Hashimoto, O., & Taguchi, H. 2005, *Proc. The 9th IAU Asian-Pacific Regional IAU Meeting 2006*, p. 338

Taguchi, H., Kinugasa, K., Indradjaja, B., Prasetyono, G.I., Setyanto,H., Malasan, H.L., Hashimoto, O., & Kurata, T. 2005, *Proc. The 9th IAU Asian-Pacific Regional IAU Meeting 2006*, p. 102

Takeda, Y. Hashimoto, O., Taguchi, H., Yoshioka, K., Takada-Hidai, M., Saito, Y., & Honda, S. 2005, *PASJ* 57, 751

Wiramihardja, S.D. 2003, *Effective Teaching and Learning of Astronomy, 25th meeting of the IAU, Special Session 4*

Section 5:

Astronomy in the Middle East and central Asia

Astronomy for the developing world
IAU Special Session no. 5, 2006
J.B. Hearnshaw and P. Martinez, eds.

© 2007 International Astronomical Union
doi:10.1017/S1743921307006916

Astronomy in Iran[*]

Yousef Sobouti

Institute for Advanced Studies in Basic Sciences, Zanjan, P.O. Box 45195-1159, Iran
email:sobouti@iasbs.ac.ir

Abstract. This paper summarizes the present situation concerning astronomy in Iran. Shiraz University introduced astronomy in the mid-1960s and established the Biruni Observatory. Astronomy is also carried out at Tabriz University and in Meshed, Zanjan, Tehran, Babol and other places. The Astronomical Society of Iran is flourishing, and is about 30 years old. Site-testing is underway for a National Observatory of Iran, which will be equipped with a 2-m class optical telescope. Iran publishes Nojum, the only astronomical monthly magazine of the Middle East, which is presently in its fifteenth year.

Keywords. Astronomy in Iran

1. Introduction

In spite of her renowned pivotal role in the advancement of astronomy on the world scale during 9th to 15th centuries, Iran's rekindled interest in modern astronomy is a recent happening. Serious attempts to introduce astronomy into university curricula and to develop it into a respectable and worthwhile field of research began in the mid 60's. The pioneer was Shiraz University, which should be credited for the first few dozens of astronomy- and astrophysics-related research papers in international journals, for training the first half a dozen of professional astronomers and for creating the Biruni Observatory. Here, I take the opportunity to acknowledge the valuable advice of Bob Koch and Ed Guinan, then of the University of Pennsylvania, in the course of the establishment of this observatory.

At present the astronomical community of Iran consists of about 65 professionals, half university faculty members and half MS and PhD students. The yearly scientific contribution of its members has, in the past three years, averaged to about 15 papers in reputable international journals, and presently has a healthy growth rate. Among the existing observational facilities, Biruni Observatory with its 51-cm Cassegrain, CCD cameras, photometers and other smaller educational telescopes, is by far the most active place. Tusi Observatory of Tabriz University has 60 and 40 cm Cassegrains, and a small solar telescope. A number of smaller observing facilities exist in Meshed, Zanjan, Tehran, Babol and other places.

The Astronomical Society of Iran (ASI), though some 30 years old, has expanded and institutionalized its activities since early 1990's. ASI sets up seasonal schools for novices, organizes annual colloquia and seminars for professionals and supports a huge body of amateur astronomers from among high school and university students. Over twenty of ASI members are also members of IAU and take active part in its events.

---

[*]This paper was presented by Dr E. Guinan, Villanova University, USA, as Dr Sobouti was unable to be present in person.

2. The National Observatory of Iran

In the past five years, astronomers of Iran have staged an intensive campaign to have a National Observatory of their own (NOI). Initial planning is for a state-of-art 2 m telescope with CCD-based instrumentation. The project is approved and will be funded by the government in the course of five years.

The site selection for NOI, however, is already in its third year and has been and is being generously funded by the government. Initially, four sites were selected for further testing from 33 candidate sites using geophysical,local and meteorological criteria. Seeing measurements were carried out at the sites with Differential Image Motion Monitors (DIMM). After the analysis of three years of DIMM measurements, the list is now shortened to the following two sites:

- Kolahbarfi Mountain in Kashan (Height: 3220 m, Long.: 51° 26′ E, Lat.: 33° 38′ N)
- Fordoo Mountain in Qom (Height: 3150 m, Long.: 50° 58′ E, Lat.: 34° 11′ N)

Further information can be found in two articles in the Persian language in the Iran J. of Physics by Nasiri *et al.* (2003) and Darudi *et al.* (2005)

3. Nojum: Iran's monthly astronomical magazine

Last, but not least, Nojum, the only astronomical monthly magazine of the Middle East, is presently in its fifteenth year. It has a good readership among both professionals and amateurs of Farsi speaking communities within the country and abroad.

References

Darudi, A. & Nasiri, S. 2005 *Iran J. Phys.* **3**, 121–128
Nasiri, S. & Abedini, A. 2003, *Iran J. Phys.* **4**, 309–315

Astronomy for the developing world
IAU Special Session no. 5, 2006
J.B. Hearnshaw and P. Martinez, eds.

© 2007 International Astronomical Union
doi:10.1017/S1743921307006928

Astronomy in the former Soviet Union 15 years after the breakup of the USSR

Nikolai G. Bochkarev[1]

[1]Eurasian Astronomical Soc. (EAAS) and Sternberg Astronomical Inst. (SAI),
Moscow, Russia
email: boch@sai.msu.ru

Abstract. During the immediate post-Soviet period, the main infrastructure of astronomy over the territory of the former Soviet Union (FSU) was saved, in spite of dramatic decreases in financial support. Overall the situation for FSU astronomy is now stable. In Latvia, the 32-m radio-dish is in working order. This allows it to participate in VLBI programmes. In Russia, all three 32-metre radio dishes of the QUASAR VLBI system are operational, as well as the 2-m telescope with a high-resolution spectrograph (up to resolution $R \simeq 500\,000$) and the horizontal solar telescope ($R = 320\,000$) of the Russian-Ukrainian Observatory on Peak Terskol (Caucasus, altitude 3100 m). However the situation with the observatory itself is worrying, because of the regional authorities' attempt to privatize its infrastructure.

The process of equipping a number of Commonwealth of Independent States (CIS) (including Russian) observatories with CCD-cameras is in progress. To solve staff problems, Kazakhstan, Tajikistan and Uzbekistan have begun to prepare national specialists in astronomy, and the Baltic States, Armenia, Azerbaijan, Georgia, Russia, and Ukraine continue to prepare astronomers.

Teaching of astronomy at schools is obligatory only in the Ukraine and partially in the Baltic states. To maintain a "common astronomical area", the Eurasian Astronomical Society (EAAS) continues its programme of reduced-price subscription to Russian-language astronomical journals and magazines in the territory of FSU, the organization of international conferences and Olympiads for school students, and lectures for school teachers and planetarium lecturers, etc.

Telescopes in Russia and other CIS territories permit to monitor an object more then 12 hours and can be used in global monitoring programmes. The Central Asian sites have some of the very best astro-climates in the world. They are similar to (or a little better than) the well known Chilean sites (median seeing 0.7″, very high fraction of clear nights, no light pollution and no high wind). It is imperative that these sites be protected and intensively used by the international astronomical community.

Keywords. Astronomy, former Soviet Union

---

1. General features of the development of astronomy in the USSR

- The "Golden Age" at the beginning of space-flights epoch.
- Equipment gradually getting out of date: large telescopes (specially, the optical ones), IR, sub-mm instruments, light receivers (mainly multi-channel), computers.
- Adherence to the "classic" fields of research.
- The world's largest experience in small-telescopes usage.
- Predominance of wide-field researches.
- Distinguished science-organizers and theoreticians.
- A high level of comprehension, analysis, interpretation of the results.

2. The situation in different regions

2.1. *The Baltic states*

• Most prosperous economically but scientific budget is less than 0.5% of GNP (compared with 2–4% in developed countries).

• Most alien from CIS politically – EC members.

• About 30 astronomers in each country working in 2–3 organizations.

• Contacts between astronomers of the Baltic States are still rare: shared journal "Baltic Astronomy" (created after the Soviet disintegration).

• Cooperation mainly with Western Europe.

• Fields of research mainly the same as under Soviet Union, with gradually increasing deviations.

• The internet and the main journals are accessible.

• Every country prepares astronomers of its own; the problem of retention of young scientists is not solved but the situation is improving progressively.

• Astronomical societies in every state are uniting professionals and amateurs.

• National language books and magazines for amateurs are being published.

• Astronomy is taught in schools (within physics courses), and manuals are available.

2.1.1. *Estonia*

• Financial base of science: 0.9 million euros (0.5% GNP): 10% via national grants, 20% via infrastructure development program.

• The Academy of Science as a national association of research institutions has been abolished.

• Astronomers: at Astronomical Observ. of Tartu University (Toravere, former IAAP), at the universities of Tartu and Tallinn.

• The staff reduced by a factor of 2 (Sweden experts were attracted used); ?40 researchers now.

• Salary: Euro 500–1000 per month.

• Equipment and instrumentation mostly inherited from SU: (150-cm telescope + CCD camera, a new one to be installed soon); real possibility to use the ESO telescopes and those at Canary Islands has opened and is taken advantage.

• The main fields of research remain the same: stellar physics and extragalactic astronomy (interpretation and theory), WR star observations (on Canary Islands now).

• Cooperation mainly with European countries (the most extensive one with Sweden) and Baltic States, at a lower level, with Russia and Finland.

• The problem of the young scientists is on the way of being solved.

2.1.2. *Latvia*

• Astronomers at: the Inst. of Astronomy (Univ. of Latvia, Riga), and Ventspils International Radioastronomy Centre (VIRC, Ventspils College, near Ventspils).

• Main fields of research: continuing: stellar astronomy, SLR, GPS and time service, ISM; growing: radio-astronomy; decreasing: solar astronomy.

• Collaboration: with western Europe (specially, Sweden), Baltic States, Russia. Basic equipment and infrastructure:

• Baldone Astronomical Observ. (is now part of Inst. of Astronomy): 0.8-m Schmidt tel. with small amateur-class CCD (new CCD is under installation); solar radio tel. RT-10 dismounted.

• Satellite laser station of Inst. Astronomy (at Riga) keeps world level of observations.

• VIRC: Radio telescope RT-32 is rebuilt out of a communication antenna left by the Russian Army in non-working order. Now it is the largest radio-telescope in north-west Europe. It is almost ready for VLBI work. Repairing of RT-16 (near RT 32) is now starting.

2.1.3. *Lithuania*

• Astronomers at: Inst. of Theoretical Physics and Astronomy (ITPA) of Vilnius Univ., Astronomical Observ. of Vilnius Univ. and State Inst. of Physics.
• Fields of research: continuing: Galactic structure, interstellar absorption, open clusters in the Galaxy, stars; new: asteroids (NEO), galaxies. For the last 10 years investigations of galaxies are successfully carried on at the Institute of Physics.
• Instrumentation: Moletai Observatory of ITPA: precision photometry, measurement of radial velocity of stars, CCD-photometry of stars and asteroids are carried out using 165-cm and 63-cm reflectors, 35-cm Maksutov meniscus telescope.
• Observations at Canary Islands (NOT), USA (Arizona), Turkey (Antalia), and the Subaru telescope are being intensified.
• Small-scale cooperation with CIS (with Russia and probably Uzbekistan).
• A planetarium was recently built in Vilnius (the only functioning one in all the Baltic States).

2.2. *East European CIS*

2.2.1. *Belarus*

• Still no professional astronomy.
• Several physicists at different pedagogical universities sometimes publish articles at the border of physics and astronomy.
• Planetarium continues to work in Minsk.
• There are some advanced amateur astronomers that carry out observations of variable stars. Some of them have publications in refereed journals. The amateurs cooperate with Ukrainian and Russian astronomers.

2.2.2. *Moldova*

• Very feeble professional astronomy: 60-cm telescope of Kishinev Univ. in working condition, but variable stars observers Smykov and Shakun are elderly and presently not active.
• Formerly active theoreticians: Chernobai died in 2004 and Gaina is not in full health.
• EAAS member Sholokhov teaches astronomy at Tiraspol Univ., where, moreover, an Astronomy Chair of Odessa Natl. Univ. has been recently opened (Andrievsky).
• Collaboration: with Romania and (a little) with Ukraine.

2.2.3. *Ukraine*

• About 1000 astronomers (1 per 50 000 citizens) work in 15 main astronomical observatories, institutes or departments (AO/AI/AD) of the National Academy of Science (NAS) or of National Universities (NU): in Kyiv: Main Astron. Observ. (MAO) NAS, AO & AD NU and Intl. (Ukraine-Russia) Centre for Astronomy, Medical and Ecological Res. (ICAMER); in Kharkiv: Inst. of Radio Astronomy (IRA NAS), AO & AD NU; in Nikolaev: (Mykolaev): AO NAS & AO NU; in Crimea: Crimean Astrophysical Observ. (CrAO) and former USSR Centre of Far Space Communication (near Yevpatoria, now a branch of IRA); Poltava: Gravimetrical Observ. NAS; AO NUs in Odessa, in Lviv, and in Uzhgorod – Crimean Lab. of Sternberg Astronomical Inst. (Moscow, Russia).
• Salary: $100–300 per month.

- Researches cover all the fields of modern astronomy and astrophysics.
- The first private astronomical observatory for professional basic research was created several years ago.
- A state programm for financial support of existing unique equipment is being started. Main instrumentation:
- Radio: Radio-telescopes: UTR-2 near Kharkiv (IRA); RT-70 near Yevpatoria (IRA branch); RT-22 in Katsievely (in Crimea, CrAO branch); radio-interferometer URAN (for $\lambda = 8 - 10$ m, baseline ~ 1000 km) consists of UTR and URAN-1 (IRA, Kharkiv) + URAN-4 (Odessa branch of IRA), + URAN-2 (Lviv Phys.-Tech. Inst.) + URAN-3 (at Poltava).
- Optics: in CrAO: 2.6-m, 1.25-m (for polarimetry), 1.22-m, large solar telescope, etc.; at Peak Terskol Observ.(ICAMER: Caucasus, Russia): 2-m telescope + échelle spectrographs (R up to 500 000) + professional CCD and horizontal solar telescope + spectrograph ($R = 320 000$); 1-m telescope at Mt Koshka (CrAO branch: Simeiz; Crimea); robotic meridian circle (AO NAS, Nikolaev).
- Gamma-rays: Cherenkov gamma-telescope at CrAO. Publications, societies, teaching and popularization:
- There are national scientific (four titles) and amateur-addressed periodicals in astronomy, both continued from the Soviet times and newly started.
- Internet, scientific literature accessible (with limitations in some places).
- Astronomical Societies: professional: Ukrainian Astronomical Assoc. (UAA), Odessa AS and Lviv AS; many amateur organizations.
- Astronomy is obligatory in schools; textbooks exist; conferences of teachers of astronomy are held on a regular basis.
- Planetariums: seven working ones; public lecture-hall at Odessa NU (the only one in CIS).
- Weekly TV-programmes on astronomy (in Odessa).
- Astronomy summer schools for young astronomers and for school teachers.
Some of the existing problems:
- At universities and astronomical observatories, financial support has been reduced (by 30% during 2005–6).
- Large age gap of staff at CrAO: serious danger that a very large experience in many fields of research in astrophysics may be lost. Main reason: no living quarters for young specialists.
- Peak Terskol Observ. (ICAMER): dangerous conflict with local (KBR) Russian authorities: the authorities are making attempts to privatize the infrastructure (hotel for observers, etc.).
- Attempts of buildings erection on the property of AO NU in the central part of Kyiv.
- Photo plate archives: no money to keep it in proper order, or to e-archive it.

2.3. Trans-Caucasus CIS

- Number of astronomers: several tens in each state.
- Besides one leading astronomical institution there are others related to astronomy.
- Internet is accessible, but with limitations and breaks; astronomical literature accessible, but with strict limitation (financial problems).
- National Universities prepare astronomers, but the lack of young specialists is an acute problem and is not solved.
- National professional Astronomical Societies were created during the last 15 years.

- The countries have passed through civil wars, deep economic crisis, breaks in power supply; close contacts of astronomical institutes' officers with the state leaders helped astronomy survive.
- Main optical instruments are saved and in working order.
- Researcher salary: $70–200 per month in Azerbaijan, $30–60 in Armenia and Georgia.

2.3.1. *Armenia*

- During the period of the war and blockade the Armenian Diaspora had helped (the Ajastan Foundation, headed by astronomer V. Petrosian); the Byurakan Astrophysical Observ. (BAO) director A. Petrosian held for some time the post of the Minister of Science.
- Energy deficit problem was solved after reviving the Metsamor Atomic Power station.
- The Armenian Astronomical Soc. (headed by A. Mikaelian) joins astronomers of Armenian nationality from all over the world (including, e.g., Terzian from Cornell Univ., USA).
- The journal "Astrofizika" (Astrophysics) is published in Russian and in English.
- Astronomers work at BAO NAS (Byurakan); Inst. for Radio Measurements (NIIRI NAS, Yerevan, dir. P. Herouni, Inst. of Space Astrophysics (Garni, G. Gurzadian), Yerevan Univ.
- The Cosmic Ray Station of Mt Aragats (3200 m) is functioning.
- The National Virtual Observ. is in planning, to be based on supercomputer of NAS and BAO photo-archive.
- Collaboration: with institutions in USA, Germany, France, Italy, Japan, Russia, etc.
- Scientific conferences in astronomy and Astronomical Olympiads for school students are held.
- In 2006 a summer school and scientific conference in honour of BAO's 60th anniversary was held.
- In 2007 Armenia is to host JENAM-2007 (Joint European and Natl. Astronomical Meeting).

Main instrumentation:
- BAO (1500 m, Mt Aragats slope, light pollution from Yerevan): 2.6-m telescope works with professional class CCD + modern spectrograph "SCORPIO" (up to 24 mag.) and 3D-spectrograph "VARG"; 1-m Schmidt telescope (in working condition, not in use); 0.5-m Schmidt telescope works; a radio-telescope is not in working condition.
- NIIRI testing area (about 1700 m, Mt Aragats slope): radio-optical telescope (54 m/2.6 m), not adjusted, no cooled receivers, no active specialists.
- Byurakan Station of St Petersburg State Univ. (at BAO): preserved, but does not operate now because of financial constraints. Property of Russia.

2.3.2. *Azerbaijan*

- Until 1997 there was no e-mail service, nor phone communication, so nothing was known about the fate of astronomy. In 1998 contacts were renewed.
- Significant "brain drain" to Turkey and Israel took place in early 1990s.
- Astronomers work at Shemakha Astrophysical Observ. NAS (ShAO, near Shemakha; office in Baku); at Nakhichevan highland station (created in 2003 on the base of Batabad Solar Observ.); at Scientific Centre "Caspian" (Baku); at Inst. Physics NAS (Baku).
- Main fields of research: Sun, comets, planets, Ap stars, variable stars, theoretical astrophysics.

- Since 1998 astronomical conferences and schools are held on a regular basis.
- Light pollution problem at ShAO has arisen but is kept under control.
- No national grants, but a number of astronomers have international grants.
- A new astronomical journal is being started this year (in English).

Instrumentation:

- ShAO: 2-m telescope was inactive for seven years, now at work; there are two amateur class CCD and a professional class one, three échelle spectrographs. Mirrors need aluminizing.
- Batabad Observ.: functions; the coronagraph has been damaged, now under repair.
- Paraga astrometry station (Paraga, Rep. Nakhichevan – an autonomous region of Azerbaijan, but geographically within Armenia) of Pulkovo Observ. (Russia): everything preserved. Mirrors need aluminizing. The station does not operate now.

2.3.3. *Georgia*

Situation in astronomy in Georgia is worse than in other Trans-Caucasus countries.

- Astronomers work at Abastumani Astrophysical Observ. (AbAO, Mt Kanabili, 1700 m), at AbAO Tbilisi Lab., Tbilisi National Univ.
- AbAO is an international standard of pure atmosphere, good seeing (average $\sim 1''$).
- In 1990s AbAO could survive, in a great measure thanks to Prof. J. Lominadze, then a member of the Central Election Commission.
- AbAO staff apartments: heated by metal chimneys. Breaks in power supply continue.
- Electronic communication is difficult.
- The problem of attracting young scientists remains completely unsolved.
- All the instrumentation (stellar, solar and atmospheric observatory facilities) are preserved, but need at least mirror aluminizing, change of electronic equipment at AZT-11 (1.25-m). Observations are carried out mainly by pure enthusiasm, (most actively by Omar Kurtanidze, the observatory director since May 2006): at 1.25 m + amateur-class CCD, 70-cm + amateur-class CCD, 40-cm + electro-photometer.

2.4. *Central Asia*

- Sites with the best astro-climate all over the former USSR territory (and generally, the best in Eurasia and all the eastern hemisphere of the Earth). They are similar to (or a little better than) Chilean astronomical sites: median seeing $0.7''$, clear sky, no light pollution and no high wind. These sites must be protected and intensively used by all the astronomical community!
- Issues of political stability.
- Salaries for scientists are low: from about \$10–20 (Turkmenistan) to \$20–60 (Uzbekistan).
- Internet and communications are available in the main cities, but not at the observatories.
- Attraction of young scientists is not a critical problem (reasons are different, see below).
- No astronomy in schools; only one planetarium (in Tashkent, Uzbekistan).

2.4.1. *Kyrgyzstan*

- No professional astronomy currently, nor in foreseeable future.
- The city Osh, is a transit point on the way to East Pamirs, including Shorbulak Observ. (4350 m), but in the post-Soviet epoch this is not in use, and has been inspected only once.

2.4.2. *Tajikistan*

During the civil war there was no salary for scientists for one year. Many astronomers emigrated to Germany, Russia, Ukraine, or went into business.

Presently:

- Good contacts of astronomers with the Government.
- 27 researchers (9 young ones) + probationers and post-graduate students.
- Astronomers work at Inst. of Astronomy (IA) of NAS at Dushanbe and its branch, Guissar Observ.; Tajik State Natl. Univ. (Dushanbe) has recently created a Dept. of Astronomy.
- Fields of research: comets, asteroids, meteors, variable stars, star formation regions, structure of galaxies, seismic-ionosphere effects, geostationary satellites observations.
- Astronomers welcome international collaboration.
- Very good astro-climate at mountain observatories (see below for details).
- Free access to internet and to main astronomical journals exists for about the last three years (a result of international help).
- There are small state grants for researchers; astronomers have one international research grant.
- Tajik Astronomical Soc. (scientists + pedagogical university teachers) and EAAS branch exist.

Equipment:

- Mirrors of all refractors need re-aluminizing.
- Guissar Observ. of IA NAS (about 30 km west of Dushanbe along the valley): the wide-field camera VAU-75 is "alive", but not used; 0.7-m and 0.4-m telescopes: photoelectric photometry.
- Mt Sanglokh Observ. of IA NAS (2300 m): median seeing is a little better than on Mt Maidanak: $0.6 - 0.7''$): the road and bridges are restored; is used now for meteor observations only; 1-m and 0.6-m telescopes are "alive" but not in use.
- Shorbulak Observ. (4350 m, on a hill on East Pamirs plateau), good site for sub-millimetre astronomy (1–2 mm of precipitate water); former branch of Pulkovo Observ. (Russia), now belonging to IA NAS: 70-cm telescope, in 2001 was in order, but not in use.
- Ak-Arkhar Cosmic Ray station (East Pamirs, 4200 m): does not work (?).

2.4.3. *Turkmenistan*

- Academy of Science has been abolished; the Phys.-Tech. Inst. that used to host several astronomical teams (solar investigations, meteors, radioastronomy) is closed; the State does not need basic research (including physics and astronomy). The fall of a meteorite (a chondrite of a mass of about 1 ton) named after Turkmenbashi could not awake any interest towards astronomy.
- S. Mukhamednazarov is a member of the President Council on Science. This permits keeping the main optical telescopes in working state for use by visitors.
- Internet and e-mail resources are strictly limited and under severe government control.

Equipment:

- Optical Astronomical instrumentation, including 1-m (focus = 2 m) telescope and 0.8-m telescope of AO NU of Odessa (Ukraine) at Mt Dushak-Erekdag (2300 m, very good astro-climate) are kept in working order (for climatological and ecological needs), but are used only by visitors from abroad (Dorokhov & Dorokhova from Odessa, Ukraine).

- The radio telescope RT-12 and metal components of RT-32 (for the initial six dishes Soviet-era version of "Quasar" VLBI system) are probably lost.
- Solar Observ. (at Vanovskoe, near Firyuza) does not work (no longer in existence?).

2.4.4. Uzbekistan

- Astronomers work at AI NAS (Tashkent + Kitab), at the Astronomical Depts. of Tashkent, Samarkand and Karshi universities.
- There are young observing assistants but mostly without astronomical education.
- Salary: $20–60 per month. Additional earnings are necessary.
- Financial support of science: only grants for institutions, no basic support; AI NAS has seven national grants of about $ 1000/year/person (for salary, travels, equipment, tax, etc.).
- The present director of AI NAS Sh. Ehgamberdiev held for several years the post of the President Councilor on Science.
- AI NAS has custom privileges for equipment importing; Maidanak Observ. has special financial support as a unique scientific object; the total AI NAS budget has increased several times up to $50 000–100 000/year.
- Space Agency has been created and is working.
- Fields of research as under USSR: astrometry, solar physics, helio-seismology, variable stars; physics of galaxies, with the addition in recent years of international collaborative AGN monitoring.
- Cooperation: mainly with South Korea, Russia, Ukraine, West Europe, USA.
- A planetarium is in operation.
- No astronomy at schools.
Equipment and infrastructure:
- Kitab: latitude station (branch of AI NAS, in valley near Kitab): classic astrometric equipment and GPS.
- Mt Maidanak Observ. (branch of AI NAS, 2600 m, three peaks several km from each other, including Kurganak, with no telescope yet): 1.5-m telescope + professional CCD (2000 × 800 pixels) for imaging and photometry; 1-m telescope + amateur class CCD; five 60-cm and one 48-cm telescopes (photoelectric photometers + two amateur class CCDs);
- Helio-seismology station of AI NAS: moved from Mt Kumbel (2300 m, near Mt Big Chimgan) to Parkent (already at work?).
- Military facilities for satellite laser location (Mt Maidanak, 2600 m): two 1.1-m quick-rotating telescopes (in working order, but not used since 2004).
- Plato Suffa (2500 m): the site for Russian-Uzbek mm range RT-70: temporarily shut down.
- Samarkand Univ. station (on the outskirts of the city): 0.48-m telescope is installed, and a 1-m is under construction.

2.5. Kazakhstan and Russia

2.5.1. Kazakhstan

- The Academy of Science as an independent institution has been abolished and replaced with scientific or science-technical centres joining, under three research institutes.
- National Space Research Program (SRP) is carried on since 2005 in cooperation with Russia.
- There is one astronomical institution: the Fessenkov Astrophysical Inst. (API). Now API is a part of Astrophysical Research Centre (other parts are National Space Research Inst. and Ionosphere Research Inst.).

- There are about 30 astronomers: in API and Almaty (Alma-Ata) State Univ. (ASU); two astronomers are still members of Sternberg Astronomical Inst. (Moscow, Russia)
- Staff researcher salary: $150–300 per month.
- Astronomy students are trained at ASU and in Lomonosov Moscow State Univ.
- The young scientists problem is not still solved. Several graduates are on API staff.
- The main fields of research are as in the Soviet epoch: dynamics of gravitating systems, spectrophotometry, polarimetry, ISM, AGN (observations and theory); solar investigations are reduced to solar cosmic rays propagation; atmosphere optics research is stopped. Satellite monitoring and theory of their motion are open in frame of SRP.
- Collaboration: with Russia; on a small level with Germany, UK and USA.
- There is a branch of EAAS.
- Internet and e-mail facilities: at API main building since 2005, before it, home addresses were used. Mostly no e-communications with API observatories yet.
- API Bulletin is published (in Russian).
- No astronomy in schools. Equipment and infrastructure:
- API main site at Kamenskoe Plato (suburbs of Almaty, 1500 m): used mainly 70-cm refractor (with 3-cascade image tube and amateur class CCD); 70-cm Maksutov meniscus telescope and other instruments in working order.
- Tian-Shan Observ. of API (2700 m, about 40 km from Almaty, former branch of Sternberg Astron. Inst., Moscow): place with very good atmospheric transparency, but bad seeing; used for photoelectric photometry; the road is in poor condition; two 1-m telescopes (one in near working order, another in non-working order; neither is used: no light receivers); 48-cm telescope with photoelectric photometer and amateur CCD is in regular use; solar instruments out of order; the lab. building and the living quarters are in excellent condition, supported on commercial basis ("scientific tourism"). Nearby former API Solar Observ. is not functioning.
- High Altitude station of Ionosphere Research Inst. (very close to API Tian-Shan Observ.): 10-m rotating dish for ionosphere radio-observatory was functioning; now under repair for SRP.
- Cosmic Ray station (3340 m, near API Tian-Shan Observ.) of Lebedev Physical Inst. RAS (Moscow, Russia) is functioning.
- Assy-Turgen Observ. of API (2800 m plateau, about 100 km from Almaty): electricity from diesel engine; 1-m telescope (+ spectropolarimeter) is used by visiting API observers; empty tall dome for 1.5-m telescope (erected to the end of 1980s) needs complete rebuilding.

2.5.2. *Russia – several topics only*

- Russian astronomers have lost real access to the main part of the FSU observatories.
- Practically all the astronomical infrastructure Russia has inherited from the USSR has been preserved. Many places are presently equipped with new light receivers. New equipment: In post-Soviet epoch new astronomical instruments and even new observatories equipped with modern facilities have been created:
- Radio-interferometer network "Quazar" consists of three radio-telescopes RT-32 located in new sites: at Svetloe near St Petersburg; near Zelenchukskaya (North Caucasus); at Badary valley, (Rep. Buryatia, Sayany mountains, Siberia);
- RT-64 at Kalyazino;
- Ukraine-Russian Observ. at Peak Terskol (3120 m) equipped by 2-m telescope and large horizontal solar telescope (both with high resolution échelle spectrographs), modern hotel and telecommunication facilities;
- 1.7-m IR telescope of ISTP (Irkutsk) at Mondy Observ. (Rep. Buryatia);

- 1.5-m telescope of Kazan SU in Antalia (Turkey).
- E-mail and internet service accessible everywhere.
- Facilities for optical and radio observations cover eight time zones and could be used for international programmes of 24-h monitoring of astronomical objects (with other CIS astronomical observatories).
- Astronomer preparation system is being "upgraded": Departments of astronomy are newly open in Tomsk and Stavropol State universities, many students can get a good training at SAO RAS.
- There is a system of financial support for active scientists: RFBR grants, Scientific School Support Program, Government research grants for young doctors of science, etc.
- Main problem. The government frequently does not recognize the significance of basic science and takes measures towards destroying the existing basic science infrastructure (as well as towards weakening the system of preparation of young scientists). Up to now it has been possible to "dampen" most of such attempts, but government pressure remains strong. Therefore the fate of fundamental science (astronomy, in particular) in Russia remains unclear. The same is true in respect to Russian Academy of Sciences and the universities.

3. Conclusions

- Overall, astronomy in all of the former Soviet territory has managed to preserve much of its former infrastructure, human potential and "common astronomical area" (partially shared by some former communist countries).
- All this has been achieved in spite of policies of our countries and thanks to the active position of the "leaders" of our national astronomical communities, and, to EAAS activity.
- Astronomy has suffered less damage compared with other basic sciences.
- In many fields the degradation process is now giving way to modest steps towards upgrading the technical base. However the rate of development is far below the world level.
- CIS astronomers seek counterparts for more effective usage of the available infrastructure. Main goals include: a progress in 24-h monitoring programmes; and increasing usage of unique world-class astro-climate of astronomical sites in Central Asia (Mt Maidanak, Mt Sanglokh, etc.).

Astronomy for the developing world
IAU Special Session no. 5, 2006
J.B. Hearnshaw and P. Martinez, eds.

© 2007 International Astronomical Union
doi:10.1017/S174392130700693X

Relativistic astrophysics and cosmology in Uzbekistan

Bobomurat J. Ahmedov[1,2,3], Roustam M. Zalaletdinov[1,4], Zafar Ya. Turakulov[1,2,3], Salakhutdin N. Nuritdinov[2,3] and Karomat T. Mirtadjieva[2,3]

[1]Institute of Nuclear Physics, Ulughbek, Tashkent 702132, Uzbekistan

[2]Ulugh Beg Astronomical Institute, Astronomicheskaya 33, Tashkent 700052, Uzbekistan

[3]National University of Uzbekistan Physics Faculty, Tashkent, 700174, Uzbekistan

[4]Department of Mathematics, Statistics and Computer Science, St Francis Xavier University, Antigonish, NS, Canada B2G 2W5

email: ahmedov@astrin.uzsci.net, nurit@astrin.uzsci.net, mkt@astrin.uzsci.net

Abstract. The theoretical results obtained in Uzbekistan in the field of relativistic astrophysics and cosmology are presented. In particular electrostatic plasma modes along the open field lines of a rotating neutron star and Goldreich-Julian charge density in general relativity are analyzed for the rotating and oscillating magnetized neutron stars. The impact that stellar oscillations of different type (radial, toroidal and spheroidal ones) have on electric and magnetic fields external to a relativistic magnetized star has been investigated. A study of the dynamical evolution and the number of stellar encounters in globular clusters with a central black hole is presented. Perturbation features and instabilities of the large-scale oscillations on the background of the non-linearly pulsating isotropic and isotropic Ω-models are studied. The non-stationary dispersion equation of the sectorial perturbations for the general case and the results of certain oscillation mode analysis are given. The model composed as the linear superposition of two other models was constructed and the stability of this model is studied. In a cosmological setting the theory of macroscopic gravity as a large-distance scale generalization of general relativity has been developed. Exact cosmological solutions to the equations of macroscopic gravity for a flat spatially homogeneous, isotropic space-time are found. The gravitational correlation terms in the averaged Einstein equations have the form of spatial curvature, dark matter and dark energy (cosmological constant) with particular equations of state for each correlation regime. Interpretation of these cosmological models to explain the observed large-scale structure of the accelerating Universe with a significant amount of the nonluminous (dark) matter is discussed.

Keywords. Theoretical astronomy, astronomy in Uzbekistan, rotating oscillating neutron stars, general relativity, dark matter, dark energy

1. Introduction

Theoretical Astrophysics is the subject which has got an essential development in Uzbekistan during the last decade, especially through the newly established collaborations with advanced western and eastern institutions. Our regional collaboration is supported by the Abdus Salam International Center for Theoretical Physics (ICTP, Trieste, Italy), Academy of Sciences for Developing World (TWAS, Trieste, Italy) and the Inter-University Centre for Astronomy and Astrophysics (IUCAA, Pune, India) in the framework of BIPTUN (Bangladesh - India - Pakistan - Turkey - Uzbekistan) Network on Relativistic Astrophysics and Cosmology. Another important scientific collaboration is with western partners, mainly at the International School for Advanced Studies (SISSA, Trieste, Italy), International Center for Relativistic Astrophysics (ICRA, Rome-Pescara,

Italy), Dalhousie University (Halifax, Canada), St Francis Xavier University (Antigonish, Canada) and at the ICTP. Local scientific activity in Theoretical Astrophysics in Uzbekistan is partly supported through the Affiliation Scheme (ICAC-83) of the ICTP. Scientific collaboration of the Department of Galactic Astronomy and Cosmology in UBAI is with Kharkov State University (Ukraine), Sternberg Astronomical Institute (SAI) at Moscow State University, St Petersburg State University, the Astrophysical Institute (AIP, Potsdam) and international funding is mainly through Centre of Science and Technology in the Ukraine.

Research on Theoretical Astrophysics in Uzbekistan is concentrated in Ulugh Beg Astronomical Institute (UBAI) and Institute of Nuclear Physics (INP) of the Uzbekistan Academy of Sciences, and in the National University of Uzbekistan. UBAI is one of the oldest astronomical institutions of the Former Soviet Union. It was founded in 1873 and is the oldest one in the Central Asian area (where Uzbekistan lies), which has absolute maximum of clear sky time for the whole Eurasian continent. This makes the area particularly important for optical astronomical observations. As a result of the site-testing expeditions organized by UBAI and SAI in the early 1970s, Maidanak mountain (2700m) located 120 km south of the famous historical city of Samarkand, was selected for an observatory. In August 1996 a seeing monitoring at Mt Maidanak was started with Differential Image Motion Monitor of ESO, designed by M. Sarazin and used for Paranal and La Silla site-testing. After one year the results of the seeing measurements showed a very high quality of seeing conditions at Mt Maidanak.

Scientific projects in UBAI are related to the theoretical and observational research in galaxies, photometric observation of eclipsing binaries, observational studies of young stars, the study of solar activities etc., and two projects are related to theoretical astrophysics. These are F2.2.01 "Gravitational lenses and collapsing galaxies" and F2.2.06 "Dynamics of gravitating systems and electromagnetic fields around compact objects" where encounters in globular clusters and galaxies with a central black hole and electromagnetic fields and waves around magnetized neutron stars are investigated.

The Institute of Nuclear Physics (INP) in Tashkent is the biggest one in Central Asia. It was founded in year 1956. It has one project related to theoretical astrophysics: F2.1.09 entitled "Investigation of equations of gravitation and electrodynamics in relativistic astrophysics and cosmology".

Several scientists deliver lectures at the Astronomy Department of the National University of Uzbekistan, which was founded in the year 1918. The Astronomy Department of the University trains Bachelors, Masters and PhD students. The scientific activity of the Department members are in the fields of formation and evolution of elliptical galaxies, physics of globular clusters, dynamics of galaxy clusters, physics of quasars, and close binary systems. In particular the influence of the third star on close binary system is investigated there.

In this paper, the theoretical results obtained in Uzbekistan in the field of relativistic astrophysics of compact objects and relativistic cosmology will be presented. In particular electrostatic plasma modes along the open field lines of a rotating neutron star and Goldreich-Julian charge density in general relativity are analyzed for rotating and oscillating neutron stars. For a steady-state Goldreich-Julian charge density the usual plasma oscillation along the field lines are found; the plasma frequency resembles the gravitational redshift close to the Schwarzschild radius. The equations that describe the electromagnetic processes in a plasma surrounding a neutron star are obtained by using the general relativistic form of Maxwell's equations in a geometry of a slow-rotating gravitational object. A new mechanism of the generation of azimuthal current under the gravito-magnetic effect on radial current in a plasma around a neutron star is predicted.

The impact that stellar oscillations have on electric and magnetic fields external to a relativistic magnetized star has been investigated. The solution of the general relativistic Maxwell equations both in the vicinity of the stellar surface and far from it has been found. The general relativistic energy loss through electromagnetic radiation for different types (radial, toroidal and spheroidal) of oscillations of relativistic magnetized stars has been calculated. Analytic solutions of Maxwell's equations in the internal and external background space-time of a slowly rotating misaligned magnetized neutron star have been obtained. A formula for magneto-dipolar radiation of rotating stars derived in general theory gives a correction for electromagnetic energy loss of 2–6 times, depending on the compactness of the neutron star.

In a cosmological setting, the theory of macroscopic gravity is a large-distance scale generalization of general relativity. The averaged Einstein equations of macroscopic gravity are modified by the gravitational correlation tensor terms as compared with the Einstein equations of general relativity, the tensor satisfying an additional system of equations. Exact cosmological solutions to the equations of macroscopic gravity for a flat ($k = 0$) spatially homogeneous, isotropic space-time are found. The gravitational correlation terms in the averaged Einstein equations have the form of spatial curvature, dark matter and dark energy (cosmological constant) with particular equations of state for each correlation regime. The interpretation of these cosmological models to explain the observed large-scale structure of the accelerating Universe with a significant amount of the nonluminous (dark) matter is discussed.

In the following two sections we will provide examples of theoretical research carried out in our groups.

2. Collapsing models of early evolution stages of self-gravitating systems: non-linear spherical and disk-like models

One of the urgent problems of present-day astrophysics, the galactic and extra-galactic astronomy, is to study the early non-stationary evolution stages and development processes of galaxy large-scale structure (Binney & Merrifield 1998) of galaxies and other self-gravitating systems. This problem calls for complex numerical calculations with modern computers, or constructing non-linear non-stationary models as well as studying the problem of their stability. The second method seems more preferable, as at the numerical experiments it is difficult to reveal some non-linear effects and to take notice of gravitational instabilities that originate in proper time.

The non-stationary phase-pulsating models, which are adequate for the investigation of the instabilities of early evolution stages of galaxies and their superclusters, have been constructed (Nuritdinov 1983, Nuritdinov 1991). In particular, we obtained the models that generalize the Einstein and Kamm equilibrium models for spherical self-gravitating systems in the non-stationary case.

The phase density of spherical self-gravitating system in the generalized Einstein model is

$$\Psi_1(r, v_r, v_\perp, t) = \rho(t)[2\pi v_b(t)]^{-1}\delta(v_r - v_a)\delta(v_\perp - v_b)\chi(R_0\Pi - r)\left[1 + \Omega\frac{v_\perp}{v_b}\sin\theta\sin\eta\right].$$
$$(2.1)$$

The phase density of spherical self-gravitating system in the generalized Kamm's model is

$$\Psi_2(r, v_r, v_\perp, t) = \rho\Pi^4/(\pi^2 R^2\Omega_0^2)\cdot f^{-1/2}\cdot\chi(f)\left[1 + \Omega\frac{r\cdot v_r}{\Omega_0 R_0^2}\sin\theta\sin\eta\right], \qquad (2.2)$$

where variables are

$$f = \left(1 - r^2/R^2\right) \cdot \left(\Omega_0^2 R^2/\Pi^4 - v_\perp^2\right) - (v_r - v_a)^2, \quad v_a = \frac{\Omega_0 \lambda r \sin \psi}{\sqrt{1 - \lambda^2 \Pi^2}}, \quad v_b = \frac{r \cdot \Omega_0}{\Pi^2}, \quad (2.3)$$

and

$$\Pi\left(\psi\right) = \frac{1 + \lambda \cos \psi}{1 - \lambda^2}, \quad \sin \theta = \frac{\sqrt{x^2 + y^2}}{r}, \quad \eta = \arctan\left(\frac{v_\varphi}{v_\theta}\right). \quad (2.4)$$

Here $\rho(t)$ is the matter density in physical space and it does not depend on r; δ and χ are the symbols of Dirac and Heaviside functions; v_φ and v_θ are the azimuthal and meridian components of velocity $v_\perp$; $\Pi(\psi)$ is the expansion factor of sphere; the variable ψ is connected with t by correlation $\Omega_0 t = (1 - \lambda^2)^{-3/2}(\psi + \lambda \sin \psi)$; Ω_0 is the frequency of star revolution in its orbit of the stationary system; $\lambda = 1 - (2T/|U|)_0$ is the pulsation amplitude, where $(2T/|U|)_0$ is the virial ratio with $t = 0$, T and U being the system kinetic and potential energies, respectively. Ω is the angular velocity of the system rotation.

Real evolution of collapsing galaxies is described by the function $\Psi + \Delta\Psi_i$ which should satisfy the system of Jeans-Poisson equations. The above-mentioned models are applicable to early evolution stages of elliptical E and lens-like S galaxies. For galaxy superclusters which are Zeldovich pancakes, the construction of disk-like non-linear non-equilibrium models is required. In this case the isotropic model for phase density

$$\Psi_{Is} = \frac{\sigma_0}{2\pi \, \Pi\sqrt{1 - \Omega^2}} \left[\frac{1 - \Omega^2}{\Pi^2}\left(1 - \frac{r^2}{\Pi^2}\right) - (v_r - v_a)^2 - (v_\perp - v_b)^2\right]^{-1/2} \quad (2.5)$$

is constructed.

This model is non-stationary generalization of the Zeldovich-Bisnovatiy-Kogan equilibrium disk and on this basis a number of new non-stationary models for early evolution stage of self-gravitating disk-like systems with the anisotropic velocity diagrams (Nuritdinov 1992, Mirtadjieva, 2003) is constructed:

$$\Psi_A = \int\limits_{-1}^{+1} \Psi_{Is} A(\Omega') d\Omega' . \quad (2.6)$$

For $A(\Omega') = (2/\pi)\left(1 - \Omega'^2\right)^{1/2}(1 + \Omega \cdot \Omega')$

$$\Psi_A^1 = \frac{\sigma_0}{\pi}\left(1 + \Omega'\frac{[x v_y - y v_x]}{R_0^2}\right)\chi(D) \quad (2.7)$$

For $A(\Omega') = 1/2$

$$\Psi_A^2 = \alpha \cdot F(\beta, k) , \quad (2.8)$$

where $F(\beta, k)$ is spatial function in the form of the Jacobi elliptical integral.

For $A(\Omega') = (8/3\pi) \cdot (1 - \Omega'^2)^{3/2}$

$$\Psi_A^3 = \frac{2\sigma_0}{3\pi}\left\{1 + \frac{r^2}{\Pi^2} + \Pi^2\left[(v_r - v_a)^2 + v_\perp^2\right] - 3r^2 v_\perp^2\right\}\chi(D) . \quad (2.9)$$

For $A(\Omega') = \Omega'^4 \cdot (1 - \Omega'^2)^{1/2}$

$$\Psi_A^4 = \frac{\sigma_0}{\pi}\left(3D^2 + 24D \cdot r^2 v_\perp^2 + 8r^4 v_\perp^4\right)\chi(D) . \quad (2.10)$$

The method of constructing non-linear models is based on the linear superposition of any two above-stated models. By this way we can obtain a new anisotropic model with the compound nature since their gravitational potentials and surface densities coincide:

$$\Psi(\vec{r}, \vec{v}, \Omega_1, \Omega_2, \lambda, \nu, t) = (1 - \nu) \cdot \Psi_i(\vec{r}, \vec{v}, \Omega_1, \nu, t) + \nu \cdot \Psi_j(\vec{r}, \vec{v}, \Omega_2, \nu, t) \qquad (2.11)$$

here ν is the parameter of superposition.

This method allows us to construct the multiple parameter models which describe conditions of systems being close to real existing early stages of its evolution. On the other hand, by this way we construct a model which covers intermediate conditions between two discrete non-stationary configurations.

3. Averaging problem in cosmology and macroscopic gravity

Macroscopic gravity is a non-perturbative geometrical approach proposed in Zalalet-dinov (1992) - Zalaletdinov (2003) to resolve the averaging problem of general relativity by its reformulation in a broader context as the problem of the macroscopic description of gravitation. The classical physical phenomena are known to possess two levels of description: the microscopic description by discrete matter models and the macroscopic description by the continuous matter models. Transition from a microscopic picture to a macroscopic one is accomplished by means of a space-time or ensemble averaging procedure. Lorentz' theory of electrons and Maxwell's electrodynamics are known examples of a microscopic theory and its macroscopic theory

$$F^{\mu\nu}_{,\nu} = \tfrac{4\pi}{c} j^\mu = 4\pi \sum_i q_i u^\mu(t_i) \quad \rightarrow \langle \text{avg} \rangle \rightarrow \quad H^{\mu\nu}_{,\nu} = \tfrac{4\pi}{c} \langle j \rangle^\mu = \tfrac{4\pi}{c}(J^\mu - c P^{\mu\nu}_{,\nu})$$

$$F_{[\alpha\beta,\gamma]} = 0 \qquad \rightarrow \langle \text{avg} \rangle \rightarrow \quad \langle F \rangle_{[\alpha\beta,\gamma]} = 0, \ H^{\mu\nu} = \langle F \rangle^{\mu\nu} + 4\pi P^{\mu\nu}$$

$$(3.1)$$

where avg is averaging.

For general relativity, such a task is much more complicated because of (1) the necessity to define covariant space, space-time volume or statistical averages on Riemannian space-times, (2) the Riemannian geometry of space-time, the nonlinear structure of the field operator of the Einstein equations and necessity to deal with gravitational field correlators, and (3) the problem of construction of models of smoothed, continuously distributed self-gravitating media $T^{(hydro)}_{\alpha\beta} = \langle T^{(micro)}_{\alpha\beta} \rangle$. The Einstein equations themselves can be shown to be insufficient to be consistently averaged out.

$$\langle R_{\alpha\beta} \rangle - \frac{1}{2} \langle g_{\alpha\beta} g^{\mu\nu} R_{\mu\nu} \rangle = -\kappa \langle T^{(micro)}_{\alpha\beta} \rangle \Rightarrow \langle R_{\alpha\beta} \rangle - \frac{1}{2} \langle g_{\alpha\beta} \rangle \langle g^{\mu\nu} \rangle \langle R_{\mu\nu} \rangle + C_{\alpha\beta}$$

$$= -\kappa \langle T^{(micro)}_{\alpha\beta} \rangle. \qquad (3.2)$$

They become a definition of the correlation function $C_{\alpha\beta}$ unless this object is defined from outside the averaged Einstein equations. In macroscopic gravity provides a formalism for a derivation of the averaged Einstein's equations which take the form

$$\bar{g}^{\alpha\epsilon} M_{\epsilon\beta} - \frac{1}{2} \delta^\alpha_\beta \bar{g}^{\mu\nu} M_{\mu\nu} = -\kappa \langle T^{\alpha(micro)}_\beta \rangle + (Z^\alpha{}_{\mu\nu\beta} - \frac{1}{2} \delta^\alpha_\beta Q_{\mu\nu}) \bar{g}^{\mu\nu} \qquad (3.3)$$

where $\bar{g}^{\mu\nu}$ is the averaged metric, $M_{\mu\nu}$ is the Ricci tensor of the Riemannian curvature $M^\alpha{}_{\beta\gamma\delta}$ which stands for the induction tensor, $Z^\alpha{}_{\mu\nu\beta} - \frac{1}{2} \delta^\alpha_\beta Q_{\mu\nu}$ is a correlation tensor constructed from the correlation connection tensor,

$$Z^\alpha{}_{\beta\gamma}{}^\mu{}_{\nu\sigma} \equiv Z^\alpha{}_{\beta[\gamma}{}^\mu{}_{\nu\sigma]} = \langle \Gamma^\alpha{}_{\beta[\gamma} \Gamma^\mu{}_{\nu\sigma]} \rangle - \langle \Gamma^\alpha{}_{\beta[\gamma} \rangle \langle \Gamma^\mu{}_{\nu\sigma]} \rangle \qquad (3.4)$$

where angular brackets denote a space-time volume averaging (Zalaletdinov (1992) - Zalaletdinov (1993)). There correlation tensors and another non-Riemannian curvature tensor $R^\alpha{}_{\beta\gamma\delta}$ which stand for polarization tensors and average field tensors satisfy a set of non-linear partial differential equations.

An exact cosmological solution to the equations of macroscopic gravity has been obtained by Coley, Pelavas & Zalaletdinov (2005) for a flat spatially homogeneous, isotropic macroscopic space-time given by the Robertson-Walker line element

$$ds^2 = -dt^2 + a^2(t)(dx^2 + dy^2 + dz^2) \tag{3.5}$$

for the constant macroscopic gravitational correlation tensor

$$Z^\alpha{}_{\beta\gamma}{}^\mu{}_{\nu\sigma} = \text{const} \tag{3.6}$$

which means that the macroscopic gravitational correlations are assumed to remain unchanged in time and space. The macroscopic gravity equations can be solved to show that there is finally only one remaining independent component $Z^3{}_{23}{}^3{}_{32}$ determined through an integration constant.

The equation of state for the macroscopic gravitational correlation field

$$p_{grav} = p_{grav}(\rho_{grav}) = -\frac{1}{3}\rho_{grav} \tag{3.7}$$

is determined by the equations of macroscopic gravity and the averaged Einstein equations with an averaged matter distribution taken to be a perfect fluid read

$$\left(\frac{\dot{a}}{a}\right)^2 = \frac{\kappa\rho_{mat}}{3} + \frac{\varepsilon}{3a^2}, \tag{3.8}$$

$$2\frac{\ddot{a}}{a} + \left(\frac{\dot{a}}{a}\right)^2 = -\kappa p_{mat} + \frac{\varepsilon}{3a^2}, \tag{3.9}$$

with a given equation of state $p_{mat} = (\rho_{mat})$, where $\varepsilon/\kappa a^2 = \rho_{grav}$ is the macroscopic gravitational correlation energy density and $-\varepsilon/3\kappa a^2 = p_{grav}$ is the isotropic pressure of macroscopic gravitational correlation field. They look similar to the Einstein equations of general relativity for either a closed or an open spatially homogeneous, isotropic Friedman-Lemaitre-Robertson-Walker space-time, but they do have different mathematical and physical, and therefore, cosmological content since

$$\frac{\varepsilon}{3} = \frac{\kappa\rho_{grav}a^2}{3} \neq -k \tag{3.10}$$

in general due to the presence of macroscopic gravitational correlation terms of a spatial curvature term $\varepsilon/3a^2$. This macroscopic gravitational correlation terms is a candidate for a "dark" cosmological agent with the following physical properties.

4. Conclusions

We may conclude for the results presented in sections 2 and 3. Early stages of galaxy formation and its large scale structures are investigated by studying instabilities of the models specified with respect to structural perturbations (Nuritdinov *et al.* 1997, Nuritdinov *et al.* 2003).

A criterion of elliptical galaxy formation is found which reads that the initial total kinetic energy of a non-rotating system must not exceed 4,2% of the total initial potential energy. In this process the Jeans instability mechanism for almost radial motion acts. Besides, the process of elliptical galaxy formation goes in the same way if the percentage specified lies in the interval (7–19%), but in this case another instability mechanism acts which has an oscillatory and resonance character. When the rotation effect is included, the percentage mentioned above grows.

A new kind of instability is discovered for pulsating spherical and disk-like galaxies, which has a combinatory-resonance character.

It is found for the first time that, among the modes studied, the increment of instability of pear-shaped perturbations is apparently larger than that of all the remaining modes, including the bar mode.

A formation criterion is found also for SB galaxies: the initial value of the total kinetic energy must not exceed 10,4% of the initial potential energy of a non-rotating system.

Instabilities of small scale modes are studied and their applications are specified.

Thus, the physics of instability of non-linear non-equilibrium models, which describe the early stages of the evolution of galaxies and their systems, differs significantly from that of the corresponding equilibrium configurations.

The dark spatial curvature term has the following physical properties as a cosmological agent:

(A) it interacts only gravitationally with the macroscopic gravitational field,

(B) it does not interact directly with the energy-momentum tensor of matter,

(C) it exhibits a negative pressure $p_{grav} = -\frac{1}{3}\rho_{grav}$ which tends to accelerate the Universe when $\rho_{grav} > 0$.

Only if one requires $12Z^3{}_{23}{}^3{}_{32} = -\varepsilon$ to be

$$\varepsilon = -3k \tag{4.1}$$

then the macroscopic (averaged) Einstein equations become exactly the Einstein equations of general relativity for either a closed or an open spatially homogeneous, isotropic space-time for the macroscopic geometry of a flat spatially homogeneous, isotropic space-time.

This exact solution of the macroscopic gravity equations exhibits a very non-trivial phenomenon from the point of view of the general-relativistic cosmology: the macroscopic (averaged) cosmological evolution in a flat Universe is governed by the dynamical evolution equations for either a closed or an open Universe depending on the sign of the macroscopic energy density ρ_{grav} with a dark spatial curvature term $\kappa\rho_{grav}/3$. From the observational point of view such a cosmological model provides a possibility to formulate a new paradigm to reconsider the standard cosmological interpretation and treatment of the observational data. Indeed, this macroscopic cosmological model has the geometry of a flat spatially homogeneous, isotropic space-time. Therefore, all measurements and data are to be considered and designed for this geometry. The dynamical interpretation of the data obtained should be considered and treated for the cosmological evolution of either a closed or an open spatially homogeneous, isotropic space-time.

Acknowledgements

This research is supported in part by the UzFFR (project 01-06) and projects F2.1.09, F2.2.01, F2.2.06 and A13-226 of the UzCST. BJA acknowledges the partial financial support from NATO through the reintegration grant EAP.RIG.981259. RMZ acknowledges the partial financial support by the RCD Grant from NSERC of Canada and the W.F. James Chair research grant at St Francis Xavier University.

References

Binney, J., Merrifild, M. 1998, *Galactic Astronomy*, Princeton University Press, 1998

Nuritdinov, S.N. , 1983 *Sov. Astronomy*, v. 27, p. 24

Nuritdinov, S.N. , 1991 *Sov. Astronomy*, v. 35, p. 377

Nuritdinov, S.N. , 1992 *Astron. Tsir.*, N. 1553, p. 9

Nuritdinov, S.N., Gaynullina, E.R., Mirtadajieva, K.T. , 1997 *Central regions of the Galaxy and gaalxies*, IAU Symposium 184, Kyoto, p. 49

Nuritdinov, S.N., Mirtadajieva, K.T., Tadjibaev, I.U. ,2003 *ASPS*, v. 316, p. 377

Mirtadajieva, K.T. ,2003 *Uzbek Journal of Physics.*, v. 5, p. 223

R.M. Zalaletdinov, Gen. Rel. Grav. **24** (1992) 1015.

R.M. Zalaletdinov, Gen. Rel. Grav. **25** (1993) 673.

R.M. Zalaletdinov, Bull. Astr. Soc. India **25** (1997) 401; *gr-qc*/9703016, 11 p.

R.M. Zalaletdinov, Ann. European Acad. Sci. (2003) 344.

A.A. Coley, N. Pelavas, and R.M. Zalaletdinov, Phys. Rev. Lett. **95** (2005) 151102.

Bobomurat Ahmedov

Astronomy for the developing world
IAU Special Session no. 5, 2006
J.B. Hearnshaw and P. Martinez, eds.

© 2007 International Astronomical Union
doi:10.1017/S1743921307006941

The astronomical observatory "Khurel Togoot" of Mongolia

Damdin Batmunkh[1]

[1]Research Centre of Astronomy and Geophysics, Mongolian Academy of Sciences,
Ulaanbaatar, Mongolia
email: btmnh_d@yahoo.com

Abstract. In this paper the basic researches, telescopes and devices of the Khurel Togoot astronomical observatory, which was founded during the International Geophysical Year, are briefly described. Our astronomical observatory is located on Bogd Mountain near the capital city Ulaanbaatar. Almost 50 years of scientific work has been carried out there. In particular, astrometric researches, GPS, solar researches and observations of minor planets are conducted. Now these scientific researches basically are maintained and extended, with the introduction of modern technology. As an example of the data received by our solar telescope 'Coronagraph', some solar images will be shown. Recently we equipped this telescope with a CCD camera. Because of the transformation of the economy in Mongolia, there are at present difficulties with the preparation of young professional astronomers and with the purchase of new astronomical equipment.

Keywords. Khurel Togoot Observatory, Mongolian astronomy, solar physics

1. Introduction

The astronomical observatory ($\phi = 47°51'50''$, $\lambda = 107°03'02''$) of Mongolia is located on the "Bogd" mountain, about 15 km south-east of the capital city Ulaanbaatar. The words "Khurel Togoot" (bronze cauldron) are the name of the site. The astronomical observatory was founded during the first International Geophysical Year by the initiative of our astronomer S. Ninjbadgar in close cooperation with scientists from Russia and Germany.

In 1957, the construction of the main buildings (Figs. 1, 2, 3) had begun and some telescopes produced by Carl Zeiss, Germany were ordered to develop the following research activities: determination of time and latitude by astronomical observations; observation of near-Earth artificial satellites; observation of solar active phenomena; recording and study of earthquakes; observation and investigations of the telluric magnetic field and its variations; study of impact of atmospheric turbulence on seeing quality ("astro-climate").

Now these scientific directions basically are continued and expanded and are based on modern information technology. During the period of the socialist system collapse, astronomy and other fundamental sciences in our country were in a difficult situation, because of insufficient financial support.

Presently, the restoration of the fundamental sciences has begun, though gradually. In 1996, the Research Centre of Astronomy and Geophysics of the Mongolian Academy of Sciences was founded. Our astronomical observatory has become the part of it.

Now there are 20 research workers in the staff of our observatory; the majority of them were educated and trained in Russia. Some our scientific results were published in Russian and Mongolian journals. In our library, we have copies of astronomical books and journals, published in Russia and East Europe 15-30 years ago. Now we use freely

Figure 1. l: The main building of the Astronomical Observatory "Khurel Togoot"

Figure 2. r: Buildings of the Coronagraph and permanent GPS station (altitude 1608 m)

Figure 3. l: Buildings of the coudé refractor and the meridian circle.

Figure 4. r: The Coronagraph telescope with Hα Halle filter

accessible via Internet scientific papers, in particular, from the NASA ADS database. In the condition of economic reconstruction in Mongolia, we are facing difficulties in training young professional astronomers and in acquisition new astronomical instruments.

2. Instrumentation

Our observatory is equipped with the following instruments: Zenith and Meridian Circle for definition of latitude and time series, the Coronagraph telescope for observations of solar active phenomena and solar corona, permanent GPS (IGS) station, coudé refractor and 45-cm Meade Schmidt Cassegrain telescope. In the Figures 4–6 some telescopes are shown, they are used now for the scientific and educational purposes.

3. Research activities

The astronomical observatory (AO) had and has been conducting observations and research in the following fields:

(a) **Astrophysics**: Since 1964 the study of solar active phenomena has been started. The following researches were carried out:

- Theoretical calculation of line formation in the atmosphere of the Sun,
- Development of computation methods for radiative transfer problems,

Figure 5. Meridian circle for time service

Figure 6. Zenith telescope for definition of latitude

- Physics of formation of spectral lines of H and CaII in the chromosphere and in the prominences of the Sun,
- Observing a flares and filament ; prominence,
- Observations of the coronal lines 530.3nm 637.4nm near prominence and study of the correlation between prominence and physical conditions of the corona,
- Emission of the solar "cold" corona.

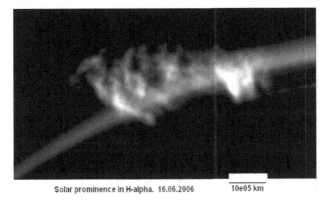

Solar prominence in H-alpha. 16.06.2006   10e05 km

Figure 7. Solar prominence in Hα, 16 Jun 2006

(b) **Astrometry**: The astrometric continuous latitude and time determination by observation of accurate star position have been performed at the observatory since 1958.
By definition of the local latitude and longitude, and their variations our AO were contributing to the international program for Earth's polar motion study and Earth Rotation
Service.

Since 2002 our AO has begun the observation of minor planets (asteroids) and meteors
by "MEADE" telescope with mounted CCD and its astrometrical processing.

(c) **Satellite geodesy**: Since 1967, the satellite observation group of our AO had been
involved in global satellite geodesy and geodynamics projects based on satellite photographic methods using the AFU-75 and FAS cameras, and a first generation satellite laser
ranging system. The staff has participated in a number of international observational
campaigns and primary data processing within the "Intercosmos" programme, including
determination of the Earth's primary parameters, study of the Earth's atmosphere, establishment of a high accurate geodetic network in Mongolia by the balloon-triangulation
method.

Since 1995 AO conducts GPS research for geodynamics and geodetic purposes and its
aim is focussed on the establishment of geocentric geodetic network, study of dynamics
and kinematics of Mongolian tectonics thus contribute to the Asian deformation model.
Currently the AO maintains and operates four permanent GPS stations almost evenly
distributed around the territory of Mongolia and has more than 15 points for campaign
style GPS measurements.

4. Examples of solar images

As an example of our results I present here the last images of the solar active phenomena observed by the coronagraph. Last year we bought an Apogee CCD camera U4
for the coronagraph. The Hα-filter with half bandwidth of 0.5 Åis used. We obtained
excellent images of solar prominences, sunspots and active regions.

5. Conclusions

Considering still good condition of our telescopes and available scientific staff we conclude that:

• Our telescopes can be used successfully for the scientific and educational purposes.
It is clear from the solar images obtained with the Coronagraph.

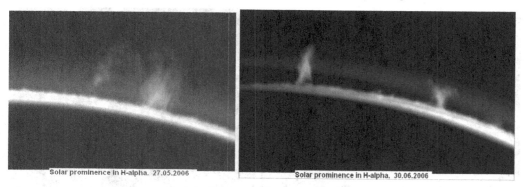

Solar prominence in H-alpha. 27.05.2006

Solar prominence in H-alpha. 30.06.2006

Figure 8. left: Solar prominence and chromosphere

Figure 9. right: Same as Fig. 8

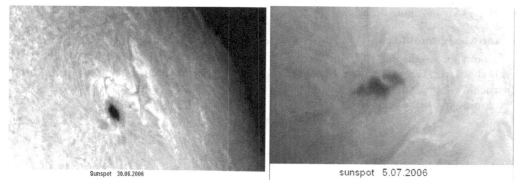

Sunspot 30.06.2006

sunspot 5.07.2006

Figure 10. a: Big sunspot and filament; b: Big sunspot in fig. 10a after 5 days

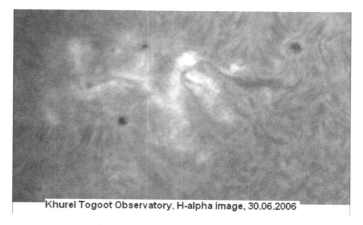

Khurel Togoot Observatory, H-alpha image, 30.06.2006

Figure 11. Solar active region

- All data and results obtained during 1964-2006 years have great scientific value.
- The further improvements of telescopes and the supply of them with modern devices are required.
- Training of young astronomers at foreign universities is very actual.
- It is necessary to expand scientific collaboration with the international astronomical organizations.

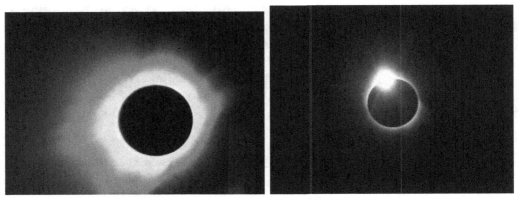

Figure 12. left: Green solar corona, at total solar eclipse, 29.03.2006. Photo by D. Batmunkh, Ch. Lkhagvajav and G. Davaakhuu. From Terskol Observatory, Russia; right: Diamond ring, otherwise same as figure on left.

Acknowledgement

The author would like to thank Prof. Oddbjorn Engvold for financial assistance to attend the IAU General Assembly in Prague. I express thanks also to Prof. John Hearnshaw for the invitation to participate in the special session 'Astronomy for the Developing World' of the IAU.

Astronomy for the developing world
IAU Special Session no. 5, 2006
J.B. Hearnshaw and P. Martinez, eds.

© 2007 International Astronomical Union
doi:10.1017/S1743921307006953

Astronomical education in Armenia

Hayk A. Harutyunian[1]

[1]Byurakan Astrophysical Observatory, Armenia
email: hhayk@bao.sci.am

Abstract. Astronomy pupils in Armenia get their first ideas on astronomy at elementary schools. Astronomy as a distinct subject is taught at all secondary schools in the country. Teaching is conducted according to a unified program elaborated jointly by professional astronomers and astronomy teachers. Unfortunately only one hour per week is allotted for teaching astronomy, which obviously is not enough workload to hire specialized astronomy teachers at every school, and at many schools this subject is tutored by non-specialists. Many schools partly compensate this lack of teachers by organizing visits to the Byurakan Astrophysical Observatory (BAO) for pupils, where they also attend short lectures on astronomy. In some schools optional training in astronomy is organized by amateurs, for the purpose of a deeper understanding in astronomy.

During recent years annual competitions for revealing gifted pupils in astronomy have been organized. These competitions have three rounds, namely, in schools, in districts and the final round is, as a rule, held at BAO. The national winners successfully participate in and win prestigious prizes at international astronomical Olympiads as well.

At Yerevan State University (YSU) there is a department for astrophysics, which was set up in 1946 and is operating to date. This department trains specialists for a career in astrophysics. Only one or two students graduate from this department yearly at present, while in the 1980s a dozen specialists were trained every year. BAO serves as the scientific base for the students of YSU as well, and a number of staff members from BAO conduct special courses for YSU students. YSU provides a Master's degree in astrophysics, and BAO is granting a Doctor's (PhD) degree since the 1970s.

Keywords. Astronomy education in Armenia, Byurakan Astrophysical Observatory

1. Introduction and historical background

Astronomical knowledge in Armenia has a rather long history. The first astronomical petroglyphs found in Armenia date to the V-IV millennia B.C. The ruins of the Metsamor ancient observatory are dated to the III-II millennia B.C. Written documents are rare, though one can find some astronomical ideas and knowledge in the "History of Armenia" by Movses Khorenatsy of the fifth century A.D. More systematically, astronomical ideas in Armenian had been given for the first time by Anania Shirakatsy of the seventh century. His manuscripts on astronomy, geography and mathematics served as textbooks for school pupils during centuries and one can study these books in the Matenadaran – the Institute of Ancient Manuscripts in Yerevan, the capital of Armenia, where many manuscripts of famous ancient Greeks are preserved as well.

2. Astronomy education in Armenian high schools

Astronomy as a separate subject was included into the secondary school program after 1920 (to the best of our knowledge) when the Soviet Armenian Republic was formed. All the Soviet republics have been implementing then the same educational programme approved by the Soviet Ministry of Education. During the second half of 20th century the astronomy textbook by Prof. Vorontsov-Velyaminov was the only mandatory course for all the pupils. In the 1970s an attempt was made to prepare another textbook in Armenia, taking into account the existence of rather high level professional astronomy in the country, and numerous good specialists available were ready to write a new textbook. Unfortunately this textbook has been in use for a few years only and the old one replaced it again very soon.

This situation changed dramatically after the Soviet Union collapsed, and every Soviet republic as a new independent country became free to carry out and implement its own programme, and any textbook could be adopted by the corresponding Ministry. Of course, such a change was welcomed with pleasure, not only by the astronomical community, and it is worth noting that with a few advantages gained as a result of independence, these countries including Armenia lost already or are losing some of the previous ones.

Turning to the present situation, one should mention that the first ideas in astronomy that pupils in Armenia get at elementary schools are owing to special hours prescribed for the introduction of elementary knowledge on the structure of the surrounding world. During about 10 hours, they learn about the geometric shape of the Earth, its rotation around its own axis and around the Sun, receive some information on the solar system structure and on other planets. The programme is prepared for 10-11 year old children to provide them with the most general information on the subject of astronomy.

At present astronomy as a distinct mandatory subject is taught at all secondary schools in the country. Teaching is conducted according to the unified programme elaborated jointly by professional astronomers and astronomy teachers and finally recommended by the Ministry of Education for all secondary schools. Alternative programmes also could be implemented through the Ministry of Education if any school justifies its necessity, and several schools with deeper education in the field of physics and mathematics use this opportunity.

However both the secondary and elementary schools suffer from the shortage of astronomy teachers, which is negatively affecting the education quality. Since only an hour per week is allotted for teaching the subject of astronomy, which obviously is not a high enough workload to hire specialized astronomy teachers at every school. Hence this course is taught very often by physicists or other specialists. For a decade in the 1980s, this problem was partially solved using the Yerevan planetarium facilities, with professional lecturers invited from Byurakan Astrophysical Observatory (BAO) and Yerevan State University (YSU). These specialists had elaborated and have been implementing a series of lectures to cover all the main topics of the astronomy course taught in the secondary schools. Unfortunately since the very beginning of the 1990s, the planetarium ceased its operation and there are negligible chances for its reopening in the near future.

Many schools partly compensate this lack of qualified teachers organizing visits to the Byurakan observatory for pupils where they can watch telescopes and also attend short lectures on astronomy, its role and purposes. At present the guide group consists of professional astrophysicists and they receive several thousands of guests per year, included among them nearly 80 per cent of school pupils. Though there is a special gallery in the building of the 2.6-m telescope for receiving visitors, where they attend a lecture and watch the largest BAO telescope, the observatory administration is seeking for a funding to build a new pavilion equipped with a special hall for lectures, as well as two small telescopes for observations etc. In a few schools facultative training is organized by the astronomy amateurs, for the purpose of a deeper learning of astronomy.

At the present date, the most serious danger for astronomical education is related, strange though it may seem, with the process of integration of the country's educational system into the European framework. This trend threatens that astronomy might be dropped from the list of mandatory subjects. In such a case, the level of astronomy knowledge would go down very rapidly. The majority of pupils who are presently learning astronomy are more or less doing so owing to the demands of the educational programme and they gain an acceptable knowledge because of that. But only a small part of them will follow their natural intellectual curiosity to learn astronomy after such a decision to make astronomy no longer mandatory. Evidently this process is going on to the detriment of knowledge in astronomy which maybe will slow down but survive only because the Byurakan observatory is well recognized by the population of the country.

3. Byurakan Astrophysical Observatory - BAO

BAO – the main center for astrophysical researches in Armenia was established in 1946 on the initiative of Professor Victor Ambartsumian (IAU President 1961-64; ICSU President 1968-72) who became the first director of the observatory and was occupying this position up to 1988. The main research directions were determined by him as well. The first studies at the Byurakan Astrophysical Observatory were related with the instability phenomena taking place in the universe and their role in the evolution of cosmic objects. This trend became the main characteristic of the research activity in Byurakan and almost all the well-known results obtained at BAO are related to this scientific direction.

First scientific results were announced in 1947 owing to the studies of a new type of stellar systems – stellar associations. Using the dynamical features stellar associations it was shown that these systems were not older than tens of millions of years, proving thus that star-forming processes are going on at present in our Galaxy. Proving the existence of these young stars, Ambartsumian actually made the universe a living organism for the first time. Though this idea met with undisguised skepticism by the astronomical community then, but not any astronomer at present has doubts concerning the existence of young or newborn stars in our Galaxy and other galaxies.

Another very fruitful idea put forward in the Byurakan observatory concerning the activity in galactic nuclei was rejected by the world astronomical community in the beginning of 1960s. However later on, when many new active galaxies have been revealed and, particularly, the Markarian survey was implemented and his catalogue of UV-excess galaxies was completed, a new very wide and rapidly developing scientific direction in extragalactic astronomy was established called AGN studies.

Undoubtedly the fact of existence of the rather famous professional observatory in the country with its scientific achievements, with modern scientific facilities including the 1-m Schmidt telescope and another 2.6-m telescope, being unique ones in the region, and a series of smaller telescopes has had a very strong influence on the educational process. However, it should be mentioned that for a long time there existed an essential gap between professional astronomy and the general educational level in astronomy. Until recently, no amateur astronomy existed in the country in any sense. Now the situation is slightly different, although it needs a long way of improvement still.

For decades, competitions on knowledge of various subjects have been organized for pupils in Armenia, and such competitions made them study these subjects more intensively. Unfortunately astronomy has not been included. Now this problem is solved. During recent years, annual competitions for revealing gifted pupils in astronomy are also organized by the Ministry of Education in collaboration with BAO. These competitions have three rounds, namely, in schools, in districts and the final one as a rule is held at BAO. The country's winners successfully participate and win prestigious prizes in the international astronomical Olympiads as well. In September 2005, an astronomical school-competition was organized at Byurakan where pupils from Russia took part as well. The Armenian Ministry of Education with BAO intend to apply for organizing an international Olympiad in Armenia in 2008 – the year when V. Ambartsumian's centenary will be celebrated in Armenia.

4. Astronomy at Yerevan State University - YSU

At YSU a department for astrophysics was set up by Prof V. Ambartsumian in 1945 and is operating to date. This department trains specialists for a career in astrophysics. Only one or two students graduate from this department yearly at present, while in the 1980s a dozen students were trained every year. Actually this department was established for a very definite purpose – to train the scientific staff for the future Armenian astrophysical observatory. The observatory was set up in the following year and BAO since then and up to now serves as a scientific base for the students of YSU as well. Mainly the BAO staff members conduct special courses for the students of the mentioned department.

YSU provides a master's degree in astrophysics, and BAO is granting a doctor's (PhD) degree since the 1970s. It is worth noting that many post-graduate students from various countries of the former Soviet Union,, as well as from abroad, received their PhD degrees in astrophysics at the Byurakan Astrophysical Observatory. The teachers working at present in the secondary schools of the republic are graduated mainly from the Armenian State Pedagogical University (ASPU) and a smaller part of them are from YSU. The BAO makes plans for organizing with the Ministry of Education at least short term regular courses for the astronomy teachers aimed at increasing their professional skills.

One of the main goals of the recently organized Armenian Astronomical Society (ArAS, 2001) is supporting astronomical education, as well as the propagation of astronomical knowledge within the population in the country. The Society intends to use public lectures as well as opportunities given by TV channels and newspapers for achieving their goals in this particular field.

In 2004 BAO renewed summer training for the students from YSU at the observatory, which was interrupted in the mid-1990s. In 2005 was organized a summer school for the best students of YSU physical faculty, which was aimed at helping them for further choosing of their specialization. Lectures on various subjects were organized for them and also they participated in a competition to show their knowledge. BAO and YSU agreed to organize such schools annually. This year it will be combined with the First Byurakan International School already announced for young astronomers and to be held in the period 22-31 August 2006.

Astronomy for the developing world
IAU Special Session no. 5, 2006
J.B. Hearnshaw and P. Martinez, eds.

© 2007 International Astronomical Union
doi:10.1017/S1743921307006965

Suffa Radio Observatory in Uzbekistan: progress and radio-seeing research plans

Alisher Hojaev[1]†, G.I. Shanin[2] and Yu.N. Artyomenko[3]

[1]Ulugh Beg Astronomical Institute, Center for Space Research, Uzbek Academy of Sciences, Tashkent, Uzbekistan
email: ash@astrin.uzsci.net

[2]Radio-observatory RT-70, Center for Space Research, Uzbek Academy of Sciences, Tashkent, Uzbekistan
email: suffa@mail.ru

[3]Astro-Space Center, Lebedev Physical Institute, Russian Academy of Sciences, Moscow, Russia
email: feli@asc.rssi.ru

Abstract. The Suffa International Radio Observatory to be completed in coming years and the new radio astro-climate (seeing) research proposals for radio weather forecasting are described.

Keywords. Astronomy in Uzbekistan, telescopes, instrumentation: adaptive optics, radio continuum, lines: general, submillimetre, site testing, atmospheric effects

1. Introduction

Radio-astronomy is a rapidly developing and very promising branch of modern astrophysics. The SKA, ALMA, the space radio telescope and other prominent projects of coming years will open new horizons in the study of the universe in unexplored ranges and with much higher spatial resolution than in other spectral bands and with higher sensitivity than before.

One of the large-scale radio-astronomy facilities is the complex of the International Radio Astronomy Observatory being created on the Suffa plateau (Uzbekistan) in close collaboration with Russia (Kardashev, 1992). It should be a basic part of the Earth-Space VLBI system (Kardashev *et al.*, 1995). With the main instrument similar to the GBT at NRAO it will be one of the main basic elements of the global radio interferometry network as well. The Suffa project has been mentioned as one of the most prominent for the new century astronomy in the millennium overview paper (Trimble, 2001).

2. General Information

The location of the created observatory is shown in Fig. 1. The project itself was started up in 1984 and since 1992 project was suspended after the Soviet Union disappeared. In 1995 the agreement was signed between the governments of Russia and Uzbekistan establishing 'The International Radio Astronomy Observatory "Suffa"' (IRAOS). This agreement provides a legal basis for the continuation of the construction of the 70-m Suffa radio-telescope which will belong to the IRAOS. The duration of the agreement is 99 years. The agreement is open for joining by other states, international organizations, national scientific institutions or personalities from third states.

There are at least three forms of joining to this project:

† Present address: UBAI, Astronomicheskaya 33, Tashkent 700052, UZB

Figure 1. A mountain view of observatory on Suffa plateau. The location map and the telescope horizon shielding diagram are inserted in the left and top right, respectively.

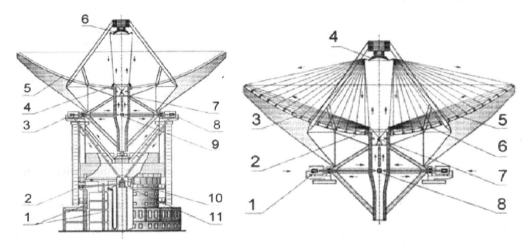

Figure 2. Sketch-section of RT-70 radio-telescope (left) and its primary mirror shape fitting system (right). Numbering in the left panel: 1 - azimuth master; 2, 3, 8 - laser position measuring devices; 4 - basic platform; 5 - primary mirror; 6 - counter-reflector; 7 - periscope mirror; 9 - guide-builder(for precise pointing); 10 - monitoring device (controller); 11 - digital feedback sensor. Numbering in the right panel: 1 - laser position measuring devices; 2 - basic platform; 3 - primary mirror; 4 - counter-reflector; 5 - reference devices; 6 - electric jacks for reflecting panels; 7 - periscope mirror; 8 - guide-builder.

(*a*) the participation in completing the complex. An immediate participation in completing the installation is encouraged and will give the preferences in future activities;

(*b*) the participation in providing facilities by last generation receivers and other modern equipment. This might be started with participation in the new site testing project (see Chapter *Radio Astro Climate. New Project* of this report);

(*c*) scientific collaboration with co-financing of operation of the observatory.

Quite recently (2001) the firm decision on completing the project has been endorsed by our governments, and Russia will invest for these; therefore the project's layouts have been considerably modernized and updated in order to build up the state-of-art instrument. The telescope operation start time is planned for 2009.

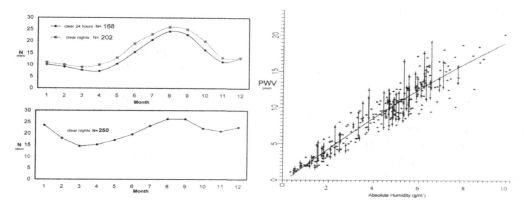

Figure 3. Left panel: distribution of clear time in 1981-1991 – general cloudiness (top left) and lower atmosphere cloudiness (bottom left). Right panel : correlation between absolute humidity and precipitable water vapour (PWV)

The purpose of the RT-70 radio-telescope should be in the following areas:

(*a*) Interferometry mode observations

(1.1) Earth-Space Very Long Base Interferometry(ESVLBI) with spaceborne radio-telescopes: (1.1.1) Radioastron ($D = 10$ m), (1.1.2) Millimetron/Submillimetron ($D = 12$ m) and (1.1.3) VSOP-2 at the angle resolution 10^3 times better than ground-based VLBI existed;

(1.2) Ground-based VLBI within the existing networks;

and

(*b*) Autonomous mode observations of the astrophysical objects and SETI targets. One should note there are no radio facilities or their creation projects in Central Asia and in its close vicinity, especially large ones, which makes the Suffa radio telescope an unique in the region. Thus and so the advantageous geographical location of the telescope will also make it indispensable for the global radio interferometry system.

Some main features of RT-70 radio-telescope are following. An optical arrangement is two-mirror Gregory scheme (Feq=571(345) m) with additional periscope mirror $D_{per} = 0.6$ m. In the left panel of Fig. 2 the sketch-section of the telescope is presented. The primary mirror is parabolic and has $D = 70$ m, $F = 21$ m, aperture angle $= 160°$. The secondary mirror(s) have ellipsoidal shape with $D = 3(5)$ m, $IF = 242$ m. The pointing accuracy is about 1 arc sec, which is achieved by using a special guide-builder and precise digital angle sensors. Guiding accuracy should be about 0.3 second of arc per second. Operational wavelengths are: in S-range (0.87÷10 mm) and in M-range (1÷10 cm) totally in the 7 (10) working bands. Total construction weight will be 4500 tonnes.

A compensation of the mirror system deformations caused by the weight and wind load, fluctuation in temperature, will be realized by homologous construction in longer wavelengths and by active adjustment of primary mirror shape in the mm-range (adaptive mirror – see Fig. 2, right panel). For this purpose special laser position measuring devices and individual electric jacks (in total 2000) will be used for each of 1188 reflecting panels (rms error < 50 microns) of the prime mirror. The accuracy of the prime mirror profile after correction should be not less than 60 microns (Shanin, 1996). The directional pattern of the radio telescope in autonomous mode observations will be 3" and the sensitivity in the mm-range up to 100 μJ which makes the facility unique.

3. Radio astro-climate

3.1. *Previous site testing*

The site location provides good seeing conditions for cm-mm range (see, e.g., Hojaev & Shanin, 1996). Site tests started in 1981. The meteorological measurements have been carried out regularly since then. Sounding balloon and pilot-balloon measurements have also been made. The transparency in different mm bands has been estimated by using the radiometer measurements of atmospheric self-radiation. Averaged annual atmospheric transmission coefficients at the zenith were derived as 0.90÷0.98 for 3.1 mm and 5.8 mm wavelengths and about 0.60 for 1.36 mm, with a maximum in the winter time. The clear time estimations were also made. In Fig. 3 we give some results of testing the observing conditions at the Suffa site.

3.2. *New project*

Now we are arranging the scientific consortium in order further to explore the Suffa site more deeply and to learn the main 'radio astro-climate' parameters by means of a new technology ('radio-seeing', radio-transparency and stability in different sub-mm, mm and cm bands, PWV , their intercorrelation and correlation with meteo-parameters) for the atmosphere modelling at the site, and we try to forecast the "radio-weather" for reliably planning the scientific schedule of the future telescope. The proposals have been prepared to submit to the international funding agencies to support the project. To carry out these experiments we need radiometers especially 225 GHz, 183 GHz, 89 GHz and automated meteorological stations. For this research any proposals and partners would be very desirable.

The 183.3 GHz heterodyne receiver radiometer (see, e.g., Hills *et al.* 2001) centered on water molecule line which absorbs the space radiation in the appropriate mm-band when it passes the atmosphere seems the best to probe for the water wapour content. Measuring at the three double-side frequency bands (1.2 GHz, 4.1 GHz and 7.6 GHz) around the centre of the line the intensity and shape of this atmospheric water line emission within the directional pattern of the instrument could be derived from the brightness temperature. Then the PWV is calculated by iteration of the water wapour density in the radiative transfer equation. Since September 1998 such a radiometer (two radiometers starting in March 1999) has been used for the long-term site test measurements at Llano de Chajnantor – 5 000 m high plateau in the Atacama Desert region of northern Chile, chosen for the ALMA project (Delgado, 2002).

4. Summary and conclusions

The large radio-telescope facility has to be completed in few years on the Suffa Plateau in Uzbekistan. The current view of the building site with parts of the facility created is presented in Fig. 4. Recently the project was considerably updated and modernized to fit present-day requirements for a first-class instrument. The site is very dry and has quite good seeing conditions, both in visual and radio ranges. A new deeper site-testing programme is being created, in order to learn more about the radio astro-climate of the site in more detail, and to prepare the ground for 'radio-weather' forecasts. The last will increase the efficiency of future observations, particularly in mm waves. We welcome any proposals and will be glad if you join us both in completing the facility complex and exploring the radio-weather.

Figure 4. Current panoramic snapshots of creating the Suffa Observatory

Acknowledgements

We would like to acknowledge our colleagues from ASC/LPI of RAS and other involved organizations who are actively working on the project and promote it. The authors also thank a referee for the useful comments.

References

Delgado, G. 2002, in: J. Vernin, Z. Benkhaldoun, & C. Munoz-Tunon (eds.), *Astronomical Site Evaluation in the Visible and Radio Range.* ASP Conference Proceedings(ISBN: 1-58381-106-0. San Francisco, Astronomical Society of the Pacific), Vol. 266, p. 246

Hills, R., Gibson, H., Richer, J., Smith, H., Belitsky, V., Booth, R., Urbain, D. 2001, *ALMA Memo* 352, pp. 16

Hojaev, A.S., Shanin, G.I. 1996, *Journal of Korean Astronomical Society* vol. 29, p. S411

Kardashev, N.S. 1992, *Suffa International Radio Astronomy Observatory* Report in NASA Headquarter, Washington

Kardashev, N.S., Andreyanov, V.V., Gvamichava, A.S., Likhachev, S.F., and Slysh, V.I. 1995, *Acta Astronautica* vol. 37, p. 271

Shanin, G.I. 1996, *Precision Radio Telescope RT70*, Abstracts of First Conference on Space Research, Technology and Conversion, Tashkent, p. 27

Trimble, V. 2001, **Future - A Year of Discovery: Astronomy Highlights of 2000** *Sky and Telescope* (February 2001), p. 50

Nikolai Bochkarev

Damdin Batmunkh

Astronomy for the developing world
IAU Special Session no. 5, 2006
J.B. Hearnshaw and P. Martinez, eds.

© 2007 International Astronomical Union
doi:10.1017/S1743921307006977

Measurement of light pollution at the Iranian National Observatory

S. Sona Hosseini[1] and Sadollah Nasiri[1,2]

[1]Zanjan University, Iran
email: s.sona.h@gmail.com

[2]IASBS, Iran
email: nasiri@iasbs.ac.ir

Abstract. The problem of light pollution became important mainly since 1960, by growth of urban development and using more artificial lights and lamps at the nighttimes. Optical telescopes share the same range of wavelengths as are used to provide illumination of roadways, buildings and automobiles. The light glow that emanates from man made pollution will scatter off the atmosphere and affects the images taken by the observatory instruments. A method of estimating the night sky brightness produced by a city of known population and distance is useful in site testing of the new observatories, as well as in studying the likely future deterioration of existing sites. Now with planning the Iranian National Observatory that will house a 2-metre telescope and on the way of the site selection project, studying the light pollution is propounded in Iran. Thus, we need a site with the least light pollution, beside other parameters, i.e. seeing, meteorological, geophysical and local parameters. The seeing parameter is being measured in our four preliminary selected sites at Qom, Kashan, Kerman and Birjand since two years ago using an out of focus Differential Image Motion Monitor. These sites are selected among 33 candidate sites by studying the meteorological data obtained from the local synoptic stations and the Meteosat. We measured and used the Walker's law to estimate the Sky brightness for three of these sites

The data obtained using an 8-inch Meade telescope with a ST7 CCD camera for above sites are consistent with the estimated values of the light pollution mentioned above.

Keywords. Light pollution, sky brightness, Iranian National Observatory

---

1. Introduction

Among 33 candidate sites at the central part of the country, 4 sites namely Kashan, Marzi, Kerman, and Birjand are selected using meteorological and geophysical data. A typical annual wind rose are shown in Fig. 1 for these four sites. To find the best observatory site at the each region, different places (shown by circular spots in Fig. 2) are examined by at least two nights of DIMM data [1] and finally the mountains Sardar in Kerman, Kolahbarfi in Kashan, Fordoo in Marzi and Mazarkahi in Birjand were selected for further site testing (shown by square spots in Fig. 2). The seeing parameter is measured for these sites and two full years of data have already been collected. Now among those four candidate sites only Marzi and Kashan are remaining, due to their seeing values.

We started to study and prepare measuring the sky brightness of these sites by one and a half years ago. Light pollution makes the level of sky brightness higher and results in less signal-to-noise ratio. With low S/N we have to build larger telescopes beside the bad effect of light pollution on them. By studying the effect of light pollution on famous telescopes we can see horrible conclusions [2].

183

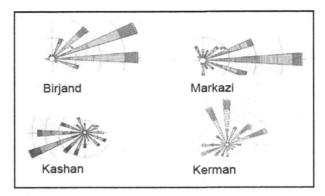

Figure 1. Histogram of wind roses of 4 candidate sites.

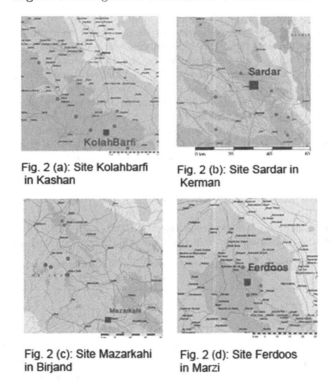

Figure 2. Locations of the four sites

We had studied many numerical and experimental methods. We had gathered sample photometric data with a SBIG ST/7E CCD and the V Johnson filter to define the sky brightness exactly.

2. A brief review on concepts of sky brightness and light pollution

Astronomy is suffering from rapidly growing environmental problems. One of these is light pollution. Urban sky glow is taking away the prime view of the stars and the universe. Many things can bring light pollution to us, such airglow, artificial sources as urban lights, zodiacal light, and solar wind as aurora and of course moon light!

The Earth's atmosphere causes the light coming from sources in an urban area to scatter, creating the halo of light visible over the city even from great distance. Even single birth sources in a dark local can be a source of local sky glow.

There are four negative factors often found with outdoor lighting that we say GLUT (GLUT = Glare + Light Trespass + Up light + Too Much Light) Fig. 3 [3].

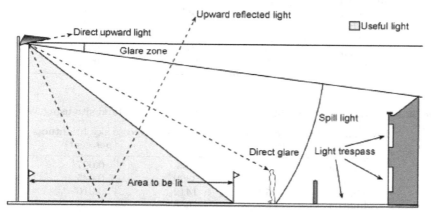

Light pollution is often caused by the way light is emitted from lighting equipment. Choosing proper equipment and carefully mounting and aiming it can make a significant difference.

Figure 3. Four negative factors found with outdoor lighting

The natural sky brightness near the zenith for V band is 21.60 (mag), for B band is 22.40 (mag) and for R band is 20.5 (mag) [4]. But we can see the measured sky brightness for observatories is less than this. The capabilities of optical observatories are continually eroded by urban growth and its harmful accompanying sky illumination. As an example percent of original value for the 4-m telescope on Kitt Peak is 93% and for the 5-m telescope at Mount Palomar is 39%! [2].

3. A brief review of previous modelling effects

A number of people have modelled light pollution in various ways. The first model made by Walker (1970) [3]. This model was a important step to start modelling the sky brightness of the sites, which other modelling base is that too. After that Treanor [4] and Berry [5] modified the model. Tomas, Modali and Roosen (1973) [6] reported calculations using a Monte Carlo method. Yocke, Hogo and Henderson [7] applied an approximate treatment of radiative transfer to a study of the effect of a proposed nuclear waste depository on the night sky brightness as seen from sites in Canyonlands National Park. As an example, Garstang (1986) has done detailed calculations for a number of observatory sites, creating maps showing how the sky glow varies at different altitudes and azimuths from each site. Burton (2000) is analyzing satellite data from the Defence Meteorological Satellite Program (DMSP; run by the U.S. Air Force) to estimate sky glow in the close vicinity of urban areas. This has the advantage of considering actual satellite data at high resolution, both spatially and in terms of intensity. However, limited consideration is given to atmospheric scattering, especially over large distances.

Table 1. Measuring excess sky brightness percentage of the Kashan site using Walker's law.

Town name	Distance from site (km)	Population	Excess sky brightness (per cent)
Marvand	12	3000	0.06
Jahak	13.2	700	0.01
Zanjan Fard	13.4	700	0.01
Gazaan	14	4000	0.05
Bon Rood	14	4000	0.054
Kom Jan	16.5	1500	0.01
Kamoo	22	3 000	0.01
Kashan	47	400 000	0.02

Table 2. Measuring excess sky brightness percentage of the Marzi site using Walker's law.

Town name	Distance from site (km)	Population	Excess sky brightness (per cent)
Bichegan	12	485	0.097
Ferdo	14	3339	0.045
Khaveh	15	1412	0.016
Voshnaveh	15	2088	0.023
Virg	18	6603	0.048
Kahak	24	7344	0.026
Delijan	27	30 000	0.079

4. Measuring of sky brightness of Iranian observatory sites

To have the largest effective aperture for the telescope we need to reduce the sky brightens as much we can. In this case, we are measuring the level of sky brightness of our candidate sites to act on the needed works. In the first step we estimated the sky brightness by Walkers law. The results are given in Tables 1 and 2.

We also measured the sky brightness of the four sites by photometry of the Landolt stars.

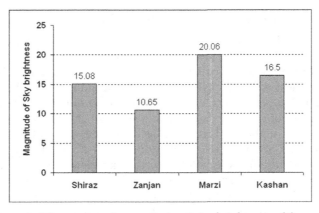

Figure 4. The measured magnitude of sky brightness of four sites.

5. Conclusion

Three years ago the Iranian Astronomical Society proposed the national observatory project that was approved by the government. Since then the site selection for this observatory started and four preliminary sites were selected by meteorological and geophysical studies. The seeing measurements immediately began and are still being continued. In addition to the different site selection parameters, the sky brightness measurement is implemented, too. This parameter is estimated for the aforementioned sites by numerical methods. The required instruments for measuring of this parameter are set up and data acquisition is currently done.

Acknowledgements

We would like to express our deep gratitude to Dr Scott W. Tear from New Mexico Tec. for his useful comments.

References

Berry, R.L., 1976, J. Roy. Astron. Soc., Canada 70, 97-115.
Cinzano, P., J., 2000, Italian Astro. Soc., 71, N. 1, 1-280.
Darudi, A., Nasiri, S., 2005, Iran. J. Phys. Research, Vol.5, No. 3, 121-128.
Garstang, R.H., 1988, The Observatory, 108.
Garstang, R.H., 1991, ASPC, 1-14.
Information Sheet 20, 1994, reported by International Dark Sky Association (IDA).
Landolt, A. U., 1983, Astro. J., 83, 3.
Tear, S. W., 2000, NASA Astrophys. Data System.
Tomas, R. W. L., Modali, S. B. and Roose, R. G., 1973, Bull. Am. Astron. Soc. 5, 391.
Treanor, P.J., 1973, The Observatory, 93, 117-120.
Walker, M.F., 1970, Publ. Astron. Soc. Pacific, 82, 672-698.
Yocke, M.A., Hogo, H. and Henderson, D., 1986, Publ. Astron. Soc. Pacific 98, 889-893.

Hayk Harutyunian

Svetlana Kolomiyets

Astronomy for the developing world
IAU Special Session no. 5, 2006
J.B. Hearnshaw and P. Martinez, eds.

© 2007 International Astronomical Union
doi:10.1017/S1743921307006989

IHY: Meteor astronomy and the New Independent States (NIS) of the Former Soviet Union

Svetlana V. Kolomiyets[1] and Vladimir V. Sidorov[2]

[1]Kharkiv National University of Radioelectronics, 14 Lenin Ave, Kharkiv, 61166, Ukraine
email: s.kolomiyets@gmail.com and kometa@kture.kharkov.ua

[2]Kazan University, 18 Kremlyovskaya str., Kazan, Tatarstan, 420008, Russia
email: vladimir.sidorov@ksu.ru

Abstract. The purpose: to emphasize, that there are some specific features of the development of science in the New Independent States (NIS) of the Former Soviet Union. These features demand enhanced attention of the organizers of the IHY. It is necessary to create effective mechanisms for the stimulation of the connection to world science of the dormant part of fundamental scientific knowledge of these countries, which has been saved up for fifty of years. Probably, the IHY is the last opportunity of rescuing the dormant part of this knowledge from full oblivion.

The method adopted is to discuss and analyse the general tendencies in science in the NIS by reference to individual cases, in particular for meteor astronomy.

Results: The features and history of the development of meteor astronomy during the existence of the Soviet Union and the subsequent period give a key to understanding of the problem. Meteor astronomy can be assumed to be a young science. It is an example of a cross-disciplinary science. It is an example of a science having a sharp rise, due to the project of the IGY and to subsequent geophysical projects. Meteor astronomy is a science directly connected with the launching of the first space satellite of the Earth and the evaluation of problems of meteoroid danger to space missions.

Commission 22 (Division III) of the IAU coordinated the development of meteor astronomy during the IGY. The known Soviet researcher of meteors V. Fedynskiy headed this Commission during four years since 1958. In the USSR numerous meteor centres were created and activated. The general management was concentrated in Moscow. Despite the close interaction under global projects of the Soviet Union with other countries, there still existed a language barrier. The language barrier, together with other reasons, has led to the creation in the USSR of a powerful meteor science, but only in the Russian language.

After the disintegration of the Soviet Union, the meteor centres have remained, but without ordinary central management. The scientific results have remained but as an isolated, inaccessible science published in English.

Conclusion: Reunification of the scientific achievements of the last few years in the NIS with international science should become the task of the IHY. Revival of the activity of the some of the centres will be useful.

Keywords. History and philosophy of astronomy, atmospheric effects, meteors, meteoroids, meteor radar techniques, space vehicles, interplanetary medium, solar-terrestrial relations

1. Introduction

A new programme of global coordinated researches on the interaction of the Sun's radiation with the Earth, planets and interplanetary space of the solar system will start in 2007. It is the fourth programme under the account from a series of the International Geophysical Years. One of the aspects of this programme is the preservation of an historical heritage of programmes of the earlier geophysical years, and, first of all, the International

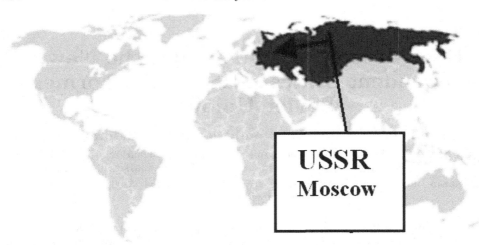

Figure 1. The territory of the USSR before the collapse, showing Moscow, the capital of the USSR, 1922–1991

Geophysical Year IGY1957. As an historical heritage of the third geophysical year from all of the received knowledge during IGY 1957, it is necessary to preserve our knowledge of meteor astronomy and meteor geophysics. It concerns knowledge of meteors which have been saved up in Russian in the Soviet Union since the time of carrying out of the third year. This knowledge in the countries of the former Soviet Union is now under the threat of full oblivion.

2. Post-Soviet space (NIS)

The Soviet Union was dissolved in 1991, and the successor states are a collection of 15 countries commonly dubbed "the Former Soviet Union". For several years after 1991 these states were commonly referred to as the "New Independent States" (NIS). As the abruptness of the fall of the Soviet Union faded, geographic-historical terms began to be used more often, such as "Eurasia" or the "Former Soviet Union". Within Russia, the former non-Russian republics have commonly been referred to collectively as the "near abroad", and the 15 successor countries together as "Post-Soviet space". Before the collapse, the Soviet Union had the most area of any territory (22 402 200 km^2) and the third highest population (299 047 571) on a global scale.

Today in the countries of the post-Soviet space, Russian as the working language of the scientific environment still continues to remain the most widespread. In these post-Soviet countries the Russian language continues for historical reasons to be the most widespread language for official communications and general understanding. For today's Russian-speaking sphere constitutes 4 per cent of all population of the Earth and 5 per cent of the land area from all the territories of the world (Table 1).

The purpose of this paper is to emphasize, that there are some specific features of the development of science in the New Independent States (NIS) of the Former Soviet Union. These features demand enhanced attention by the organizers of the IHY in general and the United Nations in particular. It is necessary to create effective mechanisms for the stimulation of the connection to world science of the dormant part of fundamental scientific knowledge of these countries (in Russian) which has been saved up for some

Table 1. Percentages of the world population and of the Earth's land surface area for some countries and for the NIS

Rank from 234	Country	Per cent population of world	Per cent area of world
1	China	20.10	1.88
2	India	16.86	0,64
3	USA	4.59	1.89
4	NIS	4.40	5.27
5	Indonesia	3.79	0.38
6	Brazil	2.89	1.67
7	Bangladesh	2.54	0.03
8	Pakistan	2.26	0.16
9	NIS minus Russia	2.21	1.92
10	Russia	2.19	3.35
28	Ukraine	0.72	0.12

fifty years, since the IGY. Probably, the International Heliophysical Year 2007 is the last opportunity of rescuing the dormant part of this knowledge from full oblivion.

The method adopted is to discuss and analyse the general tendencies in science in the NIS by reference to individual cases, in particular for meteor astronomy. Features and the problem of the development of meteor astronomy during the existence of the Soviet Union, and the subsequent period, give a key to understanding of the problem.

3. Some history of the USSR (1953-1969): the rise to power

Joseph Stalin died on March 5, 1953. In the absence of an acceptable successor, the highest Communist Party officials opted to rule the Soviet Union jointly, although a struggle for power took place behind the facade of collective leadership. Nikita Khrushchev, who won the power struggle by the mid-1950s, denounced Stalin's repression in 1956, and he eased repressive controls over the party and society. During this period, the Soviet Union continued to realize scientific and technological pioneering exploits, including to launch the first artificial satellite Sputnik 1, to launch the living being, Laika, and later, the first human being, Yuri Gagarin, into Earth orbit.

This period closely connects with the preparation and carrying out of the International Geophysical Year 1957. In the Soviet Union for a long time, beginning with the start of the Cold War in 1947, there was an initiative to catch up and overtake leading western capitalist countries, including in the domain of scientific achievements. Therefore, the preparation and carrying out of IGY 1957 in the USSR were high-quality and well organized. These activities were supervised at the highest level and have been supported by the state and regional financing. Carrying out of meteor researches in the frame of the IGY and during the subsequent geophysical programmes (the International Year of Geophysical Cooperation, the International Year of the Quiet Sun), was extraordinary successful. It has led to significant development of meteor astronomy, to the monumental boom in accumulation of knowledge on meteor astronomy, especially so in the Soviet Union and in the Russian language.

An especially important place in meteor astronomy was borrowed with a new young branch of a meteor science, connected with the radar-location of meteors. A separate interest has been that section of meteor astronomy connected with the evaluation of the meteoroid danger to space flights and the creation of a model of the distribution of meteor substances.

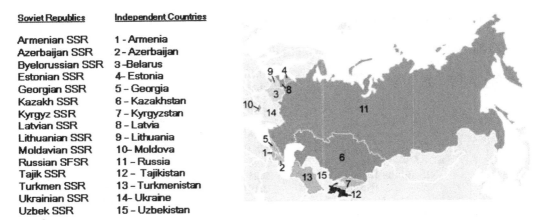

Soviet Republics	Independent Countries
Armenian SSR	1 - Armenia
Azerbaijan SSR	2 - Azerbaijan
Byelorussian SSR	3 - Belarus
Estonian SSR	4 - Estonia
Georgian SSR	5 - Georgia
Kazakh SSR	6 - Kazakhstan
Kyrgyz SSR	7 - Kyrgyzstan
Latvian SSR	8 - Latvia
Lithuanian SSR	9 - Lithuania
Moldavian SSR	10- Moldova
Russian SFSR	11 - Russia
Tajik SSR	12 - Tajikistan
Turkmen SSR	13 - Turkmenistan
Ukrainian SSR	14- Ukraine
Uzbek SSR	15 - Uzbekistan

Figure 2. List of the 15 New Independent States of the Former Soviet Union and 15 Soviet republics before the collapse

This aspect of meteor research is closely connected with the successful launch of the first artificial satellites of the Earth, and the transformation of the USSR into the first-ever space power. However, meteor science, working in Russian, has appeared to be in significant isolation from the international scientific communities. There was a linguistic barrier and also a political barrier. Besides, meteor science became a strategic science and for many interesting developments in meteor astronomy there was a ban on publication in the open press. This was especially rigid and an exacerbation during the cold war period.

4. Importance of section 5, "Ionosphere", in the programme of the International Geophysical Year

Each of the programmes of geophysical years had features connected to a history of development of human society in general and to those problems of geophysics in particular, which in turn were generated by the prevailing historical period. IGY1957 has been led in 25 years after the second year and its triumphal carrying out was promoted by a lot of favorable circumstances.

First of all, on a global scale, it was a time of economic rise and relative social well-being. The development of science and technology has achieved such a high level, such that it was possible to launch the first artificial satellite of the Earth. This has opened for mankind a new era in the conquest of space.

The IGY activities of the year 1957 were on a grand scale in every respect. The transition from studying separate aspects of geophysics to its universal study was made.

From all the broadest set of subjects of research, on which the third international geophysical year has enriched mankind with new knowledge, it is necessary to allocate, as a key, the section 5 programmes concerning the "Ionosphere". To investigate this section, it is necessary to make use of the expanded variant, which is used in all materials under the programme of each year in the Soviet Union, under titles such as "Ionosphere and meteors".

The ionosphere is the part of the atmosphere (near 60–1000 km) that is ionized by solar radiation. The meteor zone lies in the frame of the ionosphere near between 70–140 km. Edward V. Appleton was awarded in 1947 a Nobel Prize for his confirmation of the existence of the ionosphere in 1927. In 1950 L. Berker (one of the pioneers for studying of the ionosphere, the well-known American scientist) initiated to lead the third polar year in 25 years after IPY 2.

A proposal for IGY 1957 was brought before the Mixed Commission on the Ionosphere, which duly endorsed it. The Mixed Commission on the Ionosphere for IGY 1957 was formed by the International Council of Scientific Unions (ICSU), under the sponsorship of the International Union for Radioscience (URSI), with the co-operation of the International Astronomical Union (IAU) and the International Union for Geodesy and Geophysics (IUGG).

A bright example of the scale of the third year in the domain of the ionosphere is the fact, that within the second polar year, there worked six ionospheric research stations, and during the third year, work at not less than 300 ionosphere stations was planned. Such interest in the ionosphere is dictated not only by scientific geophysical problems. During the second polar year, ionosondes were necessary only for meteorology. But during the third geophysical year, new important and practical interests in the ionosphere were developed. These were connected with the wide introduction of a new radio engineering method of researches, and with modern (for that time) and most reliable means of communications by radio communications that appeared.

In 1920 the Department of Scientific and Industrial Research formed the Radio Research Board, in order to 'Direct any research of a fundamental nature that may be required, and any investigation having a civilian as well as a military interest'. The term RADAR was coined in 1941 as an acronym for Radio Detection and Ranging. This acronym, of American origin, replaced the previously used British abbreviation RDF (Radio Direction Finding) in 1942. The term has since entered the English language as a standard word 'radar', losing the capitalization in the process. The famous Sir Bernard Lovell, the British radio astronomer, and director (until 1981) and founder of the Jodrell Bank Observatory (at first the Jodrell Bank Experimental Station) did radar research and radar meteor research.

5. Phenomenon of meteor astronomy in Russia and the International Geophysical Year 1957

Many aspects of meteors have appeared extremely important for IGY 1957, both in the USSR, and in other countries, in connection with the approach of the space age and the revolutionary introduction in scientific researches of radar-tracking methods. Meteors became one of the central objects of research during IGY 1957. Meteors were a highlight in the IGY section "the Ionosphere". Their value for IGY 1957 was unique, as they were not only a subject of study, but were one of the main indirect means of studying the top layers of the atmosphere of the Earth, the circulation of the atmosphere and its parameters within the limits of the ionosphere.

Besides, meteors had one more important independent function as a means of radio communication. Meteoric communication was an alternative at times of infringements or difficulties in the use of usual radio communication, or for confidential purposes. Radio and meteors have enriched each other. With the application in meteor researches of radio engineering means and radar-tracking methods (the most advanced during IGY 1957) it began to be possible to provide meteor astronomy and meteor geophysics with an extensive observational material, not dependent on the season or time of day. It was also possible partially to automate the process of the accumulation of data, and to register weaker meteors.

The stunning file of knowledge on meteors within the limits of IGY 1957 on the volume and the importance has been received in Soviet Union.

5.1. *Special place of the Kharkiv and Kazan radar meteor centers and the development of radar methods of research on meteors in the Soviet Union*

It is necessary to note the important role of the Kharkiv and Kazan radar meteoric centres and the development of radar-tracking methods of research of meteors in Soviet Union, since preparation and participation in the project of the International Geophysical Year.

The Kazan researchers under K.V. Kostylev's manual have played a leading role in 50th years of the twentieth century in distribution of meteoric radar knowledge to the USSR and training of the future participants of the meteor program of Soviet Union to known radar methods of research of meteors in an atmosphere of the Earth and their practical realization. Subsequently, during the period with 1960 on 1986 they have created some new radar systems, which allowed to specify electrodynamic of meteoric scattering of radio wave and to execute long-term monitoring of inflow of meteoric substance on the basis of use of statistical methods of analysis of the received data?.

The Kharkiv researchers under the direction of B.L. Kashcheyev have spent the best cycle of radar observations during performance of program IGY 1957–1959, have improved the equipment for research of meteors and subsequently in 1968–1972 have created for the first time in the USSR the multipurpose automated radar system MARS for research of meteors in an atmosphere of the Earth with registration of orbits of radiometeors up to the twelfth star magnitude (one of the best and high-sensitivity in the world by the beginning of 70th years of the twentieth century). Besides for last 50 years in the Kharkiv meteor center the cycle of works on studying physics of the meteoric phenomena in an atmosphere of the Earth and on research of orbital properties of an interplanetary meteoric complex is executed in view of factors of selectivity on the basis of own long-term lines radar observations of meteors.

6. Some history of the USSR (1970-1991) and of the NIS: the recession

However, since the 1970s, the growth rate of meteor knowledge had slowed substantially. Extensive economic development, based on vast inputs of materials and labour, was no longer possible; yet the productivity of Soviet assets remained low compared with other major industrialized countries.

Besides, the period 1979-1985 is sometimes referred to as the "Second Cold War". It was marked with a change in the Western policy of detente and more confrontation against the Soviets. The meteor science in Russian basically remained an unknown item for meteor international science in English. Prior to its collapse, the Soviet Union had the largest centrally directed economy in the world. The government established its economic priorities through central planning. With disintegration of the Soviet Union and formation of the 15 new independent states, the situation with meteor astronomy became catastrophic.

Meteor astronomy in the 14 newly formed countries could probably have been developed. However, the problems were that the scientists found themselves without the usual central Moscow management system, which included the Committee on comets and meteors of Astronomical Advice of the USSR, and the Section of Interdepartmental Geophysical Committee at the Academy of Sciences of the USSR. In addition, a stable military was lacking and the former Soviet space program was also an impetus for meteor astronomy. In particular there was no motivation to reactivate the Soviet scientific structures that had previously demanded that meteor astronomy be pursued.

7. The developing world and the NIS

The terms 'First World', 'Second World', and 'Third World' can be used to divide the nations of Earth into three broad categories. At first, the term Third World had a political status during the Cold War. Today the term Third World is frequently used to denote nations with a low UN Human Development Index (HDI), independently of their political status.

The Human Development Index (HDI) is a comparative measure of poverty, literacy, education, science, life expectancy, childbirth, and other factors for countries worldwide. It is used to distinguish whether the country is a developed, developing, or an under-developed country, and also to measure the impact of economic policies on quality of life. An HDI below 0.5 is considered to represent low development and 30 of the 32 countries in that category are located in Africa, with the exceptions of Haiti and Yemen.

There are three country in Europe from the NIS group with a low HDI: Georgia (100), Azerbaijan (101) and Moldova (115). An HDI 0.8 or more is considered to represent high development. This includes countries of northern and western Europe, North America, the Southern Cone, the East Asian Tigers, Japan, Australia, New Zealand, Israel, Kuwait and the UAE.

There are two human poverty indexes for developing countries (HPI-1) and for selected OECD countries (HPI-2). (OECD is the Organization for Economic Co-operation and Development.) The last report of the United Nations (UN), 2003, only has a ranking for 17 of the 21 OECD countries with the highest Human Development Index. They are the United States (HP2-15.8), Ireland, United Kingdom, Australia, Belgium, Canada, Italy, Japan, Spain, France, Luxembourg, Germany, Denmark, Netherlands, Finland, Norway, Sweden (HP2-6.5). The following four from 21 OECD countries that are not on this list are Iceland, Switzerland, Austria, and New Zealand. There is no objective definition of Third World or "Third World country" but the use of the term remains common. The term Third World is also disliked, as it may imply the false notion that those countries are not a part of the global economic system. In general, Third World countries are not as industrialized or technologically advanced as OECD countries, and therefore in academia, the more politically correct term to use is "developing nation" or "developing world".

Countries that have more advanced economies than developing nations but haven't yet gained the level of those in the First World are grouped under the term Newly Industrialized Countries or NICs (e.g. China, India, Mexico or South Africa). In Table 2 we can see estimates of population and its dynamic for the most populous ten countries and for the NIS. Low and even negative values of parameters of an increase in population for the countries of the post-Soviet space reflect the fact, that these developing countries have serious problems, such as economic, and others. These problems are characteristic of the transition periods of states, which change their political system. However these usual transitional problems and the destroying of the old settled connections in the countries of the post-Soviet space are adverse for development in these countries of all sciences in general, and astronomy, in particular.

8. NIS problems and the UN outreach programme

The UN report for 2005 shows that, in general, the HDI for countries around the world is improving, with two major exceptions: Post-Soviet states (NIS), and Sub-Saharan Africa, both of which show a steady decline. Worsening education, economies, and mortality rates have contributed to HDI declines amongst countries in the first group, while

Table 2. Current world population (selected and ranked from 232)

rank	country	area (sq.km)	population estimate at 2006-07-01	yearly growth (%)
	World	510,072,000	6,525,170,300	1.14
1.	China	9,596,960	1,313,973,700	0.59
2.	India	3,287,590	1,095,352,000	1.38
3.	USA	9,631,418	298,444,200	0.91
4.	Indonesia	1,919,440	245,452,700	1.41
5.	Brazil	8,511,965	188,078,200	1.04
6.	Pakistan	803,940	165,803,600	2.09
7.	Bangladesh	144,000	147,365,400	2.09
8.	Russia	17,075,200	142,893,500	-0.37
9.	Nigeria	923,768	131,859,700	2.38
10.	Japan	377,835	127,463,600	0.02
26.	Ukraine	603,700	46,710,800	-0.60
42.	Uzbekistan	447,400	27,307,100	1.70
77.	Belarus	207,600	10,293,000	-0.06
92.	Azerbaijan	86,600	7,961,600	0.66
96.	Tajikistan	143,100	7,320,800	2.19
111.	Kyrgyzstan	198,500	5,213,900	1.32
113.	Turkmenistan	488,100	5,042,900	1.83
115.	Georgia	69,700	4,661,500	-0.34
120.	Moldova	33,843	4,466,700	0.28
130.	Lithuania	65,200	3,585,900	-0.30
136.	Armenia	29,800	2,976,400	-0.19
142.	Latvia	64,589	2,274,700	-0.67
151.	Estonia	45,226	1,324,300	-0.64

Human Immunodeficiency Virus/Acquired Immune Deficiency Syndrome (HIV/AIDS) and concomitant mortality is the principal cause of decline in the second group.

The United Nations Development Programme (UNDP), the United Nations' global development network, is the largest multilateral source of development assistance in the world. The UNDP is on the ground in 166 countries, working on global and national development challenges. UNDP provides expert advice, training, and grant support to developing countries, with increasing emphasis on assistance to the least developed countries.

The Human Poverty Index is an indication of the standard of living in a country, and was developed by the UN. The UN considers this a better indicator than the Human Development Index, which in turn is considered a better indicator than the Gross Domestic Product. For an understanding of all the problems of the NIS, it is necessary to study different indexes (Table 3).

The UNBSSI IHY programme is deploying arrays of small, inexpensive instruments around the world, especially in developing nations, to provide global measurements of geospace and heliospheric phenomena. It also hosts annual workshops dedicated to the IHY programme through 2009.

Table 3. The Gross Domestic Product for NIS countries. GDP dollar estimates here are derived from purchasing power parity (PPP) calculations. The data are provided by the International Monetary Fund (IMF). For comparison, max. and min. GDP figures are \$12,277B (USA) and \$221m (Kiribati). Max. and min. GDP/capita are \$69,800 (Luxembourg) and \$596 (Malawi).

Country of NIS (alphabetical order)	Country code/ capital	GDP (PPP) \$m	GDP per capita, \$	Region
1. Armenia	AM/Yerevan	14,167(127th)	4,270 (115th)	Western Asia
2. Azerbaijan	AZ/ Baku	38,708 (88th)	4,601 (107th)	Western Asia
3. Belarus	BY/Minsk	75,217 (64th)	7,711 (79th)	Eastern Europe
4. Estonia	EE/Tallin	22,118 (106th)	16,414 (43th)	Northern Europe
5. Georgia	GE/Tbilisi	15,498 (122th)	3,586 (122nd)	Western Asia
6. Kazakhstan	SQ/Astana	125,522 (56th)	8,318 (70th)	South-central Asia
7. Kyrgyzstan	KG/Bishkek	10,764 (134th)	2,088 (141st)	South-central Asia
8. Latvia	LV/Riga	29,214 (95th)	12,666 (53rd)	Northern Europe
9. Lithuania	LT/Vilnius	48,493 (75th)	14,158 (49th)	Northern Europe
10. Moldova	MD/Chisinau	8,563 (141st)	2,527 (131st)	Eastern Europe
11. Russia	RU/Moscow	1,575,561 (10th)	11,041 (62nd)	Eastern Europe
12. Tajikistan	TJ/Dushanbe	8,802 (139th)	1,388 (159th)	South-central Asia
13. Turkmenistan	TM/Ashgabat	40,685 (88th)	8,098 (73rd)	South-central Asia
14. Ukraine	UA/Kiev	338,486 (28th)	7,213(86th)	Eastern Europe
15. Uzbekistan	UZ/Tashkent	50,395 (74th)	1,920 (145th)	South-central Asia

Appendix I: Addresses of organizations coordinating meteor research in the USSR during the IGY, 1957

1. Moscow, B. Gruzinskaja 10, Astronomical Council AS of the USSR, Commission on Comets and Meteors

2. Odessa, Park named Shevchenko, Astron. Observ. of Odessa Univ. (parent organization of the meteor service of the USSR during the IGY period)

3. On organizational questions: Moscow, Kaluga Highway 71, Interdepartmental Committee of IGY at Presidium AS of the USSR, working group for studying meteors

4. For sending materials: Moscow area, post Vatutenki, Scientific Research Institute of Terrestrial Magnetism, Ionosphere and the Distribution of Radiowaves (now IZMIRAN)

Appendix II: Participants of IGY program on meteor research in the USSR

Participants of program IGY 1957 on meteor research in the USSR

R - radar, Ph - photographic, V - visual

N	City	φ	λ	H m	Scientific institutes / Country / Chairs	Program, N igy
1	*Ashkhabad*	37 ° 56'	58 ° 24 '	200	Astrophysical Laboratory of the Institute of Physics and Geophysics AS **Turkmen SSR** I.A. Astapovich, Ya.F. Sadykov.	R, Ph, V N696 (C126)
2	*Kazan*	55 ° 47 '	49° 07 '	80	Astronomical observatory named Engelgardt of the Kazan University **Russian SFSR** K.V. Kostylyov.	R N233
3	Kiev	**50 ° 27 '**	30° 30 '	185	Astronomical observatory of the Kiev University **Ukrainian SSR** A.F.Bogorodskiy,	R, Ph, N320
4	Odessa	**46 ° 29 '**	30° 46 '	50	Astronomical observatory of the Odessa University **Ukrainian SSR** V.P.Tsesevich, E.N. Kramer	R, Ph, V N621
5	Stalinabad *Dushanbe*	38 ° 34 '	68° 46 '	820	Institute of Astrophysics AS **Tajik SSR** L.A. Katasev, P.B. Babadzhanov, A.M. Bakharev.	R, Ph, V N680 (C115)
6	*Tomsk*	56 ° 29 '	84° 59 '	120	Tomsk Polytechnical Institute(*faculty of Radiophysics*) **Russian SFSR** Ye.F.Fialko.	R N224
7	Kharkov	50 ° 00 '	36° 14 '	140	Kharkov Polytechnical Institute (*faculty of Radioengineering*) **Ukrainian SSR** B.L. Kashcheyev	R N358(B141)

Acknowledgements

The authors acknowledge Wikipedia, the free internet encyclopedia, for very informative articles and pictures that were used in this contribution. S.V.K. is thankful to the IAU and to the Academy of Sciences of the Czech Republic for awarding her a grant to attend the XXVIth General Assembly and, personally, to John Hearnshaw (SPS5 chair), Oddbjorn Engvold (IAU General Secretary) and Jan Palous (GA NOC Chair) for their helpful assistance. S.V.K. also thanks the organizers of Special Session 5 of the XXVIth GA for acceptance of this contribution for publication, and especially to John Hearnshaw for his active help in the preparation of this paper.

Section 6:

Astronomy in eastern Europe

Astronomy for the developing world
IAU Special Session no. 5, 2006
J.B. Hearnshaw and P. Martinez, eds.
© 2007 International Astronomical Union
doi:10.1017/S1743921307007004

Astronomy in Serbia and in Montenegro

Olga Atanacković-Vukmanović

Department of Astronomy, Faculty of Mathematics, University of Belgrade,
Studentski trg 16, 11000 Belgrade, Serbia
email: olga@matf.bg.ac.yu

Abstract. After a brief survey of the foundations, development and the present status of astronomy education at all levels, a review of research in astronomy in Serbia and in Montenegro is given.

Keywords. History of astronomy, astronomy in Serbia, astronomy in Montenegro, astronomy education, research

---

1. Introduction

Astronomy education and research in Serbia have more than a 120-year long tradition (Dimitrijević, 2001). Namely, since 1880 astronomy has been taught together with meteorology at the Great School (the University of Belgrade since 1905). The first lectures were given in 1884, when Milan Nedeljković was elected to be the suplent (supplementary lecturer) for the courses of astronomy and meteorology. In 1887 Milan Nedeljković also initiated the foundation of the first Astronomical and Meteorological Observatory in Serbia. Since then numerous changes due to a turbulent political environment, institutional transformations, reforms of education and changes in research mainstreams have taken place.

Nevertheless, the histories of astronomy education and research in Serbia remained always strongly interlinked. In this paper we shall briefly describe the most important activities regarding astronomy education and research in Serbia and in Montenegro as well, since traditionally these two countries shared a common cultural area.

2. Astronomy education

In this section we give the status of astronomy education at all levels, from elementary school to PhD studies. Also, the activities of numerous amateur astronomical societies in public astronomy education are briefly summarized.

2.1. *Elementary and secondary school education*

Astronomy makes a part of the elementary and secondary school curricula, but not as a separate subject except in some cases mentioned below.

In elementary schools astronomy topics are taught within the courses of natural history, geography and physics. Recently, certain improvements of the program of physics concerning astronomy topics were introduced.

From 1969 to 1990 astronomy was taught as a separate course in the fourth year of secondary schools (in the beginning with one class hour per week and in the 1980s with two class hours per week).

However, due to a law passed in 1990, astronomy was integrated into the fourth year physics courses (with 10 classes in general and in social-sciences-oriented secondary schools, and with 32 classes in natural-sciences-oriented schools). In order to help secondary school teachers to keep up with new achievements in astronomy and in methods of teaching astronomy, lectures on various astronomy topics are given at regular annual seminars for physics teachers.

At present astronomy is taught as a separate course only in the Mathematical High School of Belgrade and in seven high schools of other Serbian towns (Novi Sad, Niš, Kragujevac, Kruševac, Kraljevo, Valjevo and Leskovac). Astronomy is not taught as a separate subject in the high schools in Montenegro.

Many attempts have been made within the reform of elementary and secondary education to reintroduce astronomy as a separate and compulsory course, but without success until now.

Despite the situation, pupils of higher classes of elementary school and high-school students express great and permanent interest in astronomy. Recently they scored several significant results (Milogradov-Turin, 2004). Serbia and Montenegro participated at three International Astronomy Olympiads: the VIIth (2002), IXth (2004) and Xth (2005). One junior and three senior teams participated at IAOs and won one silver and five bronze medals in total.

Special emphasis should be put to the activities of the Petnica Science Center, near Valjevo, centre for talented high-school students interested in science. It organizes seven seminars per year, lasting 7–8 days with about 20 participants on average.

2.2. *University education*

Astronomy topics are taught at all five state universities in Serbia: Belgrade, Novi Sad, Niš, Kragujevac and Priština (since 2002/2003 academic year in Kosovska Mitrovica) and at the University of Montenegro (Podgorica, Kotor).

2.2.1. *Astronomy education at the universities of Serbia*

The beginning of higher university-like education in Serbia can be traced back to 1838 when *Licej* (the lyceum) was founded in Kragujevac (Milogradov-Turin, 2002; Simovljević & Milogradov-Turin, 1998). In 1841 Licej was moved to Belgrade and in 1863 it was transformed into *Velika škola* (the Great School). The traces of "physical" astronomy can be found in the curriculum of Licej in 1854/55 academic year.

Teaching of astronomy together with meteorology was introduced in 1880 and the lectures started in 1884, when Milan Nedeljković (the founder and the first director of the Astronomical and Meteorological Observatory of Belgrade) was elected to be suplent for the courses of astronomy and meteorology at the Great School.

He became professor in 1886. The year 1880 is taken to mark the foundation of the Chair of Astronomy in Belgrade, although jointly with meteorology until 1924. When the University of Belgrade was founded in 1905, the Chair of Astronomy and Meteorology remained within the Faculty of Philosophy. Milutin Milanković, the most famous Serbian astronomer of the XX century, well known for his astronomical theory of climate, was elected professor of the University of Belgrade in 1909 and remained at the post for more than four decades. He taught a number of subjects related to applied mathematics. According to new regulations of the Faculty of Philosophy introduced in 1925, astronomy was for the first time taught as a separate subject. Vojislav Mišković, who obtained his PhD in France, became professor of the University of Belgrade in 1925. A separate study group for astronomy was established in 1927.

After the foundation of the Faculty of Mathematics and Sciences of the University of Belgrade in 1947, the Chair of Celestial Mechanics and Astronomy was formed. It was separated into two chairs, the Chair of Mechanics and the Chair of Astronomy, in

1962. In the process of the latest reorganization in 1995, the Chair of Astronomy became the Department of Astronomy and remained within the Faculty of Mathematics. The University of Belgrade is still the only one in Serbia with a Department of Astronomy.

Astrophysics was introduced as an obligatory course at the Chair of Astronomy in 1958 and it developed into several courses since then. Important changes in curricula were introduced in 1961 when two separate study groups were formed: Astronomy and Astrophysics.

Until 1988 these two groups had the same curricula at the first two study years, to become completely separate programmes afterwards.

The undergraduate studies last four years. About 4–6 students graduate each year from the Department of Astronomy in Belgrade. So far 222 students have graduated from the Department of Astronomy of the University of Belgrade, 58 students received an MSc degree and 28 students a PhD degree. The first astronomy student graduated in 1936, the first MSc degree was obtained in 1968 and the first PhD degree in 1958. It is interesting to note that 41 per cent of all graduated students are women, whereas this percentage grew to 57 per cent in the last 20 years.

At the Department of Astronomy six professors and three assistants teach fifteen subjects at the study groups of Astronomy and Astrophysics and two astronomy subjects at the study groups of Physics and Mathematics.

Apart from the courses in mathematics, physics and computer sciences, the study program in Astrophysics comprises the following two-semester astronomical courses: General astronomy (I year), General astrophysics (II), Practical astrophysics (III), Astronomical data analysis (III), Theoretical astrophysics (IV), Structure and evolution of stars (IV), Radio astronomy (IV), Stellar astronomy (IV) and Methodology of teaching astronomy and the history of astronomy (IV). In addition to being entitled to do research in astrophysics, students graduated in Astrophysics are also entitled to teach physics and astronomy in secondary schools.

Apart from the courses in mathematics and computer sciences, the study programme in Astronomy comprises the following two-semester astronomical courses: General astronomy (I year), General astrophysics (II), Spherical astronomy (II), Practical astronomy (III), Astronomical data analysis (III), Theoretical astronomy (IV), Ephemeris astronomy (IV), Celestial mechanics and the motion of artificial satellites (IV) and Stellar systems (IV). In addition to being entitled to do research in astronomy, students graduated in astronomy are also entitled to teach mathematics and astronomy in secondary schools.

Many of the students graduated from the Department of Astronomy in the last ten years enrolled at PhD studies in the USA, Canada and Australia, and lately in the EU countries as well. Most of them have already continued their research and gained post-doctoral positions.

At the University of Belgrade astronomy is also taught as a compulsory one-semester course, "Fundamentals of astrophysics", for the third-year students of the Faculty of Physics (physics teachers division), a compulsory course, "Geodetic astronomy" (4th year), at the Faculty of Civil Engineering, and a one-semester elective course, "Fundamentals of astronomy", for the fourth-year students of mathematics.

Since 2002 the University of Novi Sad, the Department of Physics (Faculty of Natural Sciences) has opened an astronomy study group with several astronomy and astrophysics courses.

At the universities of Niš, Kragujevac and Priština (now situated in Kosovska Mitrovica), "Fundamentals of Astrophysics" (3rd study year), "Astrophysics and Astronomy"

(3rd study year) and "Fundamentals of Astronomy" (2nd study year), respectively, are taught as one-semester courses at the physics study groups.

2.2.2. *Astronomy education at the University of Montenegro*

Two astronomy courses are taught at the University of Montenegro: a two-semester course "Astronomical navigation" (2nd study year) at the Faculty of Maritime Studies of Kotor and a one-semester course "Geodetic astronomy" (3rd study year) at the Department of Geodesy, Faculty of Civil Engineering of Podgorica.

2.2.3. *Reform of university education*

According to the act passed in September 2005 the new European Credit Transfer System (ECTS) is to be introduced at all universities in Serbia and in Montenegro.

The Department of Physics of the University of Novi Sad was the first to introduce this system in 2002 together with opening of the new study group of Astronomy. At the University of Belgrade new study programs of Astronomy and Astrophysics are going to be introduced starting from 2006/2007 academic year. All existing two-semester courses are reorganized and divided into one-semester courses. Some new courses are introduced as well. The model 4+1 for the first two degrees (bachelor and master) is accepted. The first two study years of the third (PhD) degree feature compulsory and elective courses, whereas the third year is dedicated to the work on the PhD thesis.

2.3. *Public outreach*

Public astronomy education in Serbia and in Montenegro is realized by the way of lectures at public universities, radio and TV programmes, popular journals and books, lectures in two Planetaria (Belgrade and Novi Sad), in public observatories and fifteen amateur astronomical societies. The activities offered by amateur societies cover public observations of all major events, lectures, courses, conferences, schools and camps.

There are 14 astronomical societies in Serbia (two in Belgrade, two in Novi Sad, one in each of Valjevo, Kragujevac, Niš, Zrenjanin, Vršac, Bor, Prokuplje, Loznica, Knjaževac, Novi Pazar) and one in Montenegro (Podgorica). As there is an increased interest in astronomy among the general public, five of them were founded in the last three years.

The largest and the oldest society of amateur astronomers is the AS "Rudjer Bošković" of Belgrade, founded in 1934. The Society organizes astronomy courses each autumn and spring, Belgrade Astronomical Weekends, Summer Astronomical Meetings and Summer Schools of Astronomy, typically lasting a week. The non-profit astronomical journal "Vasiona" ("The Universe"), published by the Society, has a 54-year long tradition.

Since 1998 the largest astronomical web site in the country, Internet magazine "Astronomical magazine" (www.astronomija.co.yu) has been maintained by the AS "Lyra" of Novi Sad. Since 2003 the Society publishes "Astronomija", a paper magazine of high-quality presentation.

More details about the activities of the amateur astronomical societies can be found in Milogradov-Turin (1996, 2000, 2002) and Atanacković-Vukmanović (2005).

3. Astronomy research

Astronomy research in Serbia is for the most part performed in two astronomical institutions: the Astronomical Observatory of Belgrade and the Department of Astronomy at the Faculty of Mathematics of the University of Belgrade.

3.1. *Astronomical Observatory in Belgrade*

The Belgrade Astronomical Observatory is one of the oldest scientific institutions in Serbia. It was founded (together with the Meteorological Observatory) in 1887. Prof. Milan Nedeljković was appointed its first director. In 1924 the Observatory was divided into two separate institutions: the Astronomical Observatory and the Meteorological Observatory of Belgrade University. From 1930 to 1932 a new astronomical observatory, 6 km southeast of Belgrade's centre, at the 253-m high hill Veliki Vračar, named Zvezdara since then (zvezda=star), was built under the direction of Prof. Vojislav Mišković. The observatory has grown into a modern institution under his supervision. Prof. Mišković won the French Academy Prize in 1925 for his studies in stellar statistics, whereas the most important of his later works were related to minor planets. Among the distinguished scientists who served as directors of the Observatory over the years, let us mention Prof. Milutin Milanković, widely known for his explanation of the ice ages phenomenon and the history of the climate of Earth and other planets, who was at the head of the Astronomical Observatory from 1948 to 1951.

3.1.1. *Instruments*

The instruments procured by M. Nedeljković from Germany, on account of the First World War reparations, were mounted in 1934 and constitute still the observing basis of the Observatory. These are:

- Large Refractor - equatorial Zeiss 650/10550 mm
- Solar spectrograph Littrow type, collimator lens 200/9000 mm, grating Bausch & Lomb 600 lines/mm
- Large meridian circle Askania 190/2578 mm
- Large vertical circle Askania 190/2578 mm
- Large transit instrument Askania 190/2578 mm
- Astrograph Zeiss 160/800 mm
- Photovisual refractor Askania 135/1000 and 125/1000 mm
- Transit Instrument Bamberg 100/1000 mm
- Zenith telescope Askania 110/1287 mm

The large meridian circle was, unfortunately, burnt up in May 1999.

Recently, a Meade reflector of 40 cm diameter has been procured and is used for CCD observations of solar system objects.

A project (initiated in 1986, but not implemented at the time) of a new astronomical station of the Belgrade Astronomical Observatory in southern Serbia is being carried out. The observatory is to be situated on the mountain Vidojevica near Prokuplje at an altitude of 1155 m. In the first phase, a reflector Astro Optik ($D = 60$ cm) is to be mounted. Later on, a larger telescope is planned.

3.1.2. *Research activities*

Research activities at the Astronomical Observatory cover a wide range of topics. Starting with 1935 the services for minor planets and solar observations, for time and latitude, for double stars and for variable stars were established. Since 1960 research in astrophysics has developed. It started with the photometry and polarimetry of eruptive stars and later it was directed towards stellar and solar physics and astronomical spectroscopy in general. Nowadays, research is carried out mostly in dynamical astronomy (solar system bodies, double stars, Earth rotation), astrophysics (solar physics, close binary stars, astronomical spectroscopy, galactic astronomy, extragalactic astronomy, cosmology, astrobiology) and history of astronomy. Thirty-five of 49 staff members are researchers. They participate in eight scientific projects financed by the Ministry of Science and Environmental Protection of Serbia and in several international projects.

The researchers of the Astronomical Observatory participate in the undergraduate study programmes at the Universities of Belgrade and Novi Sad, as well as in the Master and PhD study programs of Astronomy and Astrophysics at the Belgrade University.

3.1.3. *Publications*

Since 1936 the Astronomical Observatory has published *Bulletin de l'Observatoire astronomique de Belgrade*. From No. 145 (in 1992) it appeared under the name *Bulletin astronomique de Belgrade*, after merging with *Publications of Department of Astronomy* (founded in 1969). From No. 157 (in 1998) the name was changed into *Serbian Astronomical Journal* (http://saj.matf.bg.ac.yu). Along with this main journal, since 1947 the Belgrade Observatory has published *Publications of the Astronomical Observatory of Belgrade*. All the publications of the Belgrade Observatory are distributed to about 200 scientific institutions all over the world.

3.2. *Department of Astronomy of the Faculty of Mathematics of Belgrade*

The research activities of the staff of the Department cover the following topics: Earth's rotation, dynamics of asteroids, motion of artificial satellites, stellar kinematics and dynamics, stellar structure, radiative transfer, solar and stellar atmospheres, radio astronomy, supernova remnants, active galactic nuclei and history of astronomy. All this research is carried out together with the colleagues from the Astronomical Observatory in the framework of the common projects.

The first solar radio interferometer in Serbia was constructed by the staff of the Department of Astronomy in 1960. It worked till 1966 for daily observations and later served only for teaching purposes.

Acknowledgements

I would like to express my gratitude to the IAU and to the Ministry of Science and Environmental Protection of the Republic of Serbia for financially supporting my participation at the XXVIth General Assembly of the IAU in Prague. This work is realized within the project No. 146003 "Physics of Stars and the Sun". I would like to thank Prof. J. Milogradov-Turin, Dr Z. Knežević and Mr N. Vitas for useful comments that helped to improve the manuscript.

References

Atanacković-Vukmanović, O. 2005, *Publ. Astr. Obs. Belgrade* 80, 275
Dimitrijević, M. 2001, in: A. Antov, R. Konstantinova-Antova, R. Bogdanovski & M. Tsvetkov (eds.), *Balkan Meeting of Young Astronomers, Belogradchik* 20
Milogradov-Turin, J. 1996, *Newsletter IAU Commission 46* 45, 59
Milogradov-Turin, J. 2000, *Newsletter IAU Commission 46* 52, 8
Milogradov-Turin, J. 2002, *IAU Commission 46; Newsletter Supplement; National Liaison Triennial Reports 2002*
Milogradov-Turin, J. 2002, *Publ. Astr. Obs. Belgrade* 75, 289
Milogradov-Turin, J. 2004, *Vasiona* 5, 265
Simovljevic, J. & Milogradov-Turin, J. 1998, in: N. Bokan (ed.), *125 years of the Faculty of Mathematics (in Serbian)* (Faculty of Mathematics, Belgrade)

Astronomy for the developing world
IAU Special Session no. 5, 2006
J.B. Hearnshaw and P. Martinez, eds.

© 2007 International Astronomical Union
doi:10.1017/S1743921307007016

Astronomy and astrophysics in Bosnia and Herzegovina

Aleksandra Andic[1]

[1]Queen's University Belfast, 1 University Road, Belfast BT71NN, UK
email: a.andic@qub.ac.uk

Abstract. In Bosnia and Herzegovina astronomy teaching is almost nonexistent. There are only several courses within universities and they are usually given by physicists who themselves have had only elementary courses in astrophysics. When educational outreach is in question, the situation is even more grim. On the other hand, there is a huge interest for astrophysics in the student population. There are several solutions and possibilities which I will present, together with a discussion of the main obstacles which need to be overcome.

Keywords. education, Bosnia and Herzegovina, astronomy, astrophysics

1. Introduction

Around a decade and a half ago Bosnia and Herzegovina (BiH) was an integral part of the larger state, Socialistic Federative Republic of Yugoslavia (SFRY). For decades, development in BiH was determined by its 'big brother' surroundings. In the former SFRY the major centres for astrophysics were Belgrade and Zagreb, and therefore all students interested in astronomy went to those centres for astrophysical education. Those two centres also had astrophysical research institutes and observatories in which scientific work was done.

Thanks to the social support of the former SFRY it was not hard even for the poorest students to go and pursue education they wished for in distant cities. The war and disintegration of the former SFRY left BiH citizens without the possibility of obtaining any astrophysical education in their own country. Besides, economic and various political reasons made it virtually impossible for the average student to go to the former centres to obtain the necessary education. However, Bosnia and Herzegovina was left with two major university centres; one in Sarajevo and the other in Banjaluka.

2. Universities

2.1. University of Sarajevo

The University of Sarajevo was established after the Second World War. In December 1949 the first rector was chosen. The university at that time reflected the regional needs for the educated work force and as such there was no great need for astrophysics during that time. There was a department of "mathematics and natural sciences" as the part of the College of Philosophy at that time. During 1960-1969, a separate College for Sciences was established with five departments. Nevertheless, astrophysics remained just another course.

The University of Sarajevo today has a well established Department of Physics which is part of the College for Mathematics and Science. There is no independent part which is doing any research connected with astrophysics. There are only two courses of astrophysics held by the department. The people giving those courses are Jugoslav Strahov (on

Figure 1. The entrance of the Faculty of Mathematics and Natural Sciences, College of Mathematics and Science, University of Sarajevo

a contract basis), and Dr Zalkida Hadžibegović as a lecturer. Unfortunately, searching by ADS for both persons yields zero papers in astronomy and astrophysics related journals. Likewise, direct contact with Dr Hadžibegović did not unveil any research interest in astrophysics.

2.2. *University of Banjaluka*

The university in Banjaluka was established 7 November 1975 following the initiative of the University of Sarajevo and the growing need for an educated work force. The College of Science evolved from a teachers' college established at the beginning of the University of Banjaluka. The Physics Department was established in 1996 as part of the College of Science. Since Banjaluka is in the part of BiH with the majority of Serbian ethnicity, the leaders of the university employed on a contract basis professors from Belgrade and Novi Sad universities as teaching staff for the department. In the department there was only one course in astrophysics. It lasted one semester and it was part of the general physics education.

The University of Banjaluka today has officially two courses. One of them "Astrophysics and theory of the gravitational fields" (one semester) is held by Dr Branko Dragović (contract based, permanently employed in the Institute of Physics, Belgrade, Serbia). Also the ADS a search yielded zero papers for Dr Dragović in the journals related to astronomy and astrophysics.

The only research connected with astrophysics in this department was conducted by myself, during recent times, when I was working for one semester in the department as a teaching assistant on the subject "Introduction of the Theory of the Fields", although this is not directly related to astrophysics (Andjic (2006)).

2.3. *Other universities*

During the post-war period two more universities appeared. One in the suburbs of Sarajevo and the other in Mostar. Both are new and neither has significant astronomy or astrophysics interests.

+ Information provided with the direct contact with Dr Hadžibegović and Dr Milošević.

2.4. *Mutual cooperation*

Unfortunately, mutual cooperation is divided by the same differences which were a major problem during the war. The ghosts of the war still make it impossible to establish a connection between the two major universities in BiH. A physics society for the whole of

Figure 2. The 'Banski dvori', building where the head of the University of Banjaluka with supporting administration is located

Bosnia and Herzegovina does not exist, since current leaders of both universities prefer to pay attention to the past, ignoring the future of the country and science completely. All attempts for contact by personnel of the departments or residents of the different parts of BiH is cut. As an example, my correspondence with Dr Hadžibegović ended without any explanation. Dr Hadžibegović is refusing to reply to my emails.

One of the possible reasons might be that leaders of the universities perceive me as a member of the 'other/enemy' side despite the fact that I work abroad. Even an invitation to share the grant was left unanswered. Last year I was contacted by the chairman of a Regional network of IHY, receiving information that I was the only BiH citizen who works within Astrophysics to the best of the chairman's knowledge. Despite this fact, BiH was included as the member of the regional network, although the members of Bosnian part are, at the moment, myself and six students from Banjaluka University who show some interest in astrophysics.

Again the ghosts of war prevent any interested Bosnian person who perceives myself as a member of the 'other side' to contact me about this matter.

3. Observatories and amateur societies

3.1. *Sarajevo*

The interest in astronomy has always been present in this area. The history of the first and only Bosnian observatory shows it. In the 1960s a society of BiH amateur astronomers

Figure 3. The observatory of Sarajevo, before the war in Bosnia and Herzegovina

founded the Astronomical Society of the University of Sarajevo in 1963 and built the People's Observatory, so called Mejtas, with a 17centimetre (7inch) reflector, in 1965. By 1973, amateurs had converted an old AustroHungarian fortress, Ćolina Kapa.

The observatory is located at an altitude of 1010 metres on Mount Trebević. In the late 1980s the first serious work started. The observatory started a collaboration with the University of Zagreb on projects including photometry. This continued until the start of the war. Today, the observatory is still in ruins, without any prospect of getting restored. At the moment, it exists primarily for calculations of the sunrise and sunset for the local meteorological institute.

The government of Bosnia and Herzegovina or local universities do not show any interest in restoring the observatory. The situation is reflected on the web as well, where the only detailed presentation of the Sarajevo Observatory can be found on the site made by the Croatian, Višsnjan observatory.

3.2. *Banjaluka*

In Banjaluka there is an Amateur Astronomical society, but it has only one person with an astronomical education on the board, Mr D. Krunić (bachelor degree, working in the pedagogical institute). They plan to establish a small local observatory, but they are also crippled by the lack of funding and lack of interest by government or local universities. The only activity so far has been organizing various lectures on the subject of astronomy and astrophysics, usually by inviting astrophysicists from Belgrade.

3.3. *Mutual cooperation*

Unfortunately, I could not get any information about the existence of any Amateur Astronomical society in Sarajevo. The one in Banjaluka is registered as the official BiH society and there is the possibility that the same ghosts of war are preventing any cooperation in this case as well. Also the Amateur Astronomical society in Banjaluka did not show any interest in joining the regional network of IHY.

4. The prospective students

4.1. *Banjaluka*

Besides such a poor education and the negligible chance of conducting real research, interest among students at Banjaluka university exists. During my short (one semester) stay at the university, a group of 6 physics students formed a group for intensive astrophysics learning, under my supervision. One of the students, Nemanja Voćkić, had shown enough interest and readiness for work, to be included in a small part of one of my research projects. Currently we are working on a conference poster and consequently a paper. N. Voćkić plans to include this material into his, obligatory, research work for obtaining a bachelor's degree. Cooperation itself is hard, since students do not receive any support from the university. It is not even clear if Mr Voćkić will be able successfully to defend his work, taking into account that I, as his supervisor, cannot be part of the examination committee.

4.2. *Sarajevo*

According to the information from Dr Hadžibegović and Dr Milošević, the name of the prospective student in Sarajevo is Nedim Mujć, who is currently looking for the possibility of graduate studies abroad. Unfortunately, I was not able to contact this student personally. The reasons for notresponding are unknown.

5. Conclusions

At the moment astronomy and astrophysics do not exist on a professional level in Bosnia and Herzegovina. Not a single Bosnian citizen with a Ph.D. in Astronomy and Astrophysics holds a position at any BiH university or any kind of other institution which is connected with the subject itself.

On the other hand, there is a significant interest among students regarding the subject. The interest is usually pushed aside while most of those students end up studying engineering or similar subjects. Due to the hard economic situation, no student is capable of pursuing undergraduate education within astronomy and astrophysics elsewhere. Also the government and local universities do not provide any help for perspective students in this area, although the need for a work force within this area is evident.

The solution to this problem depends heavily on the political climate and attitudes of the leaders at local universities. At the moment there is no indication that either will change for the better; thus they are providing no support of any kind in the astronomy and astrophysics development of the country. So my recommendation for students interested in the subject is to try and get a scholarship outside BiH and pursue education in astronomy and astrophysics abroad. Hopefully, in time, new leaders of local universities will realize that only persons who perform research can actually teach the appropriate subjects at university level. Hopefully this course of action will help put to rest the ghosts of war.

Acknowledgements

I would like sincerely to thank Dr Hadžibegović and Dr Milošević for all the information they provided. Also I thank Mr D. Krunić for providing interesting details about amateur societies. And above all my thanks go to Dr Z. Rajilić for helping the student N. Voćkić to overcome administrative and prejudice problems in the completion of his research work. Also I wish to thank Dr Katya Georgieva for including Bosnia and Herzegovina in the regional network and Prof. Dr John Hearnshaw for showing interest in the BiH situation, despite the grimness of it.

Olga Atanacković-Vukmanović

l to r: José Ishitsuka, Svetlana Kolomiyets and Nat Gopalsamy

Astronomy for the developing world
IAU Special Session no. 5, 2006
J.B. Hearnshaw and P. Martinez, eds.

© 2007 International Astronomical Union
doi:10.1017/S1743921307007028

Graduate programme in astrophysics in Split

Davor Krajnović[1]†

[1]Department of Physics, University of Oxford, Oxford, UK
email: dxk@astro.ox.ac.uk

Abstract. Beginning in the autumn of 2008, the first generation of astronomy master's students will start a two-year course in Astrophysics offered by the Physics Department of the University of Split, Croatia (http://fizika.pmfst.hr/astro/english/index.html). This unique master's course in south-eastern Europe, following the Bologna convention and given by astronomers from international institutions, offers a series of comprehensive lectures designed to greatly enhance students' knowledge and skills in astrophysics, and prepare them for a scientific career. An equally important aim of the course is to recognize the areas in which astronomy and astrophysics can serve as a national asset and to use them to prepare young people for real life challenges, enabling graduates to enter the modern society as a skilled and attractive work-force. In this contribution, I present an example of a successful organization of international astrophysics studies in a developing country, which aims to become a leading graduate programme in astrophysics in the broader region. I will focus on the benefits of the project showing why and in what way astronomy can be interesting for third world countries, what are the benefits for the individual students, nation and region, but also research, science and the astronomical community in general.

Keywords. Education in astronomy, astronomy in Croatia, sociology of astronomy

1. Introduction

The knowledge-based economy is founded on discoveries and innovations. The power of a nation is measured by its ability to stimulate discoveries and capacity to innovate. A nation cannot achieve this without increasing its educational levels and intellectual competencies. Moreover, for a small nation, building knowledge-based society is more than an economic necessity – it is also a strategy for preserving its culture and identity.

Astronomy, as a scientific discipline, is a valuable incubator of new ideas, discoveries and enterprises, driven by the synergy of sophisticated technologies, different natural sciences and the wish to understand our Universe. Astronomy has also a unique public appeal, which enables it to inspire young generations and stimulate their curiosity, creativity and appreciation for science.

In developing countries, astronomy is, however, often perceived as an abstract and expensive science, without visible end-products through which it is possible to quantitatively 'weigh' its usefulness. Even in developed countries, in the times of economic uncertainty, astronomy falls into the group of fundamental sciences which are the first to lose governmental monetary support. It is the duty of professional astronomers to oppose this opinion, and to demonstrate the benefits related to astronomy as a science and, especially, as an educational option for bright young people of all ages and school levels.

The Astrophysics Initiative in Dalmatia (ApID) is an attempt to do this and promote the growth and development of astrophysics in Dalmatia, a southern region of Croatia. ApID is a collaboration of projects, institutions, and individuals, from professional scientists to amateur astronomers, sharing the common interest in astronomy and

† Present address: University of Oxford, Keble Road, Oxford, OX1 3RH, UK

astrophysics. The strategy of ApID is to ensure its sustainability through a carefully designed combination of educational, research, and outreach programmes. The operational performance and strength of ApID are enhanced by the synergy of a public university (the University of Split) and a non-profit non-governmental organization (the Society znanost.org). The backbone of ApID is the Graduate Programme in Astrophysics at the Department of Physics at the University of Split (GPAS), Croatia, which is the main topic of this contribution to the Special Session 5. The main objectives of GPAS can be summarized as follows:

- to provide the graduates with know-how to continue in diverse career paths as well as to make them attractive targets for a range of employers in different branches of business, government and finances;

- to attract young people for careers in science and technology;

- to establish connections with other disciplines present at the University of Split, other Universities in Croatia and abroad in order to actively promote interdisciplinary sciences;

- to promote top level research in astronomy in Croatia, and

- to enhance the quality of existing graduate level education in the region and set an example to other educational institutions in south-eastern Europe.

Here, I outline the graduate programme focusing on aspects which differ from existing and more classical courses (Section 2). This is followed by a discussion about the benefits that such a study can bring to the students, to the region, but also to us, the astronomical community (Section 3). Finally, I conclude with highlights of the GPAS (Section 4).

2. Study astrophysics in Split

GPAS is supported by the Department of Physics of the University of Split, which is a part of the Faculty of Natural Sciences, Mathematics and Kinesiology. Recently, following the Bologna process (standardization of higher education in Europe), the Department of Physics restructured its courses now offering three-year undergraduate and two-year MSc (master's degree) graduate study programmes. The MSc in Astrophysics will be given in English, enabling students from other countries to follow the courses in Split, thereby increasing their mobility and exchange of experiences.

Students. The MSc programme in Astrophysics is open to Croatian and international students, who completed their undergraduates studies (BSc) in physics or related subject. In the modern world, it is necessary to actively work on attracting bright students. Clearly, the best undergraduate students should be encouraged to apply to the programme. Possible candidates, however, could be identified and attracted to astronomy at even younger age, among the gifted high-school or even elementary schools pupils. This means that a part in the organization of the graduate studies also includes an active outreach through which GPAS will be present in the media and news, will actively participate in the organization of local and regional scientific events for young people, and will promote the popularization of science to the general public.

Lecturers. Currently, there are no professional astronomers at the Department of Physics in Split. The expanding department will in the following years offer positions for astronomers, but the bulk of the astronomy specific teaching at GPAS will be given by visiting lecturers. Some of these lecturers are Croatian astronomers working abroad, while others are foreign scientists. The rest of the courses (non-astronomy related, but necessary for a completion of the master degree in Astrophysics) will be given by local physics faculty. The vising lecturers will not need to spend a whole semester in Split, but rather a few weeks necessary for completing their course.

Courses and Timetable. The courses are divided between required and elective courses. The aim is to give a thorough astrophysical background, which is supplemented with optional courses in general physics, computer science and humanities (history, philosophy, etc). Since large section of courses will be given by visiting lecturers, the whole educational programme is organized in blocks. One block, lasting between 2 to 4 weeks, contains the lectures as well as the exam. The emphasis of the studies on the research is evident in the fact that in all semesters students will be required to participate in scientific projects. Moreover, the last semester is completely devoted to a research project which will become the master thesis. The aim of these projects is to encourage students to tackle complex programmes and learn methods of the scientific research. The most important aspect, however, is to enable students to work in an international environment at astrophysical centres of excellence. They will have to work on projects offered by the lecturers, visiting their institutions abroad and producing publishable scientific results.

The first light. The new graduate programme will commence during the fall of 2008 when the first group of Croatian students following the Bologna process will finish their BSc.

3. Benefits of the graduate programme

The position of astronomy in early societies was very much different than it is today. First astronomical exploits were very practical, from the determination of the seasons necessary for the agriculture, which fed the stratified human societies, to predictions of terrifying eclipses and explanation of the heavens. Astronomy was a very practical and necessary system of theoretical knowledge, with deep cultural and social consequences. Last few hundred years, however, saw a reverse of the medal. Astronomy, supported by physical laws of nature, was turning towards theoretical knowledge of Universe, its structure and properties, loosing its practicality and usefulness for everyday aspects of human lives. On the other hand, younger sciences like physics, chemistry, biology and their various mixtures (medicine, engineering ...), with their technical applications, started changing the human society as well as its influence on the Earth. Astronomy did little in this 'progress'[†], becoming more and more academic and idealistic pursuit of the nature of the Universe. Today, it is often considered a fundamental science, remote and useless for 'real' life, perhaps an exercise for the mind, interesting in terms of secular human culture; an entertainment, although rather expensive, for the general public.

Astronomy in the modern world, however, is much more than that. It still makes humanistic, educational and technical contributions to our society. Rather than going through a list of recent contributions from astronomical research, I will focus on a more specific aspect of benefits that a study of astronomy can offer to the students and to the region hosting an astronomical institution.

3.1. *Benefits for the students*

It is self-evident that training in science opens doors to a scientific career. It is, however, often not clear to the general public that the same training can be very attractive to industry and different businesses as potential employers of skilled people. An objective

---

† One should, however, remember the crucial role of astronomy in navigation and, hence, in the great geographical discoveries and world trade until the recent advent of satellite navigation. The importance of astronomy for economy of the sea-faring nations is evident also in the support astronomical research was getting from the governments: both Paris and Greenwich Observatories were opened in 17th century with purpose to perfect the art of navigation, increase the maritime power of the nations and, in general, make the seafaring safer.

of GPAS is to produce world-class graduates with the necessary skills to become local and global leaders and entrepreneurs.

Industry and finances are generally not concerned about the specialization of an academic course. They are more interested in broader skills obtained by students, which can be transferred to the work place. A master's course in astronomy is actually well suited in this respect. The interdisciplinary nature of astronomy provides a full framework for illustrating to the students the unity of natural phenomena and the evolution of scientific paradigms that explain them. On a more pedagogical level within higher education, teaching astronomy to master students prepares them for a broad range of scientific disciplines ranging from purely academic, such as astrophysics, to widely applicable tasks of engineering and computer science. GPAS aims to leave a number of transferable skills to its graduates:

- mathematical, computing and modelling skills with scientific literacy
- tacking complex problems and handling incomplete and large data sets
- cooperating in international teams and projects
- preparing and executing projects (often on tight deadlines)
- presenting an account of work to colleagues and to broader audiences

It is important to stress that the above mentioned skills provide a solid base not only for a career in science, but also in different branches of business. It is the synergy of different sciences used in astronomical research that offers a broad spectrum of transferable skills and know-how for diverse career paths.

3.2. *Benefits for the region*

The region concerned in this section is primarily Dalmatia with its capital Split. On a slightly higher level it, of course, concerns Croatia as the country where the master study is organized. It is, however, easy to generalize towards a larger region of South–East Europe, but certain aspects of the graduate programme apply to the whole Europe, both to its political and geographical entities.

Modern economy of a nation depends on its ability to compete technologically with other nations. This ability directly translates in the number of technically trained people that are able to use existing technologies and develop new ones for international markets. In addition to that, the quality of environment depends on developing safe, clean industries and sources of energy. This, also, can be accomplished only by imaginative and highly trained scientists and engineers. However, even governments of developed countries recognize the steadily decreasing number of students interested in pursuing technical careers as a major concern for the future development of their nations. To overcome this downward trend it will be necessary to support those activities that stimulate young people toward scientific thinking and the development of mathematical and technical skills. The subject of astronomy is inherently interesting to young people, thereby keeping them interested in science, whilst they learn fundamentals of mathematics, statistics, physics, chemistry, etc. GPAS follows the trends of physics departments in developed counties which started offering astronomy options in their curriculum, resulting in an increased interest in bachelor and master degree programmes in astronomy. GPAS aims to provide a highly educated and scientifically literate work force required by their home country for its future development.

A big problem facing developing countries is the *brain-drain*, the exodus of highly educated individuals who go to wealthy nations capable of offering well payed or other specific jobs (such as research positions in astronomy). Croatia and the region of Split suffer from this effect as well. A graduate course in astronomy, however, can be used to reverse the *brain drain* into *brain gain* by attracting astronomy professionals (either

Croatian or foreign) to take posts in Split and promoting a modern programme which offers transferable skills to its students preparing them for different career paths.

GPAS should also be interesting to the relevant bodies in the European Union, since it is fully compatible with the European standards for high education (Bologna process), emphasizes the mobility of students between the countries, especially the neighbouring countries of Croatia, and its existence will enhance the quality of existing graduate level education in the region, as well as set an example to other educational institutions in south-eastern Europe.

3.3. *Benefits for astronomy*

As a final remark in this section, it does not hurt to stress that with a new graduate programme in astronomy the whole astronomical community also gains. An obvious benefits are the job openings and a popularization and spread of astronomy to the general public (who is paying for most of the astronomical research). Possibly the most interesting gain is opening of a new channel for introduction of young bright people into the world of top level astronomical research. The graduate programme in Split supports this aspect by linking professional scientists and students. This is achieved by bringing astronomers (lectures, see Section 2), who work at international institutes and are involved in different international projects, to Split to give lectures and offer master projects to the students. On the other hand, they will meet capable students with different backgrounds and eager to do astronomical research but coming from Croatia and neighbouring countries which do not have developed astronomical infrastructure nor are members of major international projects or organizations. To our knowledge, the astrophysics graduate course in Split is the first programme of this scale offering the link between astronomically developed western and underdeveloped south-eastern Europe.

4. Conclusions

Astronomy is a lively natural science that can offer a broad range of products to the general community: from 'pretty' pictures to an understanding of the structure of the Universe, from technological advancements to educational programmes. Many of these are applicable and useful for developing countries. In this work I focussed on the tertiary education, on the organization of a graduate course in astrophysics and on the benefits it offers to the students and the host country. Here I summarize the main points. The aims of GPAS are:

• to recognize the areas in which astronomy and astrophysics can serve as a national asset and to use them to prepare young people for real life challenges;

• to attract young people for careers in science and technology;

• to set the example of excellence to be followed by other existing higher education programmes in the region and demonstrating to the public that investing in science is the right thing to do;

• to bring the top science and technology research to Croatia, enhance the transfer of technology to Croatia through international collaborations and joint projects with international centres of excellence, and help disseminate the skills and knowledge necessary for the establishment of similar world-class centres of excellence in the region; and

• to actively participates in the process of brain-gain by attracting Croatian science diaspora and foreign scientists to actively participate in the Croatian education system through the transfer of their skills and knowledge to Croatia.

The astrophysics graduate course is fully compatible with the European educational standards, and its international orientation (curriculum, language and lecturers) and research excellence will serve as a bridge between the existing and future EU states, empowering Croatia as a future EU member.

Acknowledgements

I would like to thank Dejan Vinković, for useful discussions and careful reading of the manuscript.

References

The following list of references have been consulted during the development of this document:

'History of Astronomy', Pannekoek, A. 1961, London: Allen & Unwin, 1961,
'The decade of discovery in Astronomy and Astrophysics', National Research council, 1991, National Academy press, Washington, D.C.
'Astronomy and Astrophysics in the New Millennium', National Research Council, 2000, National Academy press, Washington, D.C.
'Bologna Process in Croatia', http://bolonjski-proces.idi.hr
'Astrophysics Initiative in Dalmatia (ApID): Business Strategy for 2006-2011', Forum of Croatian Astronomers, Split, Croatia, 2006.

Section 7:

Astronomy education in developing countries

Astronomy for the developing world
IAU Special Session no. 5, 2006
J.B. Hearnshaw and P. Martinez, eds.

© 2007 International Astronomical Union
doi:10.1017/S1743921307007041

International Schools for Young Astronomers (ISYA): a programme of the International Astronomical Union

Michèle Gerbaldi[1]

[1]Institut d'Astrophysique de Paris et Université de Paris-Sud XI, France
email: gerbaldi@iap.fr

Abstract. This paper outlines the main features of the International Schools for Young Astronomers (ISYA), a programme developed by the International Astronomical Union (IAU) in 1967. The main goal of this programme is to support astronomy in developing countries by organizing a school lasting 3 weeks for students with typically a M.Sc. degree. The context in which the ISYA were developed has changed drastically over the past 10 years. We have moved from a time when access to any large telescope was difficult and mainly organized on a national basis, to the situation nowadays where data archives are established at the same time that any major telescope, ground-based or in space, is built, and these archives are accessible from everywhere. The concept of the virtual observatory reinforces this access. However, the rapid development of information and communications technologies and the increasing penetration of internet have not yet removed all barriers to data access. The role of the ISYA is addressed in this context.

Keywords. IAU Programme: International Schools for Young Astronomers (ISYA).

---

1. Introduction

The programme *International Schools for Young Astronomers,* hereafter named ISYA, was developed by the International Astronomical Union (IAU) in 1967 under the auspices of IAU Commission 46 *Astronomy Education & Development.* We first describe the creation of this Commission in 1964 and the ISYA programme in 1967. Then we present the objectives and the organization of these Schools from 1967 to 1990. The context in which the ISYA was created has changed over the past ten years, mainly due to the information technology and communications revolution. We depict this evolution from 1992 to 2006. An assessment of the impact of the ISYA is given and the conclusion addresses the new horizon for this IAU programme.

2. IAU XIIth General Assembly in 1964

2.1. *Creation of Commission 46 on The Teaching of Astronomy*

During the XIIth IAU General Assembly, helded at Hamburg in 1964, one of the Special Meetings organized was on *The Teaching of Astronomy* (Transactions of the IAU, vol. XIIB, page 629, 1964). In the preliminary report published, M. Minnaert concluded the discussion (Transactions of the IAU, vol. XIIB, page 648, 1964) with two proposals. The first of proposal was that: "The Members of the International Astronomical Union, present at the Hamburg meeting on the Teaching of Astronomy, strongly recommend to the Executive Committee to organize a Commission of the Union on this subject."

This Commission was created as a Commission of the Executive Committee, with E. Schatzman being the first President for the period 1964–1967.

2.2. *Summer Schools for Young Astronomers*

During the same meeting, V. Kourganoff expressed his views on the "International cooperation in the domain of astronomy teaching, including the training of the astronomers" (Transactions of the IAU, vol. XIIB, page 637, 1964). As a follow up of this discussion a meeting was organized in July 1965 in Nice (France) to discuss the creation of an International School for Young Astronomers, its organization and its funding. The aim of such a School would be to give to young astronomers an intensive training in astronomy and astrophysics during a 3-month period, rather similar to the one that could be given in a university over a longer period. The students would then spend one year in an astronomical institution to receive more practical and theoretical training (Transactions of the IAU, vol. XIIIA, page XCV, 1967). A questionnaire was sent to many institutions around the world on their potential level of involvement in this project in terms of the proposed 3-month school or 1-year training programme. More than 20 supportive replies were received. At that date already 6 Schools were foreseen. The first IAU Summer School took place in Manchester (UK) in 1967 with funding by UNESCO. The General Secretary of the School was J. Kleczek; he was appointed by the IAU Executive Committee.

3. The first Summer Schools for Young Astronomers

3.1. *1967 in Manchester: the first Summer School for Young Astronomers*

During the XIIIth IAU General Assembly in 1967, in Prague, J. Kleczek reported on the first Summer School for Young Astronomers organized at Manchester University (UK) over a period of six and a half weeks (Transactions of the IAU, vol. XIIIB, page 229, 1967). Twelve students participated in this School (India: 3, Egypt: 2, Portugal: 1, USA: 1, Romania: 1, Czechoslovakia: 1, Poland:2 and Netherlands: 1). The infrastructure was offered by the host country of the School. The advantage of having a few well prepared students joining these Summer Schools for the benefit of the other students was emphasized. This remark has always remained valid. For one or two young astronomers, these early Schools were followed up by a stay of one year at another institution.

3.2. *From 1967 to 1970: four Summer Schools for Young Astronomers*

With the financial support of UNESCO, the IAU and the host countries, four International Schools for Young Astronomers were organized consecutively: Manchester (UK) in 1967, as already mentioned, then Arcetri (Italy) in 1968, Hyderabad (India) in 1969 and Córdoba (Argentina) in 1970. Reports on these early ISYA can be found in the Transactions of the IAU (vol. XIVA, page 563, 1970 and vol. XVA, page 719, 1973). The respective durations of these Schools were: 6.5 weeks, 8.5 weeks and 8 weeks for the last two. Their numbers of participants were, respectively, 12, 10, 23 and 21. The UNESCO Department of Environmental Sciences had allocated funds to the IAU for the organization of these Schools.

The comments by J. Kleczek are reproduced here from the relevant IAU Transactions (vol. XIVA, page 563, 1970):

> "In organizing the schools in Manchester and Arcetri young astronomers from various countries were bought to an observatory to work with local lecturers and instruments. At Hyderabad, on the contrary, a new scheme was tried, namely to bring foreign lecturers and experienced research workers to an Observatory which, with newly acquired telescopes, is developing its astronomical research programmes.

> *The scheme was effective and it has the advantage that the experienced astronomers can help the host Institute to plan future research programs."*

This scheme became the rule for the future ISYA and constitutes one of its defining characteristics.

4. From the 1st ISYA in 1967 to the 18th in 1990

After such a promising beginning the ISYA became a regular programme of Commission 46. The prescription to organize an ISYA was clearly defined in the *IAU Transactions* vol. XVA, page 718, 1973).

> The Commission propose to organize each year an ISYA of about 8 weeks duration for promising young scientists from developing countries and institutions.
> The purpose of these schools is to give a concentrated expert instruction and training in special topics of modern astronomy to a number of selected young astronomers or physicists with or without a graduated degree who otherwise would not have such opportunities available to them.
>
> The schools would be organized on a regional basis. They would be held at a suitably equipped Observatory in a location of good atmospheric conditions, thus allowing ample time for the practical training of the students at the telescope. The most convenient period of time for holding a school would be fixed by the host institution.
>
> The teaching staff would be supplied mainly by the host Observatory, but some outstanding specialists from other countries would be invited to teach a course during a limited period of time. The number of participating students would depend on the available teaching facilities (astronomical instruments, assistants and teaching staff).
>
> The activity of the schools would consist of regular lectures, practical training, seminars informal discussions and study hours.

Not all the above requirements could be fulfilled. Unfortunately, in 1971 the UNESCO funding stopped. In view of the importance and usefulness of the ISYA, the IAU Executive Committee decided to allocate funding which would allow the organization of one ISYA during the triennium 1970–1973, with preference for holding the School in a developing country. The Schools therefore continued with IAU support alone, but as the available IAU funding was much less than previously obtained from UNESCO, the number of participants, the number of teachers and the duration of the Schools were drastically reduced by half compared to the earlier ISYA.

Table 1 lists the 18 ISYA organized up until 1990. The relevant information was taken from the IAU Transactions. This provides information, when available, on the total number of participants (first figure in the last column), the number of foreigners (f) and the number of different nationalities (n). The ISYA in Argentina was on the theme *Physics of Solar Plasmas, the Sun and Interplanetary Medium and Solar Energy*; it consisted in fact of three parallel schools and it was also funded by the Argentinian Commission Nacional de Estudios Geoheliofisicos. From 1979 until 1990 the ISYA received partial financial support from the UNESCO via ICSU.

Table 1. List of the ISYA from 1967 to 1990

No	Date	Location	Duration (weeks)	Participants
1	1967 March	U.K., Manchester	6.5	12 (12f, 8n)
2	1968 June-July	Italy, Arcetri	8.5	10 (10f, 7n)
3	1969	India, Hyderabad	8	23 (5f, 5n)
4	1970 Oct-Nov	Argentina, Córdoba	8	21 (5n)
5	1973 July-Aug	Indonesia, Lembang	4	8 (3f, 4n)
6	1974 May	Argentina, San Miguel	4	60 (21f, 7n)
7	1975 Sept	Greece, Athens/Thera	4	74 (35f, 16n)
8	1977 Nov	Brazil, Rio	4	29
9	1978 Aug	Nigeria, Nsukka	3	28
10	1979 Sept	Spain, Tenerife	2	36 (7n)
11	1980 Sept-Oct	Yugoslavia, Hvar	3	25
12	1981 Aug-Sept	Egypt, Cairo	3	28 (9n)
13	1983 May-June	Indonesia, Lembang	3	21 (5n)
14	1986 Aug	China, Beijing	3	52 (6n)
15	1986 Sept	Portugal, Espinho	3	30 (19f, 7n)
16	1989 Aug	Cuba, Havana	2	55 (23f, 6n)
17	1990 May-June	Malaysia, Kuala Lumpur and Melaka	2.5	27 (11f)
18	1990 Sept	Morocco, Marrakesh	2.5	53

5. Objectives and organization of the ISYA

An ISYA is always oriented towards developing countries and takes place in these countries. Nevertheless, an ISYA takes place in countries and universities with a reasonably long-term interest in astronomy to sustain further development. During an ISYA there is no donation of research equipment, such as telescopes, for example.

The main goals are:
• to broaden the point of view of the students – a young astronomer should not only stick to a single, very specialized, branch of astronomical research;
• to fight against the isolation of the "lonely astronomer";
• to initiate collaboration on a larger geographical scale.

An ISYA is organized through an agreement signed between the IAU and a host university, and this is often linked to a development project, such as the establishment of a new astronomy department, the installation of a new telescope, etc.

The main financial conditions are as follows:
• the IAU pays for the travel of the faculty members and all the participants
• the host country pays for the stay of the faculty members and all the participants and provides the facilities for the school.

The duration of an ISYA is currently 3 weeks, which is the minimum time needed for the participants to become accustomed to speaking and debating in English and in public. The lecturers are asked to stay as long as possible in order for the participants feel at ease to communicate with them.

Normally, there are 8–10 lecturers, of which 3–4 come from the host institution and 4–5 are visiting foreign faculty members. The topics covered during an ISYA are chosen by the host institution in close collaboration with the Chairperson of the IAU ISYA programme.

The number of students who can participate in an ISYA depends mainly upon financial considerations. There are about 30 to 45 participants from, on average, 10 different

Table 2. List of the ISYA from 1992

No	Date	Location	Duration (weeks)	Participants
19	1992 Aug	China, Beijing and Xinglong Observatory	3	30 (17f, 12n, 9w)
20	1994 Jan	India, Pune	3	35 (25f, 13n, 11w)
21	1994 Sept	Egypt, Cairo and Kottamia Observatory	3	41 (12f, 13n, 10w)
22	1995 July	Brazil, Belo Horizonte and Serra Piedade	3	38 (19f, 11n, 15w)
23	1997 July	Iran, Zanjan	3	38 (14f, 8n, 12w)
24	1999 Aug	Romania, Bucharest	3	41 (18f, 9n, 22w)
25	2001 Jan	ChiangMai, Thailand	3	36 (17f, 9n, 6w)
26	2002 Aug	Casleo, Argentina	3	28 (14f, 9n, 10w)
27	2004 July	Al Akhawayn, Morocco	3	29 (18f, 13n, 9w)
28	2005 July-Aug	INAOE, Mexico	3	46 (20f, 10n, 18w)

countries in the same geographical area. The participants' background is typically that of a M.Sc. degree, but it ranges from fresh graduates to more experienced PhD students.

During an ISYA there are lectures as well as practical computer-oriented activities – both are considered equally important. Participants also give talks: for most of the students it is the first time that they have the opportunity to give a talk on their research, in English, in public and in front of foreign specialists.

An ISYA can be characterized by a large collage of astronomical and cultural backgrounds among the participants and the lecturers, which makes it so rich and fruitful.

6. From the 19th ISYA in 1992 to the 28th ISYA in 2005

Table 2 lists of the last ten ISYA. This provides information on the number of foreigners (f), the number of different nationalities (n) and of the number of women (w). From 1992 to 1997, Don Wentzel (USA) and Michèle Gerbaldi (France) were, respectively, the General Secretary and the Assistant General Secretary for these Schools and since then Michèle Gerbaldi (France) and Ed Guinan (USA) have been the Chairperson and the Vice-Chairperson, respectively, for this Programme Group of Commission 46.

The ISYA were still financially supported by the UNESCO (through ICSU) up until 2000. The ISYA in Argentina in 2002 took place during an economic crisis in that country. Nevertheless this ISYA was organized thanks to significant financial support given by UNESCO-Paris and the IAU. Subsequent ISYA were funded only by the IAU.

7. From 1992: a new context for the ISYA and their evolution

We are entering into a *computerized world*. Large astronomical databases now organized routinely by the space agencies and by all the major ground-based observatories. The first database to be so organized and accessible by anybody without restriction was the International Ultraviolet Explorer (IUE) Archive, created by the European Space Agency (ESA). This is the corner stone of a new context characterized by:

- the fast development of observational databases
- web access to publications in electronic form.

It should be noted that web access does not necessary imply easy access to on-line catalogues such as those at the CDS (Centre de Données Astronomiques de Strasbourg) or to the software needed for relevant data analysis.

Since 1995 the practical activities developed during the ISYA have been increasingly computer-oriented, to query the relevant databases and to develop the awareness of the participants towards the concept of "database mining" in order to overcome the general perception in a developing countries that *research in astronomy is not possible without access to a large telescope.*

The organization of any ISYA requires:
- computers and internet access;
- a large bandwidth (i.e. fast) link to query the databases
- access to electronic publications.

These points imply that an ISYA cannot be organized in a location that is too remote: hence a university or a national observatory is becoming the rule. However, we still have to face the long standing problem of the cost of the access to recent publications, whether in print or electronic form. This question has no solution on a short-term basis. What is lacking, very often, is the large Internet bandwidth and fast speed.

It should be mentioned that there are more and more "summer schools" organized by various institutions but they usually differ from ISYA in the following significant ways (which are important if we consider the goals of the ISYA):
- shorter in duration (one to two-weeks);
- more specialized towards doctoral students or recent postdocs;
- only national in some cases;
- students not fully funded.

Nowadays we also have to face to the greater mobility of students and their demand for mobility in their training years. This is integrated into the ISYA programme by having lecturers from Institutions which offer M.Sc. or PhD programmes.

The consequences of such evolution are that an ISYA cannot take place in a location that is too isolated. The last 3 ISYA were respectively organized
- in 2002 at El Leoncito (San Juan), the Argentinian National Observatory, which has excellent communication facilities as well as offering the possibility to do observations; with a 2-m telescope,
- in 2004 at Al Akhawayn University (Ifrane, Morocco), where the computing facilities were highly appreciated;
- in 2005 at INAOE in Mexico, the leading country in Central America for astronomical research where various programmes M.Sc. and/or PhD are offered in several Mexican universities.

8. Impact of the ISYA programme

We do not repeat here the detailed analysis done by D. Wentzel in 1996. We simply quote the key points which are addressed by an ISYA.

8.1. *What do the students get out of an ISYA?*

- a much broader perspective on astronomy and how science works;
- practice in asking penetrating and challenging questions;
- lecture materials and reference addresses;
- (professional) friendships;
- practice in spoken English;

Fig. 1 displays the home countries of the participants of the 28th ISYA in Mexico, at INAOE, showing how a regional network will be created, among the participants.

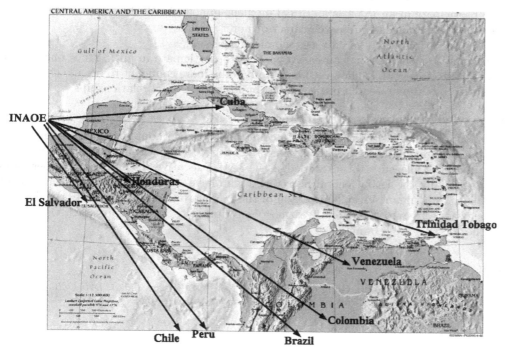

Figure 1. Location of the foreign participants at the 28th ISYA at INAOE (Mexico). (Map from PCL collection, by courtesy of the university of Texas libraries, Austin, USA)

8.2. *What does the host institution get out of an ISYA?*

- recognition
- starting a group of astronomy students
- broadening the training of astronomy students and young researchers in the host country
- exposure for the host institution and its development projects

The long-term impact of the ISYA is measured through evaluations of each ISYA done 3 to 4 years later through a questionnaire by e-mail, with a reply rate of 30% to 40%. Without a doubt, the ISYA are worthwhile for the participants and the ISYA have benefited astronomy as a whole. We quote one of many the comments received from students: *"We learned not only useful astrophysics, but also had the chance to interact with some of the more advanced researchers in the field."*

9. New horizon and conclusion

Today the *lonely astronomer* is also the one who is not associated with an international project. The development of the concept of the virtual observatory (VO) will induce, even more, the decentralization of research and will allow individuals to develop their expertise and competence in the international research arena. The objective of the ISYA is to introduce more young researchers into the international domain, but without cutting them from off their roots by, among other, offering them the possibility to start their network of scientific contacts in the context of their own national environments.

To participate in an ISYA is a critical moment in a student's life:

- a crucial cross-roads: research career or not ?
- the right time to do her/his astronomical check-up.

To become a scientist is not an easy task and to participate in an ISYA is one way to contribute to this goal. It allows students to become more confident to discuss their ideas with others, outline research projects, and build international relationships.

Acknowledgements

It should be emphasized that no ISYA could take place without all the faculty members who participate, giving so freely their time and energy to make a success of these Schools. It is my great pleasure to acknowledge the IAU Executive Committee members and more especially the IAU General Secretaries who have supported this programme over the years. Ed Guinan, Vice-Chairperson of the ISYA programme, is warmly thanked for his help as well as for the friendly atmosphere during those years in which we worked together on this programme.

References

Wentzel D., 1996, In: Eds. L. Gouguenheim, D. McNally & J. Percy, (eds.), *IAU Coll. 162, New trends in Astronomy Teaching,* Cambridge University Press, pg 27

Michèle Gerbaldi

Astronomy for the developing world
IAU Special Session no. 5, 2006
J.B. Hearnshaw and P. Martinez, eds.

© 2007 International Astronomical Union
doi:10.1017/S1743921307007053

Undergraduate and graduate programmes in astronomy for the developing world

John R. Percy

Department of Astronomy and Astrophysics
University of Toronto
Toronto ON Canada M5S 3H4
email: jpercy@utm.utoronto.ca

Abstract. In this paper, I discuss some aspects of the design of undergraduate and graduate astronomy curricula, broadly defined, for developing countries. A fundamental requirement is to develop students' ability and desire to learn, both in university and beyond. I then discuss several aspects of the curriculum: (i) The programme of coursework in astronomy and related topics such as physics and mathematics; (ii) The associated practical and project work to develop skills as well as knowledge; (iii) Linking the coursework, effectively, to various aspects of research; (iv) Development of general academic and professional skills such as oral and written communication, teaching, planning and management, and the ability to function as part of an interdisciplinary team; and (v) Orientation to the culture of the university and to the science and the profession of astronomy.

To accomplish all of these goals may seem daunting, especially as many of them are not achieved in the most affluent universities. But much can be gained by recognizing that there are well-established "best practices" in education, achieved through formal education research, reflection, and experience. Simple resources, effectively used, can be superior to the highest technology, used without careful thought. It is often best to do a few things well; "less can be more". And effective partnership, both within the local university and with the outside astronomical community, can also contribute to success.

Keywords. Astronomy education, undergraduate and graduate curricula, developing countries

1. Introduction

Under "New Initiatives for Promoting Astronomical Education and Research in Developing Countries", the sub-title of the conference on which this book is based, the training of astronomers must rank highly. Very little of my presentation will be "new" to education professionals, but it may not be widely known, appreciated, or implemented by astronomers and by university professors, most of whom are amateurs in teaching and administration, in the sense that they have received little or no pre-service or in-service training in these fields.

In this paper, I reflect on the concept of *best practices* in designing undergraduate and beginning graduate programmes and courses. Many of the principles are very general, and would apply to any university, but I shall try to express them in terms of the needs of developing countries, as far as my limited experience is able.

There are many relevant papers on astronomy education in and for developing countries in two previous IAU conference proceedings, Pasachoff & Percy (1990), and Gouguenheim *et al.* (1998), as well as in Batten (2001).

In truth, undergraduate programmes and courses at universities in the developed countries are far from perfect. The strategies for creating effective programmes and courses do not necessarily cost much money, so the developing countries have the opportunity to succeed – and exceed – by avoiding the errors that the rest of us have made.

I recognize the special challenges which face universities in the developing countries. These challenges are well described in this book, and in the proceedings of the previous conference on this subject (Batten 2001). The university (or the country!) may have only one astronomer, whose responsibilities and opportunities may be many, and who may have to teach several astronomy courses, over multiple terms, in multiple years. The astronomy programme will have to make use of other faculty members, in cognate departments, to teach other courses in the programme. Fortunately, many such faculty members have an interest in astronomy. But they must be willing to co-operate with each other. Universities tend to be compartmentalized into departments and groups which have only their own interests in mind. So negotiation, cooperation, and sharing are of utmost importance – and are usually beneficial for everyone. Education depends, more than anything else, on *people*, not equipment.

There is a general strategy for approaching this or any other education activity, which I like to express in terms of the *Hodson diagram*, since it was pictured very simply in a paper by the distinguished science educator Professor Derek Hodson (2001). We begin by defining the *objectives* of our astronomy programme and courses; we decide on a *curriculum* (which includes knowledge, skills, applications, and attitudes); we develop effective *teaching strategies* – which include much more than the lecture and the textbook. Most of all, we subject every part of this process to *assessment*, which includes *both evaluation and feedback* so that, every year, the programme and courses constantly improve.

The Hodson diagram emphasizes that we should first discuss the *objectives* of astronomy programmes and courses for the developing countries. Since there are so many aspects to astronomy, and so many applications of astronomy to public, school, and university education, to academic research, to the development of science and technology infrastructure, to communication with institutions and with the public, and to other practical matters, it follows that students in astronomy programmes should develop a broad range of knowledge, skills, and experiences. In developing countries, astronomers have to fulfil many roles.

The Hodson diagram also reminds us that educational activities such as the delivery of programmes and courses are subject to *assessment* – both evaluation and feedback – just as our scientific observations and theories are. This can be done through formal *education research*, through a more simplified process called *action research*, or even more simply, through careful *reflection*. The problem is that very few university professors and administrators have any formal training in these or other aspects of education. Normally, they begin with what they experienced themselves, as undergraduates. They may *reflect* on ways to improve the experience – which is good. But teaching and learning in higher education is a well-developed discipline, with regular conferences and refereed journals. Sometimes the papers in these conferences and journals are rather esoteric, but there are individuals such as Tim Slater in the US who are working hard to apply this research to the reality of undergraduate programmes and courses. See Slater & Adams (2003) for an excellent, inexpensive guide to teaching "Astro 101". And there is a free on-line journal *Astronomy Education Review* at http://aer.noao.edu.

The most important strategy for success is to develop students' ability and motivation to be *self-learners* – both in university and beyond. Often, students enter school or university with ability and motivation, but these are actually "turned off" by the system! One consequence, in the developed countries, is that both the strongest and the weakest students may fail in university.

The student's success thus depends, not only on their innate intelligence, but on their attitude and other academic skills. It also depends on the attitude of the instructors and their university: on whether the education of students is of high priority. In this sense, the barriers for astronomy students in developing countries are high, but not necessarily insurmountable.

2. The programme of coursework

When one thinks of an undergraduate astronomy programme, one often thinks in terms of a list of courses. At one end of the spectrum, the list is long, and includes every possible sub-topic in astronomy, physics, mathematics – and possibly other disciplines such as statistics, and computer science as well. Most or all of the courses in the programme are specified.

At the other end of the spectrum, the student may take a much broader programme, especially in the first two years; this programme may include foundation courses in astronomy, physics, and mathematics, but it may also include many courses in the humanities and social sciences. This type of programme is favoured by many liberal-arts colleges and universities in the US.

Both approaches have some merit, and I tend to prefer the middle of the spectrum. At my university, 14 of 20 courses in the programme are specified. This leaves room for courses in useful cognate subjects, or courses that develop skills in management, teamwork, and communication. Some students use this course to complete a smaller programme in a second subject that interests them.

The students who enter the astronomy graduate programme at my university must have a strong background in physics. Usually they have had at least a course or two in astronomy, and/or some astronomy research experience. Occasionally, we have students who have come from a pure physics background. In that case, we want to be sure that they have a genuine interest and motivation for astronomy.

When astronomy courses are included in an undergraduate programme, they usually include one-term courses in: The Solar System, Sun and Stars, Galaxies and Cosmology and, if possible, Practical Astronomy. If the courses do not have a significant theoretical content, then there may be an additional course in Astrophysics. In addition, it is desirable to have an upper-year research project course. And there is often an upper-year "capstone" course in current or special topics. I have seen programmes with even more astronomy courses, but I would advise on fewer courses, rather than more. This kind of programme is probably good for students who will not be proceeding to graduate work, but will be proceeding directly to astronomy-related careers (such as teaching) which require understanding of a broad range of astronomy topics.

There is ongoing discussion in the education community about whether it is better to teach as much content as possible, or to teach less material but in more depth (and more effectively) – "less is more". To some extent, the actual curriculum of courses is less important than the general and specific skills and attitudes that the students learn.

3. Practical and project work

If you think back to my comment about curriculum, you will realize that courses tend to teach *knowledge* or *content*. The practical work and projects teach *skills*. They may also teach the *applications* of the content and skills to the real world, but this can be done even better through formal or informal *internships*. Students often intern with academics by carrying out a senior thesis or research project. If their programme has a more practical bent, they may serve an internship in industry, government, schools, or other real-world setting. This can be done as a course, or through summer employment. Internships can benefit the student, the university, and the supervisor or employer, but they *must* be carefully structured, supervised and evaluated by the university if they are to be effective.

There is another approach that is becoming very popular in education circles – *community service learning* (e.g. www.communityservicelearning.ca). This means carrying out a practical project, for course credit, under the joint supervision of a faculty member and a supervisor from the community. For example: the 1000 students in first-year Engineering at the University of Toronto undertake a practical design project in the community, usually in support of worthy non-profit community organizations.

How could undergraduate astronomy students, in developing countries, serve their communities? One way would be by assisting schoolteachers in the important task of teaching astronomy and physics. But there are many other possibilities. Elsewhere in this book, Patricia Rosenzweig describes a form of community service learning in which undergraduate students assist, enthusiastically, with school outreach activities.

Most science courses have an associated *laboratory* component. This can introduce students to equipment, data, software, and other analysis tools and techniques. But there is a real danger in what are called "cookbook labs" – ones in which students follow a specified procedure to obtain a specified result. This may explain why a study in the US showed that high school students who took physics labs (presumably of the cookbook variety) in school did more *poorly* in introductory university physics than students who had not taken such labs in high school. The expensive "black box" equipment, used in the developed countries, may actually be *less* effective than simpler, more transparent equipment.

4. Linking coursework and research

Linking coursework or teaching and research is a high priority in most North American universities. An influential document is the Boyer Report (1998) entitled *Reinventing Undergraduate Education: A Blueprint for America's Research Universities*. This concept includes giving students opportunities to develop research skills, and also exposing students to modern research as part of their coursework.

In my own university department, most of my colleagues are doing exciting frontier astronomy with large telescopes, satellites, or powerful computers. Very few of them feel that they are able or willing to supervise undergraduate research. So undergraduate research supervision falls to those of us who do relatively "small-science" projects such as my research on variable stars (Percy *et al.* 2006). Yet we are able to provide students with a genuine or *authentic* (to use the current education jargon) research experience, involving real data and analysis, resulting in a conference presentation or research paper with them as co-author. They may then go on and do a more high-powered senior thesis with one of the big-telescope astronomers.

An alternative is to give students projects which are not original, and are therefore not publishable, but which involve all the steps in a typical research project. These include:

• Development and understanding of the *objectives and strategies* for carrying out the project.

• Writing a project *proposal*, and periodic *progress reports*; this is a new experience for most students.

• Critical *background reading* of print and Internet sources, from the basic level to the research level.

- Description of the *data* being used, including its origin, accuracy, and peculiarities.
- Description of the *method(s) of analysis* being used, including algorithms and software – their strengths, weaknesses, and peculiarities, and how they interact with the data.
- Testing of the analysis on *trial data*, to make sure that it produces a reasonable result.
- Analysis and interpretation of the *new data*, with appropriate output tables, graphs, and notes.
- Writing, editing, and defence of a final *project report*.
- Preparation of a *poster paper, display, or oral presentation* on the project.

The senior thesis, or major project, is a culmination of the development of the student's research knowledge and skills. While the project may be carried out one-on-one with a supervisor, it is important for the supervisor – or the course co-ordinator, if there are several students doing a senior project at the same time – to introduce some *structured learning* to the process. This structure can build upon the lower-level research skills, mentioned above.

At one time, it would have been difficult for a student in a developing country to do research at the senior undergraduate or graduate level, because of the lack of access to books, journals, and data. With the spread of the Internet, robotic telescopes, and data archives, this is gradually becoming less of a problem.

The lesson for the developing countries is thus that students can learn research by doing projects with small telescopes, or archival data that is available on-line. For instance, there are terabytes of variable star data available, with lots of science still to be "mined", along with freely-accessible software to analyze it. There are journals which do not charge an arm and a leg for page charges. Or the results can be self-published on-line.

5. General academic skills

Why, in this paper, am I putting so much emphasis on skills? One reason is that, a few years ago, a survey was done of PhD graduates of physical-science programmes in the US, 5 years after completing their education (Smith *et al.* 2002). They were asked to comment on how well their education prepared them for the job they were in. They indicated that they had been well-prepared for project design, analysis and interpretation of data, and for other aspects of problem-solving. But they had *not* been well prepared for: (i) oral and written communication; (ii) teaching, instructing, and mentoring; (iii) administering and managing projects and people; and (iv) working with individuals from different disciplines.

These are examples of general academic skills. Others are: autonomy, critical thinking, determination, enthusiasm, industriousness, initiative, teamwork, judgement, time management, and the ability to *get al.*ng with other people.

There is one highly-regarded method of instruction which can develop these skills very effectively, and that is *problem-based learning* – PBL (Albanese & Mitchell 1993). Students work in groups to solve a significant practical problem, based on research, understanding, and agreement. PBL facilitates the formation of *study groups*, which have shown to be very effective in promoting learning. An undergraduate astronomy programme also leads to the formation of a *cohort* of students who take similar courses, share common experiences, and may well become lifelong colleagues, supporters, and friends. Large universities are just now discovering the benefits of organizing students into such cohorts or *learning communities*.

6. The student experience

The current major concern of North American universities is something that goes beyond the formal curriculum – the *student experience.* Are the students satisfied with their experience? Surveys show that, especially in large universities, many students are uninterested and unengaged. Many are not motivated to learn and, even if they are, they face the barriers of large classes and untrained, uninterested instructors.

Here's where an astronomy programme in a developing country may have an advantage. It is unlikely that the astronomy, physics, and mathematics courses are large so – as long as the instructor is conscientious and genuinely interested in students – there is the potential for good learning, and good experience. How can this be encouraged? One way is by evaluating and rewarding the professors equally for their teaching, and research or professional work.

It's also important to *evaluate* the student experience by surveying students, anonymously if necessary. In my university, all *courses* are evaluated. The instructor may or may not act on the results. Programmes are not evaluated. If possible, graduates should be surveyed, both when they leave the university, and after they have been in the working world for a few years. And their comments should be taken seriously, and used to improve the programme and the experience.

7. Graduate programmes and courses

I will not comment on PhD programmes, since the structure is well-defined: the student carries out a significant original research project as an apprentice of the supervisor. Very often, the PhD is not done in the developing country; the student is sent overseas, to a more established institution, with more resources. However, the following general comments about MSc programmes have some relevance to all graduate work. And most of my comments about undergraduate programmes apply to graduate programmes as well.

In choosing students for graduate work, academic achievement is a necessary criterion, but not a sufficient one. My experience is that, as long as the student's average is B+ or greater (or the equivalent in your system), future success depends mostly on the student's general academic skills, attitudes, and motivation.

Whether entering an undergraduate programme or a graduate programme, students may have difficulty acclimatizing to the culture of the department and the university, especially if they come from another country. Formal and informal academic orientation and mentoring are absolutely essential – especially for students who come from a family or part of society without a tradition of university study – and are therefore highly recommended.

Although graduate students may have to take and pass courses – in which case, all of the above comments may apply – the essential feature of graduate education is supervised research. Just as university professors are amateurs at teaching, they are also amateurs at supervision. Students may have a supervising committee but, if the committee does not meet or take its work seriously, the student may fall by the wayside.

Graduate students should receive training in teaching, outreach, and communication, and should be encouraged to gain practical experience in these skills. These skills are essential for the students' future success. Practicing them can be a satisfying and motivating experience. And, because there are few astronomers in developing countries, it makes sense to use advanced students to contribute to teaching and outreach.

8. Becoming an astronomer – or not

The topic of this book, and this paper, assumes that the students in the undergraduate astronomy program and courses will go on to become astronomers. Based on experience in both developing and developed countries, this may not be the case. There may be insufficient positions in astronomy for all the graduates. The students may develop skills which may lead them to non-astronomical jobs in government or industry. In some ways, it is useful to have astronomers in such positions of influence! We can't let government and industry be run entirely by lawyers who are scientifically illiterate! See Tobias *et al.* (1995) for an excellent discussion of science careers.

Unfortunately, universities in the developed countries often regard these students as failures, and do not maintain contact with them. This is unfortunate for many reasons. One is because they may be future donors to the university! They may also be useful mentors and role models for students who seek careers other than astronomy.

Astronomy is unique in that it can be continued as a hobby. So-called amateur astronomers can play important roles in astronomical teaching and research, and in providing grass-roots support for astronomy (Percy & Wilson 2000). This may be especially important in developing countries. There is a tendency, in the developed countries, for astronomy departments to lose track of, and lose interest in graduates who leave astronomy. This is short-sighted – for many reasons.

9. Epilogue

It takes many years of patient but determined effort to build any new research or education structure, including an undergraduate or graduate astronomy programme; it does not happen overnight. IAU Commission 46 has a history of working with countries and institutions, on a long-term basis, to support the development of new astronomical centres. We look forward to working with you in the future.

References

Albanese, M.A. & Mitchell, S. 1993, *Academic Medicine*, 68(1), 52

Batten, A.H. 2001, *Astronomy for Developing Countries*, (San Francisco: Astronomical Society of the Pacific)

Boyer Commission 1998, *Reinventing Undergraduate Education: A Blueprint for America's Research Universities*, at:

  http://naples.cc.sunysb.edu/Pres/boyer.nsp/

Gouguenheim, L., McNally, D. & Percy, J.R., 1998, *New Trends in Astronomy Teaching*, (Cambridge: Cambridge University Press)

Hodson, D. 2001, *OISE Papers in STSE Education*, Ontario Institute for Studies in Education, at the University of Toronto, 2, 7

Pasachoff, J.M. & Percy, J.R. 1990, *The Teaching of Astronomy*, (Cambridge: Cambridge University Press)

Percy, J.R. & Wilson, J.B. 2000, *Amateur-Professional Partnerships in Astronomy*, (San Francisco: Astronomical Society of the Pacific Conference Series, 220).

Percy, J.R., Molak, A., Lund, H., Overbeek, D., Wehlau, A.F. & Williams, P.F. 2006, *Publ. Astron. Soc. Pacific*, 118, 805

Slater, T.F. & Adams, J.P. 2003, *Learner-Centered Astronomy Teaching: Strategies for Astro 101*, (Upper Saddle River NJ: Pearson Education Ltd.)

Smith, S.J. *et al.* 2002, *Am. J. Phys.*, 70 (11), 1081

Tobias, S., Aylesworth, K. & Chubin, D.F. 1995, *Rethinking Science as a Career* (Tucson AZ: The Research Corporation)

Michèle Gerbaldi

John Percy

Astronomy for the developing world
IAU Special Session no. 5, 2006
J.B. Hearnshaw and P. Martinez, eds.

© 2007 International Astronomical Union
doi:10.1017/S1743921307007065

Distance education at university level: opportunities and pitfalls

Barrie Jones

The Open University, Milton Keynes, UK
email: b.w.jones@open.ac.uk

Abstract. This paper presents an overview of distance education at university level and some of the associated challenges and pitfalls.

Keywords. Astronomy education, distance education

---

1. Introduction

For the purposes of this paper, I shall define distance education as the case where: (1) the student is remote from a "bricks and mortar" institution, a university in this case; and (2) the student studies (mainly) at home and/or in the workplace. From a university a few astronomers can reach many students who otherwise would be unable to study astronomy at university level. Students can live at home, and do not have to travel. This reduces student costs. Distance-learning students can study part-time, enabling them to finance their studies. This also applies to on-campus students.

In this paper I shall discuss what is involved in going beyond "bricks and mortar" institutions, namely the prerequisites for students, the importance of high-quality materials (electronic and printed) suitable for the distance learner, the importance of student pacing and student support, continuous assessment of students, examinations and feedback from students. Space does not permit me to discuss important issues such as curricula, developing local staff skills in distance education, and specific learning materials/course programmes/degrees.

2. Prerequisites

There must be a full specification of what the student should bring to the course. This should encompass previous knowledge, previous skills and previous courses in the institution's programme. The student must also have received advice about what courses are needed and what achievement levels are required for the target outcome: e.g. Diploma in Astrophysics, BSc in Physics and Astronomy.

3. High-quality materials

High-quality materials, suitable for distance learning are essential. These remarks apply equally to electronic and printed media. The requirements are additional to the good writing desirable for any material. To compensate for the isolation of the distance learner the materials must be complete (given the likely inaccessibility of libraries, etc.), the style must be expansive and friendly. There must be a preliminary, self-administered self-test, with advice based on the outcome (e.g. remedial material for any prospective student falling just falling short of what is needed). There should also be embedded self-tests — short stop-and-think questions in the flow, longer end-of-section questions with fully worked answers and comments. Beware of the uncritical use of lectures or outreach material made available over the Internet by some universities.

A mix of printed and electronic media is good. The latter is not essential for effective distance learning, but it greatly helps. Electronic media comprise CD-ROMs, DVDs, email, and the Internet. The cost of a computer and modem is high. In developing countries school ICT facilities often have time free of use that could be made available to distance learners. Internet cafés also offer another opportunity for the distance learner to access information. Distance-learning software should minimize the ICT skills required. The goal is to use ICT for astronomy education, not ICT education for astronomy. Avoid requiring specific commercial packages, such as Word, Excel, etc., except at the higher levels, where students are more likely to have access to such tools in their normal home or work environments.

4. Samples of material

Active learning must be encouraged. In a text, one useful device is the stop-and-think question, as follows.

"... Thus, around aphelion the body is moving slowest, and around perihelion it is moving fastest. The difference in these two speeds is larger, the greater the eccentricity.

Q What are the speeds at different positions in a circular orbit?

In a circular orbit the equal areas correspond to equal length arcs around the circle, so the body moves at a constant speed around its orbit ... "

Here is some not very helpful material. Suppose that a section has introduced the planets and their order from the Sun. Their orbits have been described as "approximately circular". Nothing else on orbits has yet appeared in the course. Then the student comes to the following material.

"... The orbits of the planets are described by Kepler's laws. There are three.
  1 Each planet's orbit is an ellipse, with the Sun at one of the foci.
  2 The radius vector from the Sun to the planet sweeps out equal areas in equal times.
  3 $P = ka^{\frac{3}{2}}$ where P is the sidereal orbital period of the planet and a is the semi-major axis of its orbit.

Question X
Using what you know about the Earth's orbit, calculate the orbital period of an asteroid in an orbit with a semimajor axis of 2.3 AU.
  .
  .
  .

(end of section)

Question X answer: 3.5 years"

The teaching here needs to be "repaired" in several ways. Will the student know what an ellipse is? Likewise, focus, radius vector, sidereal orbital period, semi-major axis?

Where's the diagram of an ellipse? What is k? Also, this is far too terse. Who was Kepler? How did he arrive at his laws? Is there going to be any explanation of these laws based on Newton's laws? And so on. Furthermore the question "Using what you know about the Earth's orbit ... " assumes that the student knows about and understands the Earth's orbit. Don't assume they know anything about the Earth's orbit. I hope a course prerequisite is simple algebra (or that this has been taught earlier in the course). The terse answer "3.5 years" is not enough for distance-learning students. You need the full working, with particular attention to units. A comment should be added on significant figures.

In addition to supportive texts, significant learning in astronomy can be accomplished via observations, without access to telescopes or laboratories, using readily available items, for example, to measure the luminosity of the Sun, with an electric light bulb (and power supply) as the most "advanced" item, or to measure the length of the sidereal day, which requires only a watch. Observational work can also be performed using virtual observatories and robotic telescopes. Regardless of how it is performed, observational work must be integrated with the rest of the course.

5. The importance of student pacing and student support

Pacing is required to prevent study-time build-up. Specify the hours needed to reach various stages of learning, with realistic ranges, based on real time, not the "university clock". Stop-and-think questions, and end-of-section questions not only aid learning, but enable the student to check that the learning outcomes associated with each stage have been reached. Continuous assessment not only provides student grades, but also provides pacing by setting submission deadlines a few times a year: some should be tutor marked assignments (TMAs), others (multiple choice) computer marked assignments (CMAs).

Student support is vital. Access to a tutor, via letter, but preferably via telephone or email, plus group meetings (tutorials) in urban areas, is desirable. Likewise, access to other students by similar means is also important, though easier in urban areas.

Continuous assessment also provides student support. As well as their grading and pacing functions, TMAs and CMAs should provide teaching, via tutor comments on TMAs, or via automated feedback on CMAs. Electronic submission and feedback could be an option, but *not* a requirement.

6. The problem of plagiarism

Plagiarism is a particular challenge for the assessment of distance-learning assignments. Take steps as follows. Avoid using questions openly available on the internet — at the very least, "version" them. Student self-help groups are to be encouraged, but plagiarism is a danger. Spot-checks on work from different students can be revealing (e.g. the curious case of the spelling mistakes common to two students' assignments points to "cut-and-paste" plagiarism!). An invigilated written examination will usually reveal if the continuous assessment grades are a fair reflection.

7. Examinations

If successful course completion is for professional advancement there needs to be an invigilated examination (which can be submitted electronically). Proof of identity with a photograph needs to be presented to the invigilator. A suitable examination venue and invigilation should be arranged as close to the student/group of students as possible. A few determined cheats might still defeat the system, but as always, they are ultimately only cheating themselves.

8. Closing the loop - student feedback

Any educational system must seek to improve itself. A crucial ingredient is feedback from students. Do not rely passively on feedback, but design a system, perhaps with some enticement, such as reduced fees on future courses.

9. Conclusion

Distance education is not an easy option for the student, but it can reach students who otherwise would have no option at all.

Barrie Jones

Astronomy for the developing world
IAU Special Session no. 5, 2006
J.B. Hearnshaw and P. Martinez, eds.

© 2007 International Astronomical Union
doi:10.1017/S1743921307007077

Projects of the Teaching and Popularization section of LIADA

Paulo Sergio Bretones[1,2] and Vladimir C. de Oliveira[2]

[1]Instituto Superior de Ciências Aplicadas, Limeira, Brazil
email: bretones@mpc.com.br

[2]Instituto de Geociências, Universidade Estadual de Campinas, Campinas, Brazil

Abstract. The goal of this work is to present an analysis of the observational projects developed by the Teaching and Popularization Section of the *Liga Ibero-Americana de Astronomía (LIADA)*. The Section is based in Brazil and counts on the support of 16 volunteer coordinators from most Latin-American countries. The observational projects are described on the home page of LIADA and aim to attract the attention of the general public, teachers and students to encourage their active participation in observational astronomy. The strategy is to circulate support material and open a discussion forum about each of the astronomical phenomena to enhance their consideration by and visibility to the public. Participants' reports are posted on the Internet forum and the web page of LIADA. We have analyzed the records of these activities and present an evaluation of the difficulties with written reports, the need of a dynamic maintenance of the home page, the establishment of a useful communications network and the visibility of the activities of LIADA.

Keywords. Astronomy education, astronomy popularization, Latin America

1. Introduction

The growing importance of scientific education in the last several decades can be seen in the quotidian presence of technical and scientific principles in modern life. However, there is a defective education in a general way in these matters. The teaching and popularization of astronomy offers many possibilities for astronomers and educators to improve scientific literacy (Percy, 1998a). Many national and international efforts have been made by means of several astronomical institutions in this regard, including, the Brazilian Astronomical Society (Sociedade Astronômica Brasileira) and IAU Commission 46.

The occurrences of astronomical phenomena such as conjunctions, oppositions, eclipses and transits present excellent opportunities to draw the attention of people to the observation of the night sky and more specifically for the teaching and popularization of these phenomena (Bretones and Oliveira, 2004a, 2004b and 2005; Pasachoff, 1998). In this context, amateur astronomers can contribute with actions that promote the teaching and popularization of astronomy in various ways (Percy, 1998b).

The purpose of this study is to present an analysis of the projects developed by the Teaching and Popularization Section (SEDA) of the Ibero-American Astronomy League (LIADA) during the period 2000 to 2005.

2. The Ibero-American League of Astronomy

The Ibero-American League of Astronomy (LIADA), was created in 1982, as a reorganization and legacy of the Latin American League of Astronomy (LIADA), founded in 1958. Among its objectives, LIADA tries to organize, conduct and facilitate the efforts of amateurs and semi-professionals in Ibero-America and to promote the works of its members. Some professional astronomers enhance the collaboration between the two groups that share our activity. With its current headquarters in Argentina and members spread over many countries, LIADA has sections in several areas of astronomy. LIADA promotes the Internet groups *foro-liada* and *ensenianza-liada*, and publishes the electronic bulletin *La Red de Observadores* (The Network of Observers) and the electronic magazine *Universo Digital* available on the LIADA website at `www.liada.net` .

Since 1992 the Teaching and Popularization Section (SEDA) of LIADA has been coordinated in Brazil and since 1998 the Section has had a website on the Internet. That website is hosted by the Morro Azul Observatory of the Instituto Superior de Ciências Aplicadas (ISCA) Faculdades in Limeira at the address `www.iscafaculdades.com.br/liada` . This is not exactly an educational site but it can be useful for anyone who works in astronomy education. The site shows the activities of the Section, lists local coordinators and their projects, and posts messages and reports of educational activities.

3. Local coordinators

Since 2001, 16 Section local coordinators have been appointed in several countries: 2001: Uruguay and Argentina; 2002: Mexico; 2003: Cuba, Guatemala and Panama; 2004: Peru, Paraguay, Chile, Venezuela, Honduras, El Salvador, Costa Rica, Spain; 2005: Colombia, Dominican Republic. The appointment of local coordinators has as its objective the decentralization of LIADA activities and promotion of activities in the various regions. Coordinators are nominated mainly as a result of their activities in astronomy education. It is thus expected that they will report on activities and projects developed in their own countries. The website shows the name, photograph, e-mail address, country and flag for each coordinator. It also lists the annual projects and reports since their appointment.

4. Projects

The present paper analyses the various observational projects (e.g. oppositions, conjunctions, eclipses and transits) that aim to attract the attention of the general public, students and teachers to encourage their participation in observational activities.

As an example for further analysis, we discuss projects related to eclipses. A particular eclipse event will have a page listing an observing project, articles and pedagogical material on eclipses, a discussion forum and links to related sites.

To initiate an observing project, the coordinator sends a general email message and posts an article which typically covers some of the following points:

• It remarks that an astronomical phenomenon, such as an eclipse, is a great opportunity for people to observe the sky;

• It invites those knowledgeable to write articles, give interviews to newspapers, radio and television, give lectures, exhibitions and observation sessions;

• It requests further, more specific articles with information about the phenomenon;

• It mentions that the starting article is just an example of the expected follow-up material;

• It invites people interested in the teaching and popularization astronomy to use the space of LIADA Forum to report their projects;

• It requests reports of the results of those projects. In some cases of specially useful information, these reports of activities are included in our teaching and popularization reports;

Table 1. Messages and reports sent by countries for the Lunar Eclipse of 8–9 November 2003.

Messages		Reports	
Guatemala	15	Argentina	5
Argentina	10	Guatemala	3
Argentina	10	Brazil	1
Mexico	1	Mexico	1
Spain	1	Uruguay	1
Total	37	Total	11

Table 2. Messages and reports of phenomena covered by LIADA members.

Phenomenon	Date	Messages	Countries	Reports	Countries
Eclipse	15-16/05/03	14	4	19	7
Mars Opposition	27/08/03	21	8	14	6
Eclipse	8-9/11/03	30	5	11	6
Eclipse	04/05/04	4	4	2	2
Transit	08/06/04	4	3	6	4
Eclipse	27/10/04	52	9	26	8
Eclipse	08/04/05	22	9	8	7

• It mentions that we are interested in stimulating initiatives of this kind and consider them as LIADA projects, and that we would like measure the outreach of our initiatives.

Since the beginning of 2004 projects launched in the forum have also had a Spanish version prepared by the Secretary of LIADA, Jorge Coghlan. For each phenomenon, the relevant messages and reports for educational purposes are selected by the coordinator and posted in the Section page. All this information is properly acknowledged and linked in the page.

Following the example of the eclipse of November 2003, an account of the messages and reports by countries was performed as shown in Table 1. Table 2 shows a list of events covered by LIADA members, excepting those related to conjunctions.

5. Discussion

In a preliminary analysis of our study period, we could verify that the space created by the LIADA web page contributed to improved organization of the information and reports on educational activities. However, the expected increase of posted messages due to the efforts of the Section was not observed.

In a general way, the reports are generally provided by the local coordinators of several countries and by active individuals that seize the opportunities as they arise. There are other members who show their ongoing activities and interest from time to time. The absence of representatives of national institutions is noticeable. There were many reports received from Argentina, which may be related to the presence of LIADA's headquarters in that country since 2000. There was also the shortage of institutions and people contributing from Brazil; a much larger number had been expected.

One problem with the posted reports was an over-emphasis on the description of instruments used in the observations, with specific details such as the aperture, eyepieces, local circumstances of the phenomenon and details or personal descriptions of the activity realized, while little or no information was provided regarding the number of people that participated, the programming of the activity, whether it was done for students, which school grade they were, in which school it was, teachers, subjects involved, etc. Such reports were not considered useful for the collection of educational data. The reasons for this are: a) The forum receives every kind of report; b) The sender of the reports is not particularly concerned with educational purposes; c) Lack of training in the area of education is not just a feature of the amateurs and institutions, but is it also the case for researchers whose main subject is not education. However, in many cases the encouragement of isolated amateurs and institutions by the Section was clearly verified and resulted in an improvement of the educational contents of these reports.

A principal result of this project is to give visibility to the well developed projects in various countries by institutions and individuals in those countries. The reports also constitute a useful data set for further analyses, for reflections and exchange of experiences, and for the establishment of a network for future projects.

The reports also shed light on the performance of local coordinators and their approach. For many of them, the reports still refer to personal experiences and are not necessarily representative of their whole countries. However, some of the reports show significant levels of activity given the astronomical development of their countries

6. Conclusion

An evaluation of the Teaching and Popularization Section activities was performed. The main positive result is that a forum for reports about astronomy education was established. However, there are still operational difficulties to maintain the page on the Internet because of a lack of staff. The question of language - translation of the projects from Portuguese into Spanish and back - if judged from the projects that have already been translated, has been practically solved.

It would be opportune to promote activities more frequently, whenever phenomena occur, such as conjunctions of the Moon with planets and bright stars, meteor showers, etc. We suggest that it would be interesting to analyze other relevant aspects with the reports available on the page. Additional work is needed to improve the information about educational methodologies relevant for the preparation of the activities and reports for the use of interested amateurs, teachers and researchers. A report guide sheet could be prepared. Finally, it would be useful to maintain continuity with local coordinators while bringing in additional countries to expand the LIADA work space, which would make an important contribution to astronomy education in the region.

References

Bretones, P.S. & de Oliveira, V.C. 2004a, In: S. Musso (ed.) *Primer Congreso de Astronomía del Centro de Estudios Astronómicos de Mar del Plata,* Centro de Estudos Astronómicos de Mar del Plata, 5 pp.

Bretones, P.S. & de Oliveira, V.C. 2004b, *Universo Digital,* vol. 53, p. 115. See also `http://www.liada.net/universo/digital/Morro%20Azul/Conferencia%20Vladimir/Artigo%2018%`

Bretones, P.S. & de Oliveira, V.C. 2005, *Boletim da Sociedade Astronômica Brasileira,* vol. 25, n. 1, p. 16

Percy, J.R. 1998a, In: L. Gouguenheim, D. McNally and J.R. Percy (eds.), *IAU Colloquium 162 - New Trends in Astronomy Teaching,* , Cambridge University Press, p. 2.

Percy, J.R. 1998b. In: L. Gouguenheim, D. McNally and J.R. Percy (eds.), *IAU Colloquium 162 - New Trends in Astronomy Teaching,* , Cambridge University Press, p. 205

Pasachoff, J.M. In: L. Gouguenheim, D. McNally and J.R. Percy (eds.), *IAU Colloquium 162 - New Trends in Astronomy Teaching,* , Cambridge University Press, p. 202

Astronomy for the developing world
IAU Special Session no. 5, 2006
J.B. Hearnshaw and P. Martinez, eds.

© 2007 International Astronomical Union
doi:10.1017/S1743921307007089

Teaching and research in astronomy using small aperture optical telescopes

Shiva Pandey †

School of Studies in Physics, Pt. Ravishankar Shukla University, Raipur, India
email: skp@iucaa.ernet.in

Abstract. Small aperture (<1m, typically 20–50 cm) optical telescopes with adequate back-end instrumentation (e.g. photometer, CCD camera and CCD spectrograph) can be used for spreading the joy and excitement of observational astronomy among postgraduate and research students in colleges and universities. On the basis of our experience over a decade of observing with small optical telescopes it has been amply demonstrated that such a facility, which any university can hope to procure and maintain, can be effectively used for teaching and research. The Physics Department of Pt. Ravishankar Shukla University at Raipur, India offers Astronomy & Astrophysics as one of the specializations of its MSc program in Physics. A set of observational exercises has been incorporated with a view to provide training in observations, analysis and interpretation of astronomical data. Observing facilities available in the department include 8"–14" aperture telescopes equipped with a photometer, CCD camera and a CCD spectrograph. A facility of this kind is ideally suited for continuous monitoring of a variety of variable stars, and thus can provide valuable data for understanding the physics of stellar variability. This is especially true for a class of variable stars known as chromospherically active stars. The stars belonging to this class have variable light curves that change from year to year in a rather strange way. A large fraction of these active stars are bright; hence the importance of small aperture telescopes for collecting much-needed photometric data. For over a decade the research activity using the 14" optical telescope has focused on photometric monitoring of well known and suspected active stars. These data, together with X-ray and radio data from archives as well as spectroscopic data obtained at Indian observatories, has led to the identification of new chromosperically active stars. This paper is aimed at sharing our experiences with the colleagues from the developing world on the usage of small optical telescopes for teaching and research with the objective of spreading the joy of astronomy among young students.

Keywords. Astronomy: teaching & research, stars: photometry, stars: active, stars: variable

---

1. Introduction

Undoubtedly, optical telescopes are amongst the most powerful tools in astronomy for probing the Universe we live in. The quest to explore the mysteries of the cosmos has led to the development of optical telescopes of bigger and bigger aperture. Optical telescopes of all sizes are useful for collecting data on a variety of objects in the Universe, and all have contributed to providing a better understanding of the Universe. It is not important what one has or can have; what is more important is what one does using what capability one has. For the purpose of this article, optical telescopes with aperture less than a metre are termed small telescopes. Development of technology for back-end instrumentation has yielded low-cost stellar photometers, CCD cameras and spectrographs. Together with the increased sophistication in telescope technology and data acquisition and reduction software packages, this has led to an enormous improvement in both the accuracy and the efficiency of observations using small optical telescopes. These features of modern small

† Visiting Associate, IUCAA, Pune, India.

optical telescopes are of great importance when one talks about introducing astronomy as a subject at various levels in schools, colleges or universities. Small optical telescopes equipped with suitable back-end instrumentation are capable of performing several tasks to enhance the teaching of astronomy. Small telescopes can also be employed for initiating research activities on suitably and carefully chosen projects on a variety of variable stars, leading to publication in reputed journals as well. In this article I will describe the activities that we have been carrying out with small optical telescopes, highlighting some research activities which have been carried out quite successfully. This has so far paid rich dividends in terms of creating manpower in astronomy. This, together with the facilities available at IUCAA and its Reference Centre in the department, has provided a big boost for promoting teaching and research in astronomy and astrophysics in our University.

2. Growth of astronomy and astrophysics in the Physics Department

Astronomy & Astrophysics (A&A) has been one of the major areas of research in the Physics Department since its establishment in 1972, but teaching in A&A started a few years later at the initiative of Professor R K Thakur, the then Head of the Physics Department, and gradually got strengthened with the addition of small optical telescopes (6" from Carl Zeiss in 1978, and a computerized 14" Celestron telescope in 1988) and stellar photometers (SSP3 and SSP5 from OPTEC). Completion of an observatory dome for housing the 14" telescope in 1991–92 sparked the beginning of observational astronomy in the department, and a new set of activities were initiated. These included a new astronomy syllabus with observational projects to supplement the theory course for the postgraduate students. The establishment of the Inter-University Centre For Astronomy and Astrophysics (IUCAA) at Pune, India in 1988 gave a big boost to our efforts. In order to find ways in which to utilize the telescopes maximally, we searched for suitable research projects for small telescopes. After a brief literature survey on variable stars, mostly done by Mr Padmakar Parihar, we selected a small sample of bright and prominent RS CVn-type active stars, and began photometric monitoring of these stars in 1994. Research projects sponsored by Indian funding agencies, such as DST, CSIR, etc., added a new dimension to our research programme by way of providing financial support for the research scholars and for maintenance of the observatory equipment. During the period 1994–2002 the facility was used for collecting good quality photometric data on several active stars. This was incorporated as a part of the thesis work of Padmakar (2000). The observational astronomy programme in the department received a severe blow when the observatory building collapsed in 2002. To make things worse, the 6" Carl-Zeiss telescope was stolen during December 2003, which brought about the collapse of the entire observational programme of the department. Fortunately, with the support from IUCAA, somehow we managed to continue our teaching programme in astronomy for the students using the small telescopes at IUCAA in Pune. A grant received from DST-FIST in 2005 helped the department to procure new telescopes and instrumentation. Since then the training programme in observational astronomy has resumed in the department.

The new observing facility in the Physics Department includes:
(i) two small telescopes, CGE800 and CGE1400, from Celestron with GPS-assisted pointing and other accessories;
(ii) a stellar photometer SSP3A from Optec;
(iii) a CCD camera ST-7XME along with spectrograph from SBIG;
(iv) software and other accessories.

The facilities are shown in Figs. 1 and 2. The department also has an adequate computing facility for data reduction and analysis, as well as a small library and internet connectivity using a VSAT facility sponsored by the University Grants Commission.

The School of Studies in Physics currently offers a two-year MSc course in physics with astronomy and astrophysics (A&A) as one of the specializations. The MSc course is split into four semesters and comprises both core and elective subjects in different areas of physics. A&A is taught as one of elective courses spread over two semesters (III & IV). The course includes topics on stellar structure and evolution, the Milky Way and other galaxies, active galaxies, general relativity and elementary cosmology. The theory course is supplemented with a project which can be either observational or theoretical. The astronomy course is quite popular in the department. Each year, about 10–12 students opt for A&A as one of their elective subjects.

It is worth mentioning here that Pt. Ravishankar Shukla University is among the few Universities in India (and the only one in the new state of Chattisgarh) where teaching and research in A&A is conducted in the Physics Department.

The mission of the Inter-University Centre for Astronomy and Astrophysics (IUCAA) is to initiate and promote teaching and research in A&A in the Indian Universities. It provides support to students, research scholars and faculty, allowing them to make use of the facilities at the Centre. Since its establishment, IUCAA has played a significant role in the sustaining the research interests of the faculty members of the department. In 1999 IUCAA created its Reference Centre in the Physics Department for the promotion of teaching and research in A&A in this region as a mark of recognition for the contribution that Department of Physics has made in the field of A&A. IUCAA provides support to short-term visitors, a small library and internet dial-up facility. The IUCAA support has become a life-line for strengthening the teaching and research activities in astronomy at the University.

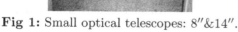

Fig 1: Small optical telescopes: 8″&14″.   **Fig 2:** Back-end instruments.

3. Activities in astronomy using small telescopes

Small optical telescopes are very useful for teaching astronomy in schools, colleges and universities. The availability of low-cost instrumentation (e.g. stellar photometers, CCD cameras and stellar spectrographs) has enhanced the capability of small telescopes in the present era of ever increasing sizes of large optical telescopes. In addition to selecting a set of observational projects to complement theoretical courses at graduate and post-graduate level, one can plan and implement research projects for which small telescopes are ideally suited (see for example Percy (1986) and Paczynski (2006)). The selection of an observing program depends on several factors such as (a) available equipment, (b) observing site, (c) time-scale of required observations, (d) experience, etc. One may seek help and advice from professional astronomers, consult the AAVSO, newsletters, etc. It turns out that photometry of well known short-period variable stars is good choice to begin with.

3.1. *Observational training projects*

Final-year students of the Department who offer Astronomy as the specialization for their MSc course carry out observational projects. For the postgraduate students, the required time-scale becomes one of the most important factors when selecting objects for them to observe. The objects for the project are chosen such that a student can carry out observations within a month, taking in to account all overheads. The observational projects include: (i) atmospheric extinction measurements, (ii) calibration of the photometer and determination of transformation coefficients, (iii) determining the H-R Diagram of nearby bright stars and star clusters like Pleiades and Hyades, and (iv) tracing light curves of various types of bright variable stars, such as Cepheids, Delta Scuti stars, eclipsing binaries like Algol and λ-tau and bright active (e.g. RS CVn) stars. During the course of the project work students are trained in observing techniques, data acquisition, reduction and analysis as well as in interpretation of the results.

3.2. *Public outreach*

Astronomy has proved to be a very effective platform for communicating science to young minds and tuning their excitement to science because of its unique quality of mass appeal. Celestial objects in the night sky have always been subject of awe and wonder to every one! This popularity of astronomy, the mother of all sciences, can be maximally utilized in attracting young students into the stream of pure science subjects. With this in mind the Physics Department started a sky gazing programme when it procured the 6" optical telescope in 1978. The programme continues to this day during the observing season and whenever special events become observable in the sky.

3.3. *Research*

Small telescopes equipped with adequate instrumentation are ideally suited for carrying out photometric observations of variable stars, and thus can provide valuable data for understanding the physics of variability. This is especially true for a class of variable stars known as chromospherically active stars, or RS Canum Venaticorum stars, after the prototype of this class identified by Hall and his group during 1972–1976. The stars belonging to this class have variable light curves, and the most puzzling feature is that their light curves change year after year in a rather strange way. The light variation is attributed to rotational modulation of dark spots, similar to sunspots, on the stellar surface, but a clear understanding of the evolution and dynamics of the dark spots remains one of the most puzzling issues in stellar physics. This bears far reaching consequences as regards the origin of the magnetic field and related dynamo problem in these stars. Continuous photometric monitoring of these active stars is therefore required to examine short-term as well long-term variations in their light curves. A large fraction of these active stars are bright enough to be monitored with small telescopes for collecting the much-needed photometric data.

With these objectives in mind we used our computer-controlled 14" Schmidt-Cassegrain Celestron to obtaining BVR photometry of several prominent RS CVn binaries. These observations were used in the theses of Padmakar Parihar and Sudhnashu Barway. These stars are found to display appreciable changes in amplitude, pulse shape, phase of minimum light as well as mean light level within a couple of rotation periods. Fig. 3. shows sample light curves of the short-term variation in the RS CVn binary V711 Tau. This is the nearest (29 pc) and brightest ($V_{max} = 5.74$) chromospherically active binary star with a short period (2.84 days) having a K1 IV primary and G5 V secondary.

In addition to regular monitoring of prominent RS CVn stars, several X-ray binaries which are suspected to be chromospherically active stars were monitored photometrically using the 16" Meade telescope at IUCAA during the period 2000–2001. Analyses of the optical photometric data in conjunction with spectroscopic, X-ray and radio continuum data of HD 61396 strongly suggested its identification as a new RS CVn-type binary (Padmakar et al., 2000a).

Likewise, Sudhnashu Barway as a part of his thesis work carried out photometric observations of five suspected variable stars selected on the basis of their intense X-ray emission as revealed from the space observatory ROSAT. Barway observed these stars during 2000–2001 using a 16" Meade telescope equipped with an SSP-3A photometer. Preliminary results indicate that four of these stars belong to the class of chromospherically active stars (Barway, 2005). Fig. 4 shows sample light curves of one of these stars, HD 39286.

These discoveries of new chromospherically active stars, apart from enriching the existing sample of active stars, are very important for understanding the underlying physical parameters like rotation, age, metallicity, etc. which are involved in generating and sustaining strong chromospheric and coronal activity in stars.

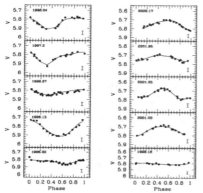

Fig 3: Short-term light variation in RS CVn star V711 Tau.

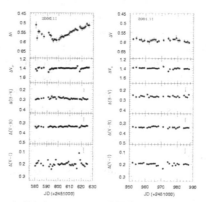

Fig 4: Light curves of HD39286 for two epochs.

4. Assessing the impact of small telescopes

If one asks whether our efforts to promote the usage of small telescopes have made any impact in the astronomy scene in India, the answer is a big 'yes' based on the feedback and compliments the Physics Department has received from astronomer colleagues in the country, specially in terms of generating manpower in astronomy. Given the constraints, and limitations of the working environments under which the department functions, the recognition earned by the department is indeed a big moral boost and encouraging for the group involved in the observational astronomy activities in the Physics Department.

Research activities using small telescopes at the department and at IUCAA have led to the following publications in journals during the period 1996–2005.

1. BVR Photometry of the three RS CVn Binary Stars: V711 tau, UX Ari and IM Peg: Padmakar and Pandey (1996a).
2. A study of long term photometric variation in the RS CVn Star V711 Tau: Padmakar and Pandey (1996b).
3. Stellar activity in the RS CVn binary star UX Arietis: Padmakar and Pandey (1999a).
4. New BVR photometry of six prominent RS CVn binary stars: Padmakar and Pandey (1999b).
5. Optical, X-ray and radio observations of HD 61396; a probable new RS CVn type binary: Padmakar et al.(2000a).
6. Study of sky conditions at Raipur and calibration of photometer-telescope system: Padmakar, Barway and Pandey (2000b).
7. HD 52452: New BVRI photometry : Barway and Pandey (2004).
8. BVR photometry of a newly identified RS CVn binary star HD 61396 : Barway, Pandey and Padmakar (2004).

So far seven students have been awarded PhD degrees for their work in the different topics in astronomy and two of them completed their thesis work using small optical telescopes. Currently, six students are working towards their PhD degree in the field of astronomy. Several students of the department have been selected for research positions in different research institutions in India as well as outside India. In addition, students are encouraged to participate in a variety of activities to sustain their interest in astronomy. The activities include seminars and lectures by visiting scientists, student participation in summer schools, vacation student programmes organized by various research institutions in the country, student seminars in the department and the Young Astronomers Meet (YAM-India), as well as the IAU International School of Young Astronomers (ISYA).

All these activities have proved extremely useful in creating a conducive atmosphere for teaching and research in the department, as well as in helping several students of the department to take up a research career in astronomy.

With the objective of spreading the culture of astronomy using small optical telescopes in other colleges and universities in India, the A&A research group at Raipur participates actively in workshops on observational astronomy, mostly sponsored by IUCAA. As a result, some colleges and university departments have acquired small telescopes and detectors and have started observational projects for their courses in astronomy. Other colleges and university departments have plans to acquire small telescopes in the near future. The Indian funding agencies are quite forthcoming in providing grants for the procurement of small optical telescopes in their bid to support the activities initiated by IUCAA.

5. Looking ahead

The future of observational astronomy using small telescopes in our University looks quite bright in view of the fact that the University administration has started the process of building a new observatory building and dome on the campus for housing the new 14" telescope. Professor Lakshman Chaturvedi, present vice-chancellor of the University, is very keen to get the work completed at the earliest opportunity. Once the permanent housing for 14" telescope is ready, it will be possible to resume our research programme on chromospherically active stars. Likewise, a new CCD spectrograph will be used to introduce projects on spectroscopic observations which were not possible earlier. The impact of our observational programme using small optical telescopes has so far been very encouraging, and we hope to continue this activity with even greater vigour in future.

Acknowledgements

I thankfully acknowledge the support received from the University, various Indian funding agencies and IUCAA. Most importantly, I acknowledge the past and present students of the Physics Department, without whom I could not have continued the observational programme. I gratefully acknowledge a travel grant received from IAU, which made possible my participation in SPS5 and the XXVIth IAU General Assembly.

References

Barway S. 2006, Ph.D. thesis, Pt. Ravishankar Shukla University, Raipur, India
Barway, S. & Pandey, S.K. 2004, *IBVS* 5553, 1, (astro-ph No. 0408351)
Barway S., Pandey S.K., Padmakar 2004, *NewA.* 10, 109, (astro-ph No. 0408177)
Paczynski, B. 2006, *Astronomical Society of the Pacific* 118, 1621
Padmakar, 2000, Ph.D. thesis, Pt. Ravishankar Shukla University, Raipur, India
Padmakar & Pandey, S.K. 1996a, *Ap&SS* 235, 337
Padmakar & Pandey, S.K. 1996b, *BASI* 24, 717
Padmakar & Pandey, S.K. 1999a, *BASI* 27, 117
Padmakar & Pandey, S.K. 1999b, *A&AS* 138, 203
Padmakar, Barway S., Pandey S.K. 2000a, *BASI* 28, 437
Padmakar, Singh K.P., Drake S.A., Pandey S.K. 2000b, *MNRAS* 314, 733
Percy, J.R. 1986, The study of variable stars using small telescopes (Cambridge University Press)

Shiva Pandey

Paulo Bretones

Jay Pasachoff

Astronomy for the developing world
IAU Special Session no. 5, 2006
J.B. Hearnshaw and P. Martinez, eds.

© 2007 International Astronomical Union
doi:10.1017/S1743921307007090

How to look for planetary transits using small telescopes and commercial CCDs in developing countries

Eder Martioli[1] and F. Jablonski[1]

[1]Divisão de Astrofísica, Instituto Nacional de Pesquisas Espaciais,
São José dos Campos, Brazil
email: eder@das.inpe.br, chico@das.inpe.br

Abstract. The main goal of this work is to have a better understanding of the problems and characteristics of photometric surveys with small-sized affordable equipment, like the one available at the Astrophysics Division/INPE, in São José dos Campos, Brazil. The use of low-cost instruments is appealing in the context of the detection of Extrasolar Planets (ESP), in the sense that many observers are available for survey and follow-up programmes. It could also make possible the inclusion of many developing countries in the search for planetary transits. We describe the data collection and analysis procedure for differential photometry of the transit of HD 209458 b, using a small telescope and a commercial CCD camera. According to the HST observations of Brown *et al.* (2001), the transit produces a box-shaped light curve with 2% depth and 184-min duration. The orbital period is ∼ 3.5 days. The equipment consists of a f/10, 11" Schmidt-Cassegrain Celestron telescope equipped with a SBIG ST7E CCD camera. Since the seeing at the campus is quite poor, we used a focal reducer to produce an effective focal ratio of about f/5, still keeping a good sampling of the PSF but with a larger field of view. The larger field of view allows the simultaneous observation of a relatively bright nearby star, suitable for differential photometry. We discuss the IRAF reduction procedures for the large number of images collected and present the results obtained in the transit of September 8, 2004

Keywords. techniques: photometric, stars: planetary systems, instrumentation: miscellaneous

1. Introduction

HD 209458 is one of the stars known to bear an Extra-Solar Planet (ESP). The detection of this planet was firstly done by Henry *et al.* (2000) using the conventional radial-velocity technique. Subsequently, Charbonneau *et al.* (2000) showed that a photometric transit can be detected in the system. The detection of photometric transits requires high-quality data since the decrease in flux is only 2% and the event spans only 3.5 hours.

The probability of a transit to happen is a few per cent, and thus many ESP could be detected photometrically. However, the difficulties in achieving the required photometric precision from ground-based observatories has kept the number of detections small.

This scenario is beginning to change with space instruments like MOST and COROT which are able to reach photometric precisions of micro-magnitudes. Similarly, ground-based photometric surveys of a large number of targets would also be effective in detecting transits. The OGLE survey, even though designed for different purposes, has detected tens of candidate planetary transits.

2. Equipment

Table 1 summarizes the instrumentation and Figure 1 shows the 11" telescope used in our work.

Table 1. Description of Instrumentation

Telescope	Celestron 11" Schmidt-Cassegrain
Accessories	focal reducer
Effective focal ratio	f/5
Filters	clear
Mount	equatorial Losmandi G11
CCD	SBIG ST7E

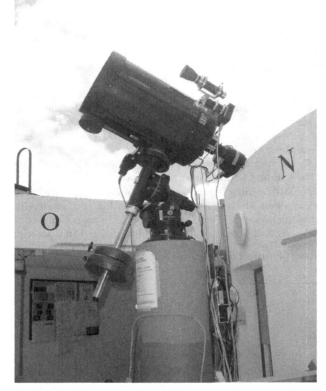

Figure 1. The 11" telescope in its mount at the main campus of INPE.

3. Data management

3.1. *Data acquisition*

The images were collected using the CCDSoft (version 5.00.126) software. Since HD 209458 is quite bright we used 3-sec integrations to be safely under the saturation limit. We added a 10-sec wait-time to the observing cycle to keep the number of images small. The CCD temperature was kept close to -10°C during the observing run. Since we did not use auto-guiding, small corrections were done manually to keep the objects more or

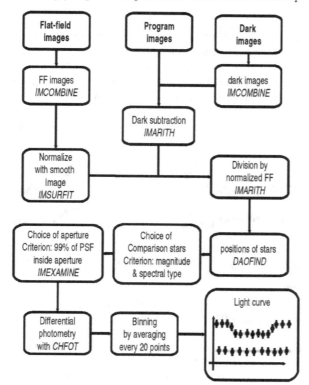

Figure 2. Flowchart showing the steps to be followed in the data reduction.

less in the same region of the detector. This is important since the focal reducer used to increase the field-of-view has strong vignetting as one goes to the borders of the field.

The usual flat-field images were taken to allow a pixel-to-pixel sensitivity correction to be carried out. Also, dark frames with the same integration time as the programme images were collected to allow the subtraction of the dark noise component.

The measurements were started one hour before the beginning of transit and lasted until the end of the event, resulting in a total of 895 CCD frames. Unfortunately, the sky conditions at the end of the event were quite poor.

3.2. *Data reduction*

The preparation of the data was carried out under IRAF and consisted of: *(i)* producing a median dark frame with imcombine; *(ii)* flat-field median frame from which the dark frame was subtracted; *(iii)* normalized flat-field from the division of combined flat-field frame from step *(ii)* by a smooth image fit to the same image with imsurfit. The *RMS* noise of the normalized flat-field frame is 0.2%. This can be considered quite good for a low-cost CCD detector.

The flux extraction was done with an IRAF script called chfot. This task automatically identifies the stars in each frame, centers precisely in each object, extracts the correspondent fluxes, calculates the magnitude differences and presents the data together with the heliocentric Julian date. A useful feature of the program is its capability to add up all the comparison stars. This is important in cases (like the present one) where the variable object is the brightest in the field-of-view.

4. Results

The original light curves that produced the binned version in Figure 3 have *RMS* errors comparable to the depth of the transit itself, namely, 2%. This makes it difficult to visualize the event. To improve visualization we combined the original points averaging them in bins containing 20 points. A side benefit of the binning operation is to get a better idea of the absolute precision of the measurements.

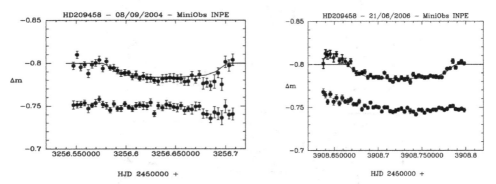

Figure 3. Light curves of the transit of HD 209458 b (top) and the sum of comparison stars (bottom) obtained in 2004 September 08 (left) and 2006 June 21st (right).

Figure 3 shows the binned light curves for HD 209458 and for the sum of the comparison stars. We superimposed the light curve of the HST data from Brown *et al.* (2001). One can see that for the two transits measured there is consistency in terms of time of beginning of the transit, its depth and duration. The photometric conditions were poorer in 2004 September 08 at the end of the run. The opposite happened in 2006 June 21st. The light curves show that there are limits even for a robust technique like differential photometry. Besides the larger extinction itself, differential extinction and differential refraction contribute to systematic trends and larger scatter in the measurements as one goes down to the horizon.

Some possible improvements given the lessons learned with the acquisition, reduction and analysis of the data can be summarized as:

- To use a larger unvignetted field-of-view to obtain more comparison stars.
- To use auto-guiding to keep the target in the same (good) area of the detector.
- To use a red bandpass filter to minimize the effects of differential extinction and differential refraction.
- To select comparison stars as close in spectral type to HD 209458 as possible.

Acknowledgements

IRAF is distributed by the National Optical Astronomy Observatories, which are operated by the Association of Universities for Research in Astronomy, Inc., under cooperative agreement with the National Science Foundation.

References

Charbonneau, D., Brown, T.M., Latham, D.W., Mayor, M. 2000, *ApJL* 529, L45-L48
Brown, T.M., Charbonneau, D., Gilliland, R.L., Noyes, R.W., Burrows, A. 2001, *ApJ* 552, 699
Henry, G.W., Marcy, G.W., Butler, R.P., Vogt, S.S. 2000, *ApJL* 529, L41-L44

Astronomy for the developing world
IAU Special Session no. 5, 2006
J.B. Hearnshaw and P. Martinez, eds.

© 2007 International Astronomical Union
doi:10.1017/S1743921307007107

Basic astronomy as part of a general higher education in the developing world

John Baruch[1], Dan Hedges[1], James Machell[1], K. Norris[2] and Chris Tallon[1]

[1]School of Informatics, University of Bradford, Bradford, U.K. BD7-1DP
email: john@telescope.org
[2]Department of Education, Bradford College, Bradford, U.K. BD7-1AY

Abstract. This paper describes a new initiative in support of the aim of Commission 46 of the IAU to develop and improve astronomy education at all levels throughout the world. This paper discusses the ideal specification of a facility to support basic astronomy within education programmes which are delivered to students who have access to the Internet. The available robotic telescopes are discussed against this specification and it is argued that the Bradford Robotic Telescope, uniquely, can support many thousands of users in the area of basic astronomy education, and the resource is free.

Access to the Internet is growing in the developing world and this is true in education programmes. This paper discusses the serious problems of delivering to large numbers of students a web based astronomy education programme supported by a robotic telescope as part of a general education. It examines the problems of this form of teaching for teachers who have little experience of working with IT and little knowledge of basic astronomy and proposes how such teachers can be supported.

The current system (`http://www.telescope.org/`) delivers astronomy education in the language, culture and traditions of England. The paper discusses the need to extend this to other languages, cultures and traditions, although for trainee teachers and undergraduates, it is argued that the current system provides a unique and valuable resource.

Keywords. Telescopes, robotic telescope, astronomy education, e-learning, robot, developing countries

1. Introduction

When Tony Blair came to power in the UK in 1997 he said that his three priorities were "Education, Education and Education". It is education and not biology that separates us from our forefathers who chased mammoths across the frozen tundra of Europe in the last ice age. It is education and knowledge that empowers us to find effective routes to deal with the threats of AIDS and global warming and differentiates between those who are terrorized by voodoo, black magic or creationist ideas and those who have the framework to understand the evolution of diseases like bird flu, the aggregation of oil reservoirs or the awe inspiring vast universe that we inhabit.

Commission 46 of the IAU seeks to further the development and improvement of astronomical education at all levels throughout the world. One of the key problems for many students around the world and many teachers is the difficulty of teaching the basics of astronomy illustrating our position in the universe and the daily, monthly and yearly rhythms that our environment is subjected to.

2. Basic astronomy education for all

The growth of cities with their high buildings, street lights and general light pollution, television, computer games and other distractions ensure that astronomy and the position of the Earth in the Universe barely enter into the consciousness of many young people. In many nations, the majority of the engineers and scientists were first brought to science by the accessibility of astronomy and the nightly spectacle of the stars. This is no longer the case and many nations particularly the developed nations are concerned about the future supply of engineers and scientists. In the 2006 House of Lords report on science teaching in schools in the UK they were most concerned about education in physics up to age 18 for the A level examination:

> *"The most problematic subject of all is physics, with the number of students opting to take the subject showing a precipitous fall since 1998. The numbers continue to fall and currently stand at less than 60 per cent of the total in the late 1980s. Moreover, the situation is far worse in some schools than these aggregated figures would suggest, as the Royal Society noted, "science take-up is strongly skewed at present, with half of all A-level entries in science coming from just 18 per cent of schools"*

The fact that few people now sit under the stars at night admiring the glorious procession of the heavens means that the stars and basic astronomy are not part of life for the majority. To deliver an understanding of the basics of astronomy with its nightly spectacle requires education. An essential core of basic education is an understanding of the place of the Earth in the Universe and the processes which give us day and night, seasons, the phases of the moon, eclipses, comets and the annual cycle of the stars.

The United Nations Convention on the rights of the Child, accepted and signed by 192 of the states of the United Nations, which is more than any of the other Human Rights Conventions, states that all young people have the right to education and that:

> *"States Parties shall promote and encourage international cooperation in matters relating to education, in particular with a view to contributing to the elimination of ignorance and illiteracy throughout the world and facilitating access to scientific and technical knowledge"*

Astronomy is a key part of that education.

3. Educating teachers is the first step to Education for All

For many developing countries this universal education even to primary level delivered free for all children as required by the Convention on the Rights of the Child is not immediately possible. They first have to educate people to the level necessary to be teachers and then to train them to be teachers. Few schools in the developing world have enough books at both primary and secondary levels and even fewer have access to the Internet. The situation in higher education and especially in teacher training appears much better. In 1998 UNESCO called for all teacher training institutions to include ICT skills within their courses, which included access to the Internet (UNESCO 1998). Tanzania is a typical developing country which in a report by UNESCO (2002) listed the problems of teacher training as:

- Low tutor competencies;
- Lack of a systematic tutors continuous professional development;
- Inadequate college management skills;
- Poor Information Communications Technology;
- Lack of competency based curriculum and curriculum materials.

The International Institute for Communications and Development, a charity funded by the Governments of Holland, Britain and Switzerland, working with developing countries had connected up 44 Tanzanian Teacher Training Colleges to the Internet, installed computers and trained their technicians by the end of 2004†. The results of this are typically indicated in Reports in 2005‡ and verbal reports in 2006 which indicated that at least the ICT aspects of teacher training had been greatly improved with Tanzanian students training to be teachers in the main Teacher Training institutions of Mzumbe and Dar Es Salaam having good access to Internet facilities. There still remain some problems associated with the cost of supplying adequate bandwidth to deliver good access. But it is probable that by this time (2007) all Tanzania's teacher training institutions are connected and effectively working on the Internet with their education programmes for pre-service teachers. It is also probable that with the lead being taken by the governments of the developed countries that the situation in Tanzania is reflected in most of the developing countries. Thus in 2007 most students training to be teachers will probably have some access to the internet during their course.

Another key problem for the training of teachers in the developing countries that was noted in the 2002 UNESCO report on Tanzania was the low tutor competencies. With the Internet available to many of the Teacher Training Institutions it is now possible to start to reap the benefits of ICT and Internet access that the UNESCO reports discussed in 1998 and 2004. The Bradford Robotic Telescope is part of that benefit.

4. The Bradford Robotic Telescope is available for all

The Bradford Robotic telescope http://www.telescope.org/ is an extensive web site for teaching and learning the basics of astronomy and the position of the Earth in the Universe. It is because it works at the basic level that it can handle an enormous number of users. The users work through the interactive e-learning programmes on the web site. These are really designed for under 16-year-old students, but there are extensive materials available to support teachers who are working at the limits of their expertise. The programmes with their range of interactive learning modules include the option for a learner to request observations to be taken by the Bradford Robotic Telescope to produce data that will support their learning.

These requests are all handled automatically by the telescope systems which schedule the observation and return the data to the user. The telescope is situated at one of the best observatory sites in the northern hemisphere. The Observatorio del Teide on the Island of Tenerife in the Canary Islands. The excellent weather in the Canary Islands means that the user will normally have their request returned within days.

The Bradford Robotic Telescope is already widely used. It has over 11 000 registered users, but the system is extensive and registration is only required for those users who wish to have data returned to them. Registration provides an email address for the system to return the data. There are about 5000 individual visitors per day to the site, where multiple hits from the same IP address and access agent on a single day are called a single visit. The telescope has completed over 20 000 user requests and returned data to the user. The telescope is widely used by developing countries with about 100 of the 140 countries of the 11 000 registered users being regarded as developing countries. The number of users from these countries varies from over 170 in India, 86 in Brazil, 48 in Mexico, to around 30 countries with a single user which include Belize and Vanuatu. It

---

† See http://www.iicd.org/projects/articles/iicdprojects.2004-11-12.6571234687
‡ See http://cs.joensuu.fi/cbe/

is not known how many of these are students in teacher training institutions but from the e-mail comments it is clear that students are one of the groups using the telescope system.

The objectives of IAU Commission 46 talk about astronomy education to all levels and it can only be assumed that this encompasses all levels of education and the complete range of our astronomical understanding. The UN Convention on the rights of the Child is more specific with a view to

"the elimination of ignorance and illiteracy throughout the world and facilitating access to scientific and technical knowledge."

5. A resource to deliver basic astronomy education

For a robotic telescope system to contribute to these objectives, especially the objectives contained in the Convention of the Rights of the Child it must be capable of supporting an education programme to deliver the basic framework of a modern understanding of our place in the Universe starting with the basic facts about the Sun and Earth, its Moon, the orbits of the planets and the local solar system. It must also be capable of delivering these to very large numbers of students. Initially, for the developing world, these students will be trainee teachers.

The basic knowledge associated with astronomy has been encapsulated in many ways, books like Astronomy for Beginners (Becan & Becan, 2004) are one area, the Web, through the English National Curriculum for Science 2006 † is another. A general search for basic astronomy will reveal hundreds of sites giving a wide range of knowledge. An understanding of astronomy starts with gravity and an understanding that the Earth is a sphere, followed by what causes day and night and the seasons, building upon these concepts opens the door to understanding so that mature students can continue themselves. It is these basic concepts that are so difficult to illustrate and which benefit greatly from the use of simulations on a computer screen and access to a telescope to confirm concepts. This is precisely the structure of the Bradford Robotic Telescope system.

The extensive web site which can be found at `http://www.telescope.org/` includes a comprehensive e-learning site with 34 basic astronomy topics which include making requests for observations with the robotic telescope to support learning and understanding. The extensive web site includes animations, models and simulations. There is an extensive archive of images and a host of sensors and web cams surrounding the robot so that the user can see the day and night time sky. Every night a video is taken of the stars rotating around the pole star and another of the stars setting on Mount Teide. Users can see for themselves that although the Sun and Moon set at different places throughout the year the stars always set in precisely the same place.

6. Detailed description

The robot telescope itself is actually four telescopes each with CCD cameras, filter sets and automatic focussing. There is the constellation camera which takes large 40 degree panoramas of the sky. The cluster camera takes three degree images of the sky showing many smaller stellar clusters and nurseries along with full images of the Moon. The galaxy camera takes 20 arc minute images of deep sky objects and the planetary camera is currently used to ensure good tracking with the galaxy camera but its pixels cover 0.7 arc seconds of sky and it is intended to upgrade it into a planetary camera to

† Available at `http://www.nc.uk.net/`

look at brighter objects which include the planets and the International Space Station taking thousands of images and selecting those which show sub arc second resolution. The only thing that the system is, as yet, unable to do is to take images of the Moon and include the surrounding stars.

This assembly of optics and data processing is able to deliver the images to support study into all areas of basic astronomy, encouraging the users to adopt the scientific method to investigate ideas and check out the limits of their veracity.

7. Compared to other robotic telescopes

There are many robot telescopes operating in the world. Most of these are remotely controlled telescopes (e.g. `http://slooh.com/`) operating over the internet with a web cam for the user to see what they are doing. These are regarded as remote telescopes rather than robotic. Robotic telescopes like the Bradford Robotic telescope operate without interference from the user. The user submits a request, the request is scheduled by the telescope and the data returned to the user. Most of the remote telescopes charge for their use a given fee for a given time. Some of these remote telescopes are very large. The Faulkes and Liverpool telescopes are 2-metre aperture telescopes which can do serious research but their field is about 5 arc minutes and so they are unable to capture the full lunar image or the panorama of the constellations which is an important starting point for astronomy education. The Bradford Robotic telescope provides the background understanding for students to graduate to using these other systems.

Another great advantage of the Bradford Robotic telescope is that it can handle very large numbers of users this is only possible because unlike any of the other robotic telescopes it provides service observing, self scheduling the requests and by working at the most basic levels there are only about 25 objects requested by the great majority of users and these can be handled every night they are visible.

At the present time the telescope is free for international users, English users are expected to subscribe to the service and provide funds to cover the running costs on `http://schools.telescope.org/`. This is a service focussed precisely on the needs of teachers in England delivering the National Curriculum to students from the age of ten up to sixteen.

It is recognized that the general service is delivered in the language culture and traditions of England. This may be fine for some countries but it clearly is not ideal for most countries of the world. It is desirable to have the facility translated into other languages and linked into other cultures. It is understood that young people will learn better if the astronomy, whilst presenting a modern scientific views, links into the mythology and cultures of their nation. It is planned to seek funding to translate the site into the major European languages incorporating the cultures associated with those languages.

8. Conclusion

The Bradford Robotic telescope is an extensive e-learning web site supported by a robotic telescope. It offers a unique resource to support the general education of teachers in developing countries. It has all the facilities required for such general education and provides extensive support for both teachers and learners.

Acknowledgements

We would like to acknowledge the support for this work from the Nuffield Foundation, PPARC, the Gatsby Trust and the Paul Instrument Fund of the Royal Society. We would also like to acknowledge the help given by Dr Faustin Kamuzora of Mzumbe University Tanzania.

References

Becan J., & Becan S. 2004, *Astronomy for Beginners,* Writers and Readers Publishing, ISBN 0863169996

UNESCO 1988, *UNESCO World Education Report,* pp. 19–20

UNESCO 2002, In: Buyela Wepukhulu (ed.), *Capacity building for Lead Teacher Training Institutions in Tanzania,* UNESCO

John Baruch

Astronomy for the developing world
IAU Special Session no. 5, 2006
J.B. Hearnshaw and P. Martinez, eds.

© 2007 International Astronomical Union
doi:10.1017/S1743921307007119

Modern facilities in astronomy education

Hayk Harutyunian and Areg Mickaelian

Byurakan Astrophysical Observatory, Armenia
email: hhayk_ast@yahoo.com

Abstract. Astronomical education is entering a new stage of development which is closely connected with the development of new technologies for communication, computing and data visualization. We discuss this evolution in the context of astronomy education in Armenia. As students spend only a short time in Byurakan Observatory for training in observations, they are not able to carry out systematic astronomical observations. Hence their training places emphasis on the use of astronomical archives and analysis of observational data obtained previously with the Byurakan telescopes and other ground-based and space telescopes. Thus, one of the aims of the Armenian Virtual Observatory is to support the training of students in this modern context.

Keywords. Astronomy education, Armenia

1. Introduction

Astronomical education in the world is entering a new stage of organization and development. This process is closely connected with the development of new technologies and technological diffusion in all the spheres of human activity, including education, enabling the use of entirely new possibilities in electronic communication, computing, visualization of observational data, new methods of data reduction, etc. Undoubtedly this process gives teachers many new favourable conditions for training their students. However, this trend also places completely new requirements on students in relation to their knowledge and skills, which must be be taken into account as well in the development of training programmes.

This progress is distinctly noticeable in Armenia and has resulted from several ongoing changes in the republic which, in their turn, dictate new approaches based mainly on technological developments. While suggesting many new possibilities, modern technologies at the same time require a higher level of knowledge. Although economic difficulties in the past decade (and the associated lack of opportunities for pursuing astronomy research) restricted the interest in astronomy among school pupils, the present young generation is gradually finding astronomy more and more attractive. Knowledge of computers and the Internet is the typical difference between present-day students and earlier generations of students. Thus, nowadays astronomy education requires heavy use of computer facilities and the Internet, just as it does in case of modern astronomical research. These facilities should be available for students as early as possible, since the sooner the students become proficient in the use of new technologies, the more efficient will be their future use of those technologies.

2. Astronomy education in Armenia

In Armenia there are sharp distinctions in the conditions in the universities and schools. Particularly noticeable differences exist in regard to access to modern technologies in the schools and universities. Only a small fraction of schools are accommodated with even the simplest modern facilities such as Internet and computers. School pupils use these facilities chiefly in Internet cafés, of which there are many in the large cities, but very few in the villages. So the students who enter university very often have entirely dissimilar starting knowledge on how to use computers and their accessories. The universities are better equipped with computers and are connected to the Internet by rather high-speed links.

During this period of rapid technological evolution astronomy educators have to confront the new educational reality presented by the realization of new technologies. This implies a certain risk of misunderstanding of the real meaning of astronomy on the part of students. Sometimes an exaggerated view of the role computers in astronomy (and in other sciences as well) becomes a danger for this science, transforming an astronomy student into the computer addict who simply dabbles with astronomical ideas or does a technician's work instead of using computers as a tool for serious scientific study. Thus, in using various technical devices one should always remember and teach the students that computers are not an end in themselves, but rather the means to facilitate mathematical or other operations for astronomical research.

In any given case a certain optimal solution should be found for training the students to use modern facilities as astronomers. Obviously the students need modern hardware such as powerful computers and Internet (including GRID technologies) as well as computing methods, usage of large astronomical databases, virtual observatories (VOs), etc. VOs, though not yet widely used, seem to be very efficient tools both for scientific research and for the training of students. The Armenian astronomy community, for example, has a unique database of the famous Markarian survey. The Digitized First Byurakan Survey (DFBS) is the digitized version of the spectroscopic survey completed for high galactic latitudes in 17,000 square degrees of the northern sky. The newly created Armenian Virtual Observatory (ArVO) contains the DFBS.

The new capabilities offered by such new facilities also indicate the need to train of a new breed of astronomy computing specialists who can drive modern research and make it much more efficient. Thus in 2005 we introduced a new graduate level course at the Yerevan State University called "Astronomical surveys, databases and virtual observatories." This course, which we believe is the first of its kind in the world, covers the modern understanding and treatment of large multi-wavelength data volumes.

As the students are not able to carry out systematic astronomical observations (they spend only a short time in Byurakan observatory for training in observations), our training emphasizes the usage of the astronomical archives and the reduction and analysis of previously obtained observational data, either from the Byurakan telescopes, or from other ground-based and space telescopes. The ArVO was thus created for such purposes as well. We are currently establishing a large data centre at the Armenian National Academy of Sciences, where a huge volume of astronomical data will be stored on the newly created ArmCluster supercomputer and students will have access to these data. A number of science projects will be carried out with these facilities, including both astronomical research and research on informatics methods.

Astronomy for the developing world
IAU Special Session no. 5, 2006
J.B. Hearnshaw and P. Martinez, eds.

© 2007 International Astronomical Union
doi:10.1017/S1743921307007120

Observing solar eclipses in the developing world

Jay M. Pasachoff

Williams College – Hopkins Observatory, Williamstown, MA 01267, USA
email: jay.m.pasachoff@williams.edu

Abstract. Astronomers have opportunities at least twice a year to use partial, annular, or total eclipses of the Sun, or planetary transits, to interest the public in astronomy through their observations. It is important to provide accurate information about the pleasures and hazards of looking toward the Sun. The International Astronomical Union helps by providing knowledgeable information from experienced eclipse observers.

Keywords. eclipses: solar, astronomy education

1. Introduction

Solar eclipses are perhaps the most dramatic event possible: when it goes dark as night in the middle of the day, people of all locations and all cultures are awed. Contemporary astronomers are even able to discern the locations of eclipses that took place thousands of years ago since ancient records sometimes record such dramatic events, allowing the Earth's changing rotation rate to be determined.

Though it is completely safe to observe the total phase of a solar eclipse, since the brightness of the solar corona is comparable to that of the full moon and therefore equally safe, the partial phases before and after totality can be hazardous to vision. On a normal day, the eye-blink reflex keeps you from staring at the Sun. In the crescent partial phases of a solar eclipse, though, the total flux of solar radiation may be too low to activate the eye-blink reflex while the surface intensity of the remaining image of the solar photosphere can damage the retina. The problem is acute when any optical aid — telescope or binoculars — is used, since retinal damage can be immediate. But the problem can occur if one stares unaided at the crescent partial phase for too long — many seconds or even minutes. Such cases are rare but they do exist. For information on safe observation of eclipses see Chou (1997, 1998), Espenak (1996), Land (2000), Pasachoff and Covington (1993), and Pasachoff (2006).

2. Safe observation of solar eclipses

A public education programme is therefore necessary to explain to people in countries over which an eclipse will be visible just why the eclipse is interesting to look at and how to observe it safely (Golub and Pasachoff, 1997; Pasachoff, 2001). We scientists also usually want to overcome local prejudices and pseudoscientific beliefs. Unfortunately, we have found students locked in basement rooms, sometimes even on the side away from the Sun, in, historically, the United States, Canada, and Australia; people refusing to eat food during an eclipse in India; and pregnant women being kept indoors during the eclipse in many countries. Yet we astronomers do have a tremendous teaching opportunity at the times of eclipses, since newspapers and sometimes public authorities tell people about the coming eclipse, and for some days in advance of an eclipse we have the attention of many people (Pasachoff, 1996a, 2001).

Accordingly, the International Astronomical Union's Working Group on Solar Eclipses advises people around the world about eclipses, including their scientific value and how to observe them safely. A Programme Group on Public Education at the Times of Solar Eclipses of the IAU's Commission on Education and Development takes advantage of the public attention to aid local astronomers and educators about eclipse safety (Pasachoff, 2002). I chair both groups. The Working Group has as other members Iraida Kim (Russia), Hiroki Kurokawa (Japan), Jagdev Singh (India), Vojtech Rusin (Slovakia) as general members; Fred Espenak (United States) and Jay Anderson (Canada) for providing predictions and the NASA Technical Publications and Website; and Michael Gill, the moderator of the Solar Eclipse Mailing List (SEML@yahoogroups.com; eclipsechaser@yahoo.com). The previous triennium also had Atila Ozguc (Turkey) as a member, because of the 2006 eclipses passing through Turkey. The 2006–2009 triennium has Jingxiu Wang (China) and C. Fang (China) as members because of the 2008 and 2009 eclipses passing through China. The Programme Group has as other members Julieta Fierro (Mexico) and Ralph Chou (Canada), a professor of optometry who has special expertise in eye safety and in solar-filter testing.

The Working Group on Eclipses' Website is http://www.eclipses.info. The site can be reached through a link on the site of the Commission on Education and Development at http://www.astronomyeducation.org, going to Description of Our Activities (Programme Groups). Eclipses.info also contains links to providers of solar-filter materials.

3. The importance of disseminating public information on eclipses

An early example of false information from an official source was in a 1970 memo from the Department of Justice in the United States, which misleadingly states, "The public must be made aware that the so-called protective devices, such as sun glasses, smoked glass, or film negatives do not protect the eye from the invisible infra-red rays which do the damage." Some film negatives, those with silver content, indeed do and did protect the eye. So warnings about not using sunglasses or smoked glass are valid, but proper credit to acceptable protective filters should be provided.

An egregious example of public miseducation took place at the 2002 eclipse in Australia, when the official health authorities in Western Australia, where only a partial eclipse would be visible, advised people not to look at the eclipse at all. They even prevented eye-protection solar filters from being sold, on the pretense that even though they were approved in the EU, they had no specific tested verification from an Australian authority. Such people do not listen to the advice of experts, even when we try to describe the truth to them. A 5-cm-high front-page headline in *The Advertiser*, saying only "DON'T LOOK," gave an inadequate message, and the subheadline, "Eclipse glasses can hurt your eyes," was misleading.

A major problem with such official misinformation is that local people soon find out that, in fact, the eclipse was not harmful and that people they know viewed it with delight. These people may then lose confidence in other official pronouncements, including health warnings about AIDS, STD's, vaccinations, and other important subjects. The negative consequences of false warnings about eclipse eye hazards can thus be severe and widespread.

4. March 2006 eclipse in Nigeria

One of the triumphs of our IAU activities took place in Nigeria during the 29 March 2006 total solar eclipse. Our colleague from there, Prof. P.N. Okeke of the Centre for Basic Space Science, was in touch with us, and we arranged for sheets of solar filter material to be provided. Since eclipse "glasses," with a quantity of solar-filter material in each of two eyes, are expensive in quantity, even at a cost of 50c or $1 per item, we have figured out how the filter material can be used much more efficiently. At the very least, individual "glasses" can be cut in half at the bridge, giving two for the price of one. But much more efficiently, only about 4 millimetres square are needed if they are properly mounted in a square or round hole in a card or cardboard. So out of a one metre square piece of filter material, aluminized Mylar or other, one can get 250 filters across or 250 squared, which equals 62 500 filters! With labour provided locally in Nigeria to mount the filter material, they were able to make and distribute 400 000 solar filters for eye safety at the partial phases of the eclipse. The Federal Ministry of Information and National Orientation of Nigeria provided an excellent poster, asserting, "Come Wednesday 29th March, 2006 Nigeria will witness an eclipse of the Sun. DO NOT BE AFRAID! Solar eclipse is a natural phenomenon which occurs when the moon comes between the Sun and the Earth. It is historical... It could be a once in a life time experience. But DO NOT WATCH with your NAKED EYES. PROTECT YOUR EYES with special filter glasses recommended by the National Space Research and Development Agency." Would that all countries would provide such information, though the distinction between partial and total phases should have been made. A newspaper article even headlined "Total eclipse of March 29, 2006 will be used to educate Nigerians."

5. Concluding remarks

The educational opportunities provided at the times of solar eclipses are not limited to the zones in which totality is eventually visible, though certainly the event is more dramatic when totality appears. Furthermore, in the path of totality, one must educate the public as to when they can observe without eye protection (or projection methods) and when it is safe to look at the Sun directly, namely, when only the corona is visible. At the 1973 total solar eclipse in Loiengalani in the northern frontier district of Kenya, the local priest notified people when it was safe to look by ringing the mission's bell.

The path from which a partial eclipse is visible is much more widespread, often covering many countries in the developing world. And about as many eclipses never pass the partial phase. Given that there are as many annular eclipses as total eclipses, each about every 18 months (Golub and Pasachoff, 1997), the addition of half as many partial eclipses means that at least twice a year there is an occasion to explain how to look at the Sun with safety. In recent years, we have also had the 2004 transit of Venus and the 1999, 2003, and 2006 transits of Mercury. In the years to come, we will not have another transit of Venus until 2012 or a transit of Mercury until 2016 (Pasachoff, 2006). Though totality occurs only about every 400 years on average at any given location on Earth, a partial phase of a solar eclipse or a planetary transit can be seen from any given spot almost yearly (Held, 2005; Pasachoff, 2003, 2006).

In sum, we have opportunities at least twice a year to use partial, annular, or total eclipses of the Sun, or planetary transits, to interest the public in astronomy through their observations. It is important to provide accurate information about the pleasures and hazards of looking toward the Sun, and our International Astronomical Union organizations try to help by providing knowledgeable information from experienced eclipse observers.

References

Chou, R. 1997, *Eye Safety During Total Eclipses,* adapted from NASA Reference Publication 1383, *Total Solar Eclipse of 1999 August 11,* p. 19.
    http://www.williams.edu/Astronomy/IAU_eclipses/eclipse_public_info.html
    http://sunearth.gsfc.nasa.gov/eclipse/SEhelp/safety2.html linked through
    http://www.eclipses.info

Chou, R. 1998, *Sky and Telescope,* February 1998, p. 36

Espenak, F. 1996, *Eye Safety During Total Eclipses,* adapted from NASA Reference Publication 1383, *Total Solar Eclipse of 1998 February 26* p. 17
    http://sunearth.gsfc.nasa.gov/eclipse/SEhelp/safety.html linked through
    http://www.eclipses.info

Golub, L., Pasachoff, J.M. 1997, *The Solar Corona,* Cambridge University Press, Cambridge, UK.

Held, W. 2005, *Eclipses 2005-2017* Floris Books, Edinburgh

Land, D. 2000, *OK, Look Directly at a Total Solar Eclipse,*
    http://www.williams.edu/Astronomy/IAU_eclipses/look_eclipse.html linked through
    http://www.eclipses.info

Pasachoff, J.M., Covington, M.A. 1993, *The Cambridge Eclipse Photography Guide,* Cambridge University Press, Cambridge UK.

Pasachoff, J.M. 1996a, In: Z. Mouradian and M. Stavinschi (Eds.) *Theoretical and Observational Problems Relating to Solar Eclipses,* NATO Advanced Research Workshop, Sinaia, Romania; 1997, Kluwer, 249-255.

Pasachoff, J.M. 1998, In: L. Gouguenheim, D. McNally, and J. R. Percy, (Eds.) *New Trends in Astronomy Teaching,* Cambridge University Press, , Cambridge UK, pp. 202-204.

Pasachoff, J.M. 2001, In: Alan H. Batten, (Ed.) *Astronomy for Developing Countries,* Astronomical Society of the Pacific, pp. 101-106; additional comments on pp. 139-140, 338-339.

Pasachoff, J.M. 2002, In: K. Bocchialini and S. Koutchmy (Eds.) *Observations et Travaux,* Vol. 53, p. 56, Institut d'Astrophysique, Paris.

Pasachoff, J.M. 2003, *The Complete Idiot's Guide to the Sun,* Alpha Books, Indianapolis.

Pasachoff, J.M. 2006, *A Field Guide to the Stars and Planets,* 4th ed, Houghton Mifflin, Boston.

Jay Pasachoff

Astronomy for the developing world
IAU Special Session no. 5, 2006
J.B. Hearnshaw and P. Martinez, eds.

© 2007 International Astronomical Union
doi:10.1017/S1743921307007132

The role of voluntary organizations in astronomy popularization: a case study of Khagol Mandal

Aniket Sule[1,5], S. Joshi[2], H. Joglekar[3], A. Deshpande[3,4], M. Naik[3] and S. Deshpande[3,5]

[1]Astrophysikalisches Institut Potsdam, Potsdam, Germany
[2]Jodrell Bank Observatory, University of Manchester, Macclesfield, UK
[3]Khagol Mandal, Mumbai, India
[4]Society for Applied Microwave Electronics Engineering and Research, Mumbai, India
[5]Homi Bhabha Center for Science Education, Mumbai, India
email: aniket.sule@gmail.com

Abstract. We present a case study of "Khagol Mandal," a voluntary organization primarily based in Mumbai, India. In the 20 years since its inception, Khagol Mandal has given more than 1000 public outreach programmes. The volunteers strive to go beyond amateur level by means of various study tours, astronomical experiments and workshops. These activities have inspired a number of students to take professional astronomy careers. With a volunteer force, probably largest in India or even south Asia, Khagol Mandal is well poised to take advantage of the archival data of large telescopes. With a little guidance from senior researchers, organizations like Khagol Mandal can provide a solution to the ever-increasing need for manpower for ancillary science from these large-scale facilities.

Keywords. Astronomy education, astronomy popularization

1. The need for voluntary organizations in astronomy popularization

In India the population density is such that it overwhelms the education and public outreach potential of most astronomical research institutions, who are hard pressed to cover all the schools and colleges in their area. Organizations of amateur astronomers can extend the outreach potential of astronomy research institutions in several ways by conducting various programmes such as exhibitions, slide shows and sky observations to popularize astronomy. At the same time, with their programmes, these organizations can attract young students and guide them to a research career in astronomy. We present the case study of an organization run by amateur astronomers by giving a detailed account of the formation and activities of the organization.

2. Khagol Mandal

Khagol Mandal (www.khagolmandal.com) is a voluntary organization, working in the field of astronomy, primarily based in Mumbai. In the regional language, Marathi, 'Khagol' refers to the celestial sphere, while 'Mandal' refers to a group of people involved in an activity. Khagol Mandal was started by a group of astronomy enthusiasts from Mumbai in 1986, the year when the Halley's comet reappeared in the sky. Twenty years down the line, Khagol Mandal operates from various units in and around Mumbai, with over 1000 members. The main unit is based in Mumbai while the other 4 units are at Badlapur, Dombivali, Thane and Nashik. These units have their own libraries dedicated

Figure 1. Response from all age groups for an exhibition by Khagol Mandal. Credit: Sameer
Kadam

to different aspects of astronomy, arrange regular lectures on various astronomical topics
for their members and train them for various public outreach activities.

3. Activities of Khagol Mandal

Khagol Mandal organizes various programmes for the general public on a regular basis,
such as sky observation programmes, exhibitions, and astronomy-related talks in schools,
colleges, clubs and community centres.

3.1. *Weekly meetings and library*

At each unit the members gather on a particular day of the week. On that day, lectures,
discussions and question-and-answer sessions are conducted. The library opens on that
day and the members can read and/or borrow the books. There are about 650 books
available at the library of the main unit, while the libraries of the other units contain
about 600 books in total.

3.2. *Sky observation programmes*

Khagol Mandal organizes regular overnight sky observation programmes at Vangani, a
sleepy village on the outskirts of Mumbai, the city where the main unit is based. Avoiding the cloud-covered monsoon months, the programmes take place once a month from
October to May. Programmes are conducted on Saturday nights which are close to the
new moon. These overnight programmes are open to the general public. The programme
consists of an introduction to telescopes, the use of constellations to find directions, identification of constellations visible with their main points high-lighted and observations
of solar system and deep sky objects through telescopes. The programme also involves

Figure 2. Model of the Sun-Earth-Moon system that demonstrates phenomena like the seasons, eclipses, precession, etc. Credit: Abhir Joshi

a slide show (see below) on one of the various astronomy topics. A question-and-answer session of a highly technical nature is always conducted during the programme where the experts from Khagol Mandal give scientific answers to the questions asked by the audience. Sometimes parallel programmes are conducted in the regional language Marathi and in English. Apart from the overnight sky observation programmes, short duration programmes are conducted as per the requirements of schools, colleges, clubs or people of residential complexes on their premises.

3.3. *Exhibitions*

Khagol Mandal has a collection of about 100 A0 and A3 sized charts on topics such as the solar system, Messier objects, and the Lonar meteor crater, which are the made by the members of Khagol Mandal. Khagol Mandal puts up these charts either in the independent programmes arranged by Khagol Mandal itself or in the programmes arranged by other science popularization bodies (see Figure 1). The exhibitions are also put up in education institutions at their request. Though the charts are self-explanatory the volunteers of Khagol Mandal are present to share the information with the viewers.

3.4. *Slide-shows*

Khagol Mandal has a collection of about 650 slides which cover a wide range of topics such as stellar evolution, the solar system, Saturn, Jupiter, extra-terrestrial life, eclipses, transits, occultations and comets. The slide shows are conducted independently or as part of other programmes of Khagol Mandal. The members conduct slide shows at various public places and at educational institutions.

Table 1. List of conferences and workshops in which volunteers of Khagol Mandal have participated, the activities they performed and the topics they covered

Year	Conference/Workshop	Activities/Topics
1987	Astronomy conference, Thane, India	General astronomy
1993	Conference on Bhaskaracharya's 800th anniversary, Mumbai, India	Mercury transit
1996	Astronomy conference, Aurangabad, India	Exhibition, sky observation, slide shows, Lonar Crater discussion
1998	Astronomy conference, Nashik, India	Exhibition, sky observation, slide shows
2002	Lonar Crater Conservation Conference	Physics of Lonar Crater, Introduction to Lonar
2003	International Conference for Science Communicators	Roles of amateurs, astro-photography and planetary transits
2005	Astronomy conference, Pune India	Exhibition and sky observation
2005	Konkan Marathi Sahitya Parishad (conference for Marathi literature)	Exhibition
2005	Astronomy conference, Goa, India	Computers in astronomy, Observational astronomy, Exhibition, sky observation, slide shows

3.5. *Basic and advance courses in astronomy*

Volunteers of Khagol Mandal conduct basic and advanced courses in astronomy almost every year. These courses are offered to the general public. The lectures are conducted by the experts of Khagol Mandal and by invited professional astronomers and physicists.

3.6. *Study tours, workshops, conferences and special event programmes*

On the occasion of rare astronomical events, Khagol Mandal organizes extensive public outreach campaigns as well as study tours, like the tours during the total solar eclipses (TSE) of 1995 and 1999, and public telescope observations at popular city locations during the Mercury and Venus transits. The Venus transit in 2002 was observed by over 23 000 people through the telescopes set up by Khagol Mandal at various locations. The annual public outreach of Khagol Mandal exceeds 15 000 people. The star party at Vangani on the day of the Leonid meteor storm in 1999 was attended by roughly 10 000 people, presumably a record in at least the south Asian region. From time to time volunteers of Khagol Mandal participate in various conferences (see Table 1).

3.7. *Publications*

Since its inception 20 years ago Khagol Mandal has been publishing a bulletin called *Khagol Varta* (which means 'Astronomy News'). This Marathi bulletin started as a quarterly magazine. It soon became bimonthly and from 2002 it became monthly. The bulletin covers the latest astronomy news and the activities of Khagol Mandal. Since 2004 Khagol Mandal has also published an English quarterly bulletin called *Vaishwik* (which means 'related to the Universe'). Table 2 lists the other publications of Khagol Mandal.

3.8. *Astronomy research*

Most of the active volunteers are students and young professionals with a basic training in scientific methodologies. A good number of members have also chosen research careers in astronomy and astrophysics for themselves. Khagol Mandal encourages its members

Table 2. List of publications of Khagol Mandal

	Year	Title
Since	1986	Khagol Warta (monthly bulletin published in Marathi)
	1986	Khagol Parichay, Khagol Prakashan, Mumbai, India, 1st edition
	1995	Totality (special bulletin on the occasion of TSE 1995)
	1999	Khagras Khagol Prakashan, Mumbai, India, 1st edition
	2001	Tarangan, Pradeep Nayak, Khagol Prakashan, Mumbai, 1st edition
	2002	Care Lonar, Kardile *et al.* Khagol Prakashan, Mumbai, 1st edition
	2002	Lonar Vivar, Khagol Prakashan, Mumbai, 1st edition
Since	2004	Vaishwik (quarterly bulletin published in English)
	2005	Khagol Parichay, Khagol Prakashan, Mumbai, India, 2nd edition
	2006	Astronomy for amateurs, Khagol Prakashan, Mumbai, India, 1st edition

to plan and participate in various projects related to astronomy. During TSE'95 and TSE'99, members participated in various experiments including spectrography of the solar corona. The meteor crater at Lonar, India is frequently visited by members to study rock and microbe samples from the area. Khagol Mandal also arranged the 'Lonar Crater Conservation Conference' in 2002 where topics like near-Earth objects, Lonar Crater: is it volcanic?, Physics of craters, Archeology of Lonar Crater, Flora and fauna of Lonar Crater, and Water of Lonar Crater were presented. Some members have been developing low-cost but efficient tools to demonstrate astronomical events, otherwise difficult to visualize (see Figure 2). With a large and semi-skilled volunteer force at its disposal, Khagol Mandal can surely undertake research projects that involve a huge amount of man-hours, if it works under the guidance of senior researchers.

4. Discussion

Khagol Mandal can share the burden of public outreach with the professional astronomical community thorough its extensive public programmes and by publishing bulletins (such as *Khagol Varta* and *Vaishwik*) and books on basic astronomy. It can provide a semi-skilled workforce of volunteers who want to get a flavour of research and are capable of simple data analysis tasks. It can attract new students to astronomy by offering them the opportunity to work on research projects as a part of their hobby. Examples of such projects include:

- Determination of the extent of the bulge and disk of Milky Way by multi-colour stellar photometry from HST archives;
- Data mining in huge sky surveys, e.g. obtaining photometric red shifts for each object in the Hubble Deep Fields;
- Extensive surveys of archives for objects like gravitational lenses and novae, which may have missed detection.

Such projects can be performed as pilot surveys for professional astronomers.

Section 8:

Promoting astronomy in developing countries through the UN, the IHY and COSPAR

Astronomy for the developing world
IAU Special Session no. 5, 2006
J.B. Hearnshaw and P. Martinez, eds.

© 2007 International Astronomical Union
doi:10.1017/S1743921307007156

The United Nations Basic Space Science Initiative: the TRIPOD concept

Masatoshi Kitamura[1], Don Wentzel[2], Arne Henden[3], Jeffrey Bennett[4], H.M.K. Al-Naimiy[5], A.M. Mathai[6], Nat Gopalswamy[7], Joseph Davila[8], Barbara Thompson[9], David Webb[10] and Hans Haubold[11] †

[1]National Astronomical Observatory, Mitaka, Tokyo 181-8588, Japan

[2]University of Maryland, College Park, MD 20742-2421, USA
email: d.wentzel@worldnet.att.ne

[3]American Assoc. of Variable Star Observers, 25 Birch Street, Cambridge, MA 02138, USA
email: arne@aavso.org

[4]3015 10th St., Boulder, CO 80304, USA
email: jeffrey.bennett@comcast.net

[5]College of Arts and Sciences, Sharjah University P.O. Box, 27272 Sharjah, UAE
email: alnaimiy2@yahoo.com

[6]Centre for Mathematical Sciences, Pala Campus, Arunapuram P.O. Box, Pala-686574, Kerala, India
email: mathai@math.mcgill.ca

[7]NASA Goddard Space Flight Center, Greenbelt, MD 20771, USA
email: gopals@ssedmail.gsfc.nasa.gov

[8]NASA Goddard Space Flight Center, Greenbelt, MD 20771, USA
email: joseph.m.davila@nasa.gov

[9]NASA Goddard Space Flight Center, Greenbelt, MD 20771, USA
email: barbara.j.thompson@nasa.gov

[10]AFRL/VSBXS and ISR, Boston College, 29 Randolph Road, Hanscom AFB, MA 01731-3010, USA
email: David.Webb.ctr@hanscom.af.mil

[11]United Nations Office for Outer Space Affairs, Vienna International Centre, A-1400 Vienna, Austria
email: hans.haubold@unvienna.org

Abstract. Since 1990, the United Nations has held an annual workshop on basic space science for the benefit of the worldwide development of astronomy. Additional to the scientific benefits of the workshops and the strengthening of international cooperation, the workshops lead to the establishment of astronomical telescope facilities through the Official Development Assistance (ODA) of Japan. Teaching material, hands-on astrophysics material, and variable star observing programmes had been developed for the operation of such astronomical telescope facilities in the university environment. This approach to astronomical telescope facility, observing programme, and teaching astronomy has become known as the basic space science TRIPOD concept. Currently, a similar TRIPOD concept is being developed for the International Heliophysical Year 2007, consisting of an instrument array, data taking and analysis, and teaching space science.

Keywords. Teaching astronomy, International Heliophysical Year 2007

† HJH presented this paper as an invited speaker on behalf of the United Nations Office for Outer Space Affairs

1. Introduction

Research and education in astronomy and astrophysics are an international enterprise and the astronomical community has long shown leadership in creating international collaborations and cooperation: Because (i) astronomy has deep roots in virtually every human culture, (ii) it helps to understand humanity's place in the vast scale of the universe, and (iii) it teaches humanity about its origins and evolution. Humanity's activity in the exploration of the universe is reflected in the history of scientific institutions, enterprises, and sensibilities. The institutions that sustain science; the moral, religious, cultural, and philosophical sensibilities of scientists themselves; and the goal of the scientific enterprise in different regions on Earth are subjects of intense study (Pyenson and Sheets-Pyenson 1999).

The Decadal Reports for the last decade of the 20th century (Bahcall, 1991) and the first decade of the 21st century (McKee and Taylor, 2001) have been prepared primarily for the North American astronomical community, however, it may have gone unnoticed that these reports also had an impact on a broader international scale, as the reports can be used, to some extent, as a guide to introduce basic space science, including astronomy and astrophysics, in nations where this field of science is still in its infancy. Attention is drawn to the world-wide-web sites at `http://www.seas.columbia.edu/~ah297/un-esa/` and `http://www.unoosa.org/oosa/en/SAP/bss/index.html`, where the TRIPOD concept is publicized on how developing nations are making efforts to introduce basic space science into research and education curricula at the university level. The concept, focusing on astronomical telescope facilities in developing nations, was born in 1990 as a collaborative effort of developing nations, the European Space Agency (ESA), the United Nations (UN), and the Government of Japan. Through annual workshops and subsequent follow-up projects, particularly the establishment of astronomical telescope facilities, this concept is gradually bearing fruitful results in the regions of Asia and the Pacific, Latin America and the Caribbean, Africa, and Western Asia (Wamsteker *et al.* 2004).

2. United Nations Office for Outer Space Affairs (UNOOSA)

In 1959, the United Nations recognized a new potential for international cooperation and formed a permanent body by establishing the Committee on the Peaceful Uses of Outer Space (COPUOS). In 1970, COPUOS formalized the UN Programme on Space Applications to strengthen cooperation in space science and technology between developing and industrialized nations.

The overall purpose of the programme "Peaceful Uses of Outer Space" is the promotion of international cooperation in the peaceful uses of outer space for economic, social and scientific development, in particular for the benefit of developing nations. The programme aims at strengthening the international legal regime governing outer space activities to improve conditions for expanding international cooperation in the peaceful uses of outer space. The implementation of the programme will strengthen efforts at the national, regional and global levels, including among entities of the United Nations system, to increase the benefits of the use of space science and technology for sustainable development.

Within the secretariat of the United Nations, the programme is implemented by the Office for Outer Space Affairs. At the inter-governmental level, the programme is implemented by the Committee on the Peaceful Uses of Outer Space, which addresses scientific and technical as well as legal and policy issues related to the peaceful uses of outer space.

The Committee was established by the General Assembly in 1959 and has two subsidiary bodies, the Legal Subcommittee and the Scientific and Technical Subcommittee. The direction of the programme is provided in the annual resolutions of the General Assembly and decisions of the Committee and its two Subcommittees.

As part of its programme of work, the Office provides secretariat services to the Committee and its subsidiary bodies and implements the United Nations Programme on Space Applications. The activities of the Programme on Space Applications are primarily designed to build the capacity of developing nations to use space applications to support their economic and social development.

In its resolution 54/68 of 6 December 1999, the United Nations General Assembly endorsed the resolution entitled "The Space Millennium: Vienna Declaration on Space and Human Development", which had been adopted by the Third United Nations Conference on the Exploration and Peaceful Uses of Outer Space (UNISPACE III), held in July 1999. Since then, the focus of the work undertaken by the Office under this programme has been to assist the Committee in the implementation of the recommendations of UNISPACE III.

In October 2004, the United Nations General Assembly reviewed the progress made in the implementation of the recommendations of UNISPACE III and, in its resolution 59/2, endorsed the Committee's Plan of Action for further implementation. The Plan of Action, contained in the report of the Committee to the Assembly for its review (A/59/174), constitutes a long-term strategy for enhancing mechanisms to develop or strengthen the use of space science and technology to support the global agendas for sustainable development. The report also provides a road map to make space tools more widely available by moving from the demonstration of the usefulness of space technology to an operational use of space-based services.

In its report, the Committee noted that in implementing the Plan of Action, the Committee could provide a bridge between users and potential providers of space-based applications and services by identifying needs of Member States and coordinating international cooperation to facilitate access to the scientific and technical systems that might meet those needs. To maximize the effectiveness of its resources, the Committee adopted a flexible mechanism, action teams, that takes advantage of partnerships among its secretariat, Governments, and inter-governmental and international non-governmental organizations to further implement the recommendations of UNISPACE III.

At its forty-ninth session, held in June 2006, the Committee had before it for its consideration the proposed Strategic Framework for the Office for Outer Space Affairs for the period 2008–2009, as contained in document (A/61/6 (Prog. 5)). The Committee agreed on the proposed strategic framework.

The expected accomplishments and the strategy reflected in the strategic framework proposed by the Office for Outer Space Affairs for the period 2008-2009 (A/61/6) are aimed at achieving increased international cooperation among Member States and international entities in the conduct of space activities for peaceful purposes and the use of space science and technology and their applications towards achieving internationally agreed sustainable development goals.

In brief, the three expected accomplishments of the Office are: (a) greater understanding, acceptance, and implementation by the international community of the legal regime established by the United Nations to govern outer space activities; (b) strengthened capacities of countries in using space science and technology and their applications in areas related, in particular, to sustainable development, and mechanisms to coordinate their space-related policy matters and space activities; and (c) increased coherence and synergy in the space-related work of entities of the United Nations system and international

space-related entities in using space science and technology and their applications as tools
to advance human development and increase overall capacity development. The establishment and operation of Regional Centres for Space Science and Technology, affiliated
to the United Nations
(http://www.unoosa.org/oosa/en/SAP/centres/index.html),
as well as workshops on basic space science
(http://www.unoosa.org/oosa/en/SAP/bss/index.html)
and the International Heliophysical Year 2007
(http://www.unoosa.org/oosa/en/SAP/bss/ihy2007/index.html)
are part of the accomplishments of the Office.

3. Official Development Assistance (ODA) of the Government of Japan

In conjunction with the workshops, and to support research and education in astronomy, the Government of Japan has donated high-grade equipment to a number of developing nations (Singapore 1987, Indonesia 1988, Thailand 1989, Sri Lanka 1995, Paraguay
1999, the Philippines 2000, Chile 2001) within the scheme of ODA of the Government
of Japan (Kitamura 1999). Here, reference is made to 45-cm high-grade astronomical
telescopes furnished with a photoelectric photometer or CCD, a spectrograph and computer equipment. After the installation of the telescope facility by the host country and
Japan, in order to operate such telescopes, young observatory staff members from the
host country have been invited by the Bisei Astronomical Observatory for education and
training, sponsored by the Japan International Cooperation Agency [JICA] (Kitamura
1999, Kogure 1999, Kitamura 2004, UN document A/AC.105/829). Similar telescope facilities, provided by the Government, were inaugurated in Honduras (1997) and Jordan
(1999).

The research and education programmes at the newly established telescope facilities
focus on time-varying phenomena of celestial objects. The 45-cm class reflecting telescope
with photoelectric photometer is able to detect celestial objects up to the 12th magnitude
and with a CCD attached up to the 15th magnitude, respectively. Such results have been
demonstrated for the light variation of the eclipsing close binary star V505 Sgr, the X-
ray binary Cyg X-1, the eclipsing part of the long-period binary ϵ Aur, the asteroid 45
Eugenia, and the eclipsing variable RT CMa (Kitamura 1999).

Also in 1990, the Government of Japan through ODA, facilitated the provision of
planetariums to developing nations (Kitamura 2004; Smith and Haubold 1992).

4. Observing with the telescopes: research

In the course of preparing the establishment of the above astronomical telescope facilities, the workshops made intense efforts to identify available material to be used in
research and education at such facilities. It was discovered that variable star observing by
photoelectric or CCD photometry can be a prelude to more advanced astronomical activity. Variable stars are those whose brightness, colour, or some other property varies with
time. If measured sufficiently carefully, almost every star turns out to be variable. The
variation may be due to geometry, such as the eclipse of one star by a companion star, or
the rotation of a spotted star, or it may be due to physical processes such as pulsation,
eruption, or explosion. Variable stars provide astronomers with essential information
about the internal structure and evolution of the stars. The most preeminent institution
in this specific field of astronomy is the American Association of Variable Star Observers.

The AAVSO co-ordinates variable star observations made by amateur and professional astronomers, compiles, processes, and publishes them, and in turn, makes them available to researchers and educators.

To facilitate the operation of variable star observing programmes and to prepare a common ground for such programmes, the AAVSO developed a rather unique package titled "Hands-On Astrophysics" which includes 45 star charts, 31 35-mm slides of five constellations, 14 prints of the Cygnus star field at seven different times, 600,000 measurements of several dozen stars, user-friendly computer programmes to analyze them, and to enter new observations into the database, an instructional video in three segments, and a very comprehensive manual for teachers and students (http://www.aavso.org/). Assuming that the telescope is properly operational, variable stars can be observed, measurements can be analyzed and sent electronically to the AAVSO.

The flexibility of the "Hands-On Astrophysics" material allows an immediate link to the teaching of astronomy or astrophysics at the university level by using the astronomy, mathematics, and computer elements of this package. It can be used as a basis to involve both the professor and the student to do real science with real observational data. After a careful exploration of "Hands-On Astrophysics" and thanks to the generous cooperation of the AAVSO, it was adopted by the above astronomical telescope facilities for their observing programmes (Mattei and Percy 1999, Percy 1991, Wamsteker *et al.* 2004).

The AAVSO is currently undertaking a massive effort to translate its basic Visual Observing Manual into Spanish and Russian and to eventually make this basic material available in the native language of any developing nation. The AAVSO is actively pursuing translations in Arabic and Chinese so as to have versions available in all the official United Nations languages.

5. Teaching astrophysics: education

Various strategies for introducing the spirit of scientific inquiry to universities, including those in developing nations, have been developed and analyzed (Wentzel 1999a). The workshops on basic space science were created to foster scientific inquiry. Organized and hosted by Governments and scientific communities, they serve the need to introduce or further develop basic space science at the university level, as well as to establish adequate facilities for pursuing a scientific field in practical terms. Such astronomical facilities are operated for the benefit of the university or research establishment, and will also make the results from these facilities available for public educational efforts. Additional to the hosting of the workshops, the Governments agreed to operate such a telescope facility in a sustained manner with the call on the international community for support and cooperation in devising respective research and educational programmes.

Organizers of the workshops have acknowledged in the past the desire of the local scientific communities to use educational material adopted and available at the local level (prepared in the local language). However, the workshops have also recommended to explore the possibility to develop educational material (additional to the above mentioned "Hands-On Astrophysics" package) which might be used by as many as possible university staff in different nations while preserving the specific cultural environment in which astronomy is being taught and the telescope is being used. A first promising step in this direction was made with the project "Astrophysics for University Physics Courses" (Wentzel 1999b, Wamsteker *et al.* 2004). This project has been highlighted at the IAU/COSPAR/UN Special Workshop on Education in Astronomy and Basic Space Science, held during the UNISPACE III Conference at the United Nations Office Vienna in 1999 (Isobe 1999). Additionally, a number of text books and CD-ROMs have been reviewed over the years which, in the view of astronomers from developing nations, are particularly useful in the research and teaching process: Bennett *et al.* (2007), for teaching purposes and the Bennett (2001) and Lang (1999, 2004) reference books for research purposes.

As part of the 15th anniversary celebrations of the Hubble Space Telescope, the European Space Agency has produced an exclusive, 83-minute DVD film, titled "Hubble – 15 Years of Discovery". The documentary also mentions the role of the Hubble Space Telescope project in facilitating some of the activities of the United nations Office for Outer Space Affairs, particularly processing of Hubble imagery as part of the education and research activities of the UN-affiliated Regional Centres for Space Science and Technology and the workshops on basic space science. The Hubble DVD was distributed world-wide, through the Office, as a unique educational tool for astronomy and astrophysics.

6. International Heliophysical Year 2007: a world-wide outreach programme

In 1957 a programme of international research, inspired by the International Polar Years of 1882-83 and 1932-33, was organized as the International Geophysical Year (IGY) to study global phenomena of the Earth and geospace. The IGY involved about 66,000 scientists from 60 nations, working at thousands of stations, from pole to pole to obtain simultaneous, global observations on Earth and in space. The fiftieth anniversary of IGY will occur in 2007. It was proposed to organize an international programme of scientific collaboration for this time period called the International Heliophysical Years (IHY) in 2007 (http://ihy2007.org/). Like IGY, and the two previous International Polar Years, the scientific objective of IHY is to study phenomena on the largest possible scale with simultaneous observations from a broad array of instruments. Unlike previous international years, today observations are routinely received from a vast armada of sophisticated instruments in space that continuously monitor solar activity, the interplanetary medium, and the Earth. These spacecraft, together with ground level observations and atmospheric probes, provide an extraordinary view of the Sun, the heliosphere, and their influence on the near-Earth environment. The IHY is a unique opportunity to study the coupled Sun-Earth system. Future basic space science workshops will focus on the preparation of IHY 2007 world-wide, particularly taking into account interests and contributions from developing nations.

Currently, in accordance with the United Nations General Assembly resolution 60/99, the Scientific and Technical Subcommittee of the UNCOPUOS is considering an agenda item on the IHY 2007 under the three-year work plan adopted at the forty-second session of the Subcommittee
(http://www.unoosa.org/oosa/en/SAP/bss/ihy2007/index.html).

A major thrust of the IHY 2007 is to deploy arrays of small, inexpensive instruments such as magnetometers, radio antennas, GPS receivers, all-sky cameras, etc. around the world to provide global measurements of ionospheric, magnetospheric, and heliospheric phenomena. This programme is implemented by collaboration between the IHY 2007 Secretariat and the United Nations Office for Outer Space Affairs. The small instrument programme consists of a partnership between instrument providers and instrument host nations. The lead scientist or engineer provides the instrumentation (or fabrication plans for instruments) in the array; the host nation provides manpower, facilities, and operational support to obtain data with the instrument, typically at a local university. In preparation of IHY 2007, this programme has been active in deploying instrumentation, developing plans for new instrumentation, and identifying educational opportunities for the host nation in association with this programme (http://ihy2007.org/observatory/observatory.shtml; UN document A/AC.105/856). Currently, a TRIPOD concept is being developed for the International Heliophysical Year 2007, consisting of an instrument array, data taking and analysis, and teaching space science.

7. Remark

In 2006, 27 November - 1 December, the Indian Institute of Astrophysics hosted the second UN/NASA Workshop on the International Heliophysical Year and Basic Space Science in Bangalore, India (http://www.iiap.res.in/ihy/). In 2007, 18-22 June, the National Astronomical Observatory of Japan, Tokyo, will host a workshop on basic space science and the International Heliophysical Year 2007 (http://solarwww.mtk.nao.ac.jp/UNBSS_Tokyo07/), co-organized by the United Nations, European Space Agency, and the National Aeronautics and Space Administration of the United States of America, and will use this opportunity to commemorate the cooperation between the Government of Japan and the United Nations, as highlighted in this article, since 1990.

References

Bahcall, J. 1991, *The Decade of Discovery in Astronomy and Astrophysics,* National Academy Press, Washington D.C.

Bennett, J. 2001, *On the Cosmic Horizon: Ten Great Mysteries for the Third Millennium Astronomy,* Addison Wesley Longman

Bennett, J., Donahue, M., Schneider, N., and Voit, M. 2007, *The Cosmic Perspective,* 4th ed. Addison Wesley Longman Inc., Menlo Park, California. CD-ROMs and a www site offering a wealth of additional material for professors and students, specifically developed for teaching astronomy with this book and upgraded on a regular basis are also available: http://www.astrospot.com/.

Haubold, H.J. 1998, *Space Policy,* 19, 215

Haubold, H.J. 2003, *Journal of Astronomical History and Heritage* 1(2):105-121

Isobe, S. 1999 *Teaching of Astronomy in Asian-Pacific Region,* Bulletin No. 15

McKee, C.F. and Taylor, Jr., J.H. 2001, *Astronomy and Astrophysics in the New Millennium,* National Academy Press, Washington D.C.

Kitamura, M. 1999, "Provision of astronomical instruments to developing countries by Japanese ODA with emphasis on research observations by donated 45-cm reflectors in Asia," in: *Proceedings Conference on Space Sciences and Technology Applications for National Development,* held at Colombo, Sri Lanka, 21-22 January 1999, Ministry of Science and Technology of Sri Lanka, pp. 147-152.

Kitamura, M. 2004. *Space Policy,* 20, 131

Kogure, T. 1999, "Stellar activity and needs for multi-site observations," in: *Proceedings Conference on Space Sciences and Technology Applications for National Development,* held at Colombo, Sri Lanka, 21-22 January 1999, Ministry of Science and Technology of Sri Lanka, pp. 124-131.

Lang, K.R. 1999, *Astrophysical Formulae, Volume I: Radiation, Gas Processes and High Energy Astrophysics, Volume II: Space, Time, Matter and Cosmology,* Springer-Verlag, Berlin

Lang, K.R. 2004, *Space Policy,* 20, 297

Mattei, J. and Percy, J.R. 1998, Eds: *Hands-On Astrophysics,* American Association of Variable Star Observers, Cambridge, MA

Percy, J.R. 1991, Ed: *Astronomy Education: Current Developments, Future Cooperation,* Astronomical Society of the Pacific Conference Series Vol. 89

Pyenson, L. and Sheets-Pyenson, S. 1999, *Servants of Nature: A History of Scientific Institutions, Enterprises, and Sensibilities,* W.W. Norton & Company, New York

Smith, D.W. and Haubold, H.J. 1992, (Eds.): *Planetarium: A Challenge for Educators,* United Nations, New York; available in Japanese, English, Spanish, and Slovak languages.

284   M. Kitamura *et al.*

UN document A/AC.105/829: Report on the Twelfth United Nations/European Space Agency Workshop on Basic Space Science, Beijing, China, 24-28 May 2004, United Nations, Vienna 2004.

UN document A/AC.105/856: Report on the United Nations/European Space Agency/National Aeronautics and Space Administration of the United States of America Workshop on the International Heliophysical Year 2007, Abu Dhabi and Al-Ain, United Arab Emirates, 20-23 November 2005, United Nations, Vienna 2005.

Wamsteker, W., Albrecht, R., and Haubold, H.J. 2004, (Eds.): *Developing Basic Space Science World-Wide: A Decade of UN/ESA Workshops*, Kluwer Academic Publishers, Dordrecht

Wentzel, D.G. 1999a, *Teaching of Astronomy in Asian-Pacific Region*, Bulletin No. 15, 4

Wentzel, D.G. 1999b, *Astrofisica para Cursos Universitarios de Fisica*, La Paz, Bolivia. English language version available from the United Nations in print and electronically at `http://www.seas.columbia.edu/~ah297/un-esa/astrophysics`; printed version also contained in Wamsteker at al. 2004.

Hans Haubold

Astronomy for the developing world
IAU Special Session no. 5, 2006
J.B. Hearnshaw and P. Martinez, eds.

© 2007 International Astronomical Union
doi:10.1017/S1743921307007168

The COSPAR Capacity-Building Workshop Programme, 2000–2007

Keith A. Arnaud[1,2] and A. Peter Willmore[3]†

[1]Laboratory for X-ray Astrophysics, NASA GSFC, Greenbelt, MD 20771, USA

[2]Department of Astronomy, University of Maryland, College Park, MD 20742 USA
email: kaa@milkyway.gsfc.nasa.gov

[3]School of Physics and Astronomy, University of Birmingham, Birmingham, B15 2TT, UK
email: apw@star.sr.bham.ac.uk

Abstract. The Committee on Space Research (COSPAR) Capacity-Building Programme introduces astronomers in the developing world to the rich resource of space research online archives. The programme consists of a series of regional workshops which each bring together about 30 developing world astronomers with around 8 teaching faculty for lectures and hands-on projects. Five workshops have been held so far with another two planned for next year and a budget which enables on average one workshop each year. Proposals for future workshops are encouraged.

Keywords. methods: data analysis

---

1. Introduction

Space missions being rare, expensive, and generally short-lived have long had a policy of archiving all their observational data to ensure that these resources are not lost. In the past these data were only available to researchers at the institutions running the space missions. However the rise of the Internet and World Wide Web has changed this. Extensive online archives and the software to use them are now available for all major space research missions, with smaller missions being served by archival research centers. In addition many missions are open world-wide to observing proposals and archival research can provide a stepping stone towards collaborations or even acquiring one's own dedicated observations.

Most users of these archives are from the US, Europe, and Japan with relatively few from the developing world. There are a number of likely reasons for this: the poor quality of internet connections in much of the developing world; a lack of awareness that such archives are freely available, along with a belief that space data are for people who build space missions; research programmes in developing countries are built around affordable ground-based facilities; cultural factors such as conservatism among graduate supervisors.

To help more astronomers in the developing world make use of space research archives COSPAR has organized a series of practical workshops centred on current, active space missions. The aims are to ensure that all participants leave with the ability to set up and use software and data at their own institutions without further support. The workshops also serve to build links between participants and lecturers, as well as among the participants.

† Paper presented by K.A.; A.P.W was the original invited speaker

Table 1. COSPAR workshops and local projects in *italics*.

2001	X-ray astronomy	Chandra, XMM-Newton	INPE, Brazil
2003	X-ray astronomy	Chandra, XMM-Newton, *Astrosat*	Udaipur, India
2004	Magnetospheric physics	Cluster, *Double-Star*	Beijing, China
2004	X-ray astronomy	Chandra, XMM-Newton, *SALT*	Durban, South Africa
2005	Space Oceanography	GOOS	CRTS, Rabat, Morocco

2. The history of the programme

The COSPAR programme of capacity-building workshops was established in 1999. The first two workshops concentrated on X-ray astronomy because this was the field of the instigator of the programme (APW) and with two major observatories, Chandra from NASA and XMM-Newton from ESA, the field is particularly dynamic at present. Following the initial success a panel to manage the programme was set up in 2004 with an annual budget large enough to fund approximately one workshop per year. Table 1 lists the workshops held. The India, China, and South Africa workshops also included discussion of a local project with applicability to the workshop theme. These projects are shown in italics in Table 1.

The workshops are held in a developing country university or space centre with good facilities and access to the internet. Participants are then drawn from the host country and its neighbours. For instance, the workshop in South Africa catered to participants from a range of sub-Saharan countries. There are usually 25-30 participants chosen by the organizing committee from applicants and about 8 lecturers, both local and foreign experts. The participants are generally senior postgraduate students, post-docs, or young faculty. All their expenses are covered by COSPAR and the co-sponsors.

COSPAR covers 50–60% of the cost, the host country 15–20% (usually US$10,000–15,000) and the rest from a range of partners, one of the most important being the IAU. Other past partners have been ICSU, UNESCO/IOC, UN/OOSA, URSI, Abdus Salam ICTP, US National Academy of Science, ESA, NASA, IFREMER, IRD, ISPRS, Medias France, UK Met Office, Univ. of Plymouth, MERCATOR OCEAN MERSEA Project.

The host country organizations provide considerable logistical support in addition to contributing to the funding and the workshops would not be possible without their enthusiastic help. For past workshops these organizations have been FAPESP Brazil, Indian Space Research Organization, Physical Research Lab. Ahmedabad, CSSTEAP India, CCSAR China, South African National Research Foundation, and CRTS Morocco.

There are currently two more workshops approved: magnetospheric physics in Sinaia, Romania in June, 2007 organized by Joachim Vogt, Thierry Dudok de Wit, and Octav Marghitu; and planetary science in Montevideo, Uruguay also in 2007 organized by Gonzalo Tancredi and Mike A'Hearn. If funds allow, there may be another applications workshop in 2007; otherwise this may be in 2008. There is also an attempt underway to organize a Chandra/XMM-Newton workshop at the Library of Alexandria in Egypt, however this will not be fully funded by COSPAR and depends on obtaining sufficient funds from other sources.

We encourage proposals for workshops. Guidelines are available from the COSPAR website (www.cosparhq.org) or by contacting one of the authors (APW).

3. Workshop structure

The available workshop time is split about equally between lectures and a data analysis project. The lectures aim to give the participants a basic understanding of the topic, including a description of the mission(s), the software, and the sorts of science that they make possible. More basic topics are sometimes necessary, for instance a series of lectures on statistics was included in the X-ray astronomy workshops.

The data analysis project is the key element of the workshop. Each participant has a dedicated PC for the duration of the workshop on which they work on data analysis under the direction of an advisor who is one of the lecturers. An attempt is made to match the project to the participant's research interests and to those of their home institution. Participants are also encouraged to form teams if they have similar interests.

The workshop closes with a poster session where each team presents a standard meeting poster on the work they have accomplished on their project. Some of these posters have been very impressive.

4. Success of the workshops

To judge the impact of the programme a questionnaire was e-mailed to the participants of the Brazil workshop two years after the event. Of the 50% who replied, all regarded the workshop as having been valuable for their careers, 86% had used Chandra or XMM-Newton data or both, 70% were still in contact with one or more of the lecturers, 70% had a publication either in press or in preparation which was based at least in part on Chandra or XMM-Newton data. All the respondents regarded the project as a key element in the success of the workshop.

There is other more anecdotal evidence for the success of the programme. Guest observer proposals for the Chandra satellite from Latin America doubled in number the year after the Brazil workshop. Following the India workshop, two institutes in Bangalore established a joint research programme in X-ray astronomy. Finally, approximately 10 graduates of the workshops attended the 2006 Beijing assembly of COSPAR and presented papers.

5. Conclusion: challenges and questions

There have been a number of challenges to be met during these workshops. The working language is English and there have been communication difficulties in a few cases. It certainly helps to have lecturers fluent in all the languages spoken by the participants. Another problem has been participant familiarity with Linux, which is usually required to run the data analysis software. Participants from countries with less well-developed scientific infrastructure are more likely to be familiar with Windows than Linux and this provides a steep learning curve before they can start doing anything useful. With the adoption of Linux systems by a number of developing country governments this problem may not be as serious in future.

Another challenge is how to follow up these workshops. It may be difficult for a participant to apply what they have learnt when they return to their home institution. The hope has been that the workshops lead to regional networking but there has been no formal structure put in place to ensure this happens. Finally, it is worth considering how this programme fits in with other developing world initiatives, some of which are described in other papers in this volume. Should COSPAR continue going it alone with these workshops or look to fit them into some larger structure?

Acknowledgements

This programme is now rolling and ≈ 135 mostly young scientists have taken part as participants. It depends crucially on: those who have funded it; the local organizers Joao Braga and Tania Sausen (Brazil), H.S.S. Sinha (India), Ji Wu (China), Arthur Hughes (South Africa) and Driss el Hadani (Morocco), and, in the cases of the Beijing and Rabat workshops, the main directors Joachim Vogt and Thierry Dudok de Wit (Beijing) and Jean-Louis Fellous and Raymond Zaharia (Rabat); the lecturers who give up two full weeks of their time for an activity that brings no direct returns in research. We thank them all.

Keith Arnaud

Astronomy for the developing world
IAU Special Session no. 5, 2006
J.B. Hearnshaw and P. Martinez, eds.

© 2007 International Astronomical Union
doi:10.1017/S174392130700717X

Globalizing space and Earth science–the International Heliophysical Year Education and Outreach Programme

M. Cristina Rabello-Soares[1], Cherilynn Morrow[2], Barbara Thompson[3] and David Webb[4]

[1]HEPL Solar Physics, Stanford University, 445 Via Palou, Stanford, CA 94305-4085, USA,
csoares@sun.stanford.edu

[2]Space Science Institute, 4750 Walnut Street, Suite 205, Boulder, Colorado 80301, USA,
morrow@SpaceScience.org

[3]Laboratory for Solar & Space Physics, NASA Goddard Space Flight Center, Solar Physics
Branch, Greenbelt, MD 20771, USA, Barbara.J.Thompson@nasa.gov

[4]ISR, Boston College, 140 Commonwealth Ave., Chestnut Hill, MA 02467 USA,
david.webb@hanscom.af.mil

Abstract. The International Heliophysical Year (IHY) in 2007 & 2008 will celebrate the 50th anniversary of the International Geophysical Year (IGY) and, following its tradition of international research collaboration, will focus on the cross-disciplinary studies of universal processes in the heliosphere.

The main goal of the IHY Education and Outreach Programme is to create more global access to exemplary resources in space and Earth science education and public outreach. By taking advantage of the IHY organization with representatives in every nation and in the partnership with the United Nations Basic Space Science Initiative (UNBSSI), we aim to promote new international partnerships. Our goal is to assist in increasing the visibility and accessibility of exemplary programmes and in the identification of formal or informal educational products that would be beneficial to improve space and Earth science knowledge in a given country; leaving a legacy of enhanced global access to resources and of world-wide connectivity between those engaged in education and public outreach efforts that are related to IHY science.

Here we describe how to participate in the IHY Education and Outreach Programme and the benefits in doing so. Emphasis will be given to the role played by developing countries; not only in selecting useful resources and helping in their translation and adaptation, but also in providing different approaches and techniques in teaching.

Keywords. education, Sun, solar-terrestrial relations, Earth, solar system, planets and satellites

1. Introduction

In 1957 a programme of international research, inspired by the International Polar Years of 1882 and 1932, was organized as the International Geophysical Year (IGY) to study global phenomena of the Earth and geospace in an unprecedented effort. The International Heliophysical Year (IHY) will celebrate the 50th anniversary of the International Geophysical Year (IGY) and, following its tradition of international research collaboration, will focus on the cross-disciplinary studies of universal processes in the heliosphere (Davila *et al.* 2004).

Nowadays, we routinely monitor the Sun, the interplanetary medium and the Earth's atmosphere. IHY represents a logical next-step from IGY, extending the studies into the heliosphere and thus including the drivers of geophysical change. The term

"heliophysical" is an extension of the term "geophysical", where the Earth, Sun & Solar System are studied as a whole domain.

2. IHY overview

International leaders have been identified throughout the world for this event and planning teams are already active in every region. Figure 1 illustrates the IHY organizational structure where the Steering Committee, Advisory Committee and Secretariat are charged with the guidance, execution and coordination of the international activities, respectively (see UNOOSA report 2006 for more information).

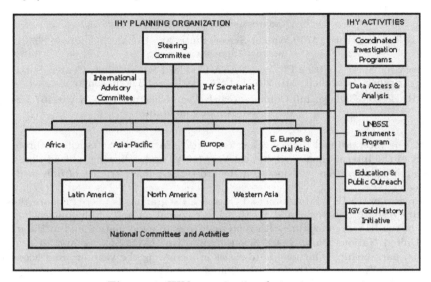

Figure 1. IHY organizational structure.

The IHY has three primary objectives:

• Advancing our understanding of the physical processes that govern the Sun, Earth and heliosphere;

• Continuing the tradition of international research and advancing the legacy on the 50th anniversary of the International Geophysical Year;

• Demonstrating the beauty, relevance and significance of space and Earth science to the world.

To achieve these goals, IHY is divided in four main components, called programmatic thrusts: Science activities, IHY/UNBSSI Observatory Development, Education and Public Outreach, and "IGY Gold" History Initiative.

IHY's scientific activities will be organized via Coordinated Investigation Programmes (CIPs) allowing cross-disciplinary studies and driving towards a more complete understanding of the universal processes that govern the Sun, Earth, planets and heliosphere (Harrison *et al.* 2005). It will involve scientists from a variety of disciplines: solar physics; planetary magnetospheres; heliosphere and cosmic rays; planetary ionospheres, thermospheres and mesospheres; climate studies; and heliobiology (i.e., study of the space environment's influence on and interaction with biological systems and processes).

The IHY/UNBSSI Observatory Development is in collaboration with the United Nations Basic Space Science Initiative (UNBSSI) of the UN Office for Outer Space Affairs (UNOOSA) and is dedicated to stimulating research on Earth and space science, specially in developing nations, through the deployment of small-instrument arrays around the world (see Gopalswamy *et al.* 2006).

The "IGY Gold" History Initiative aims to preserve the history and legacy of IGY by identifying and recognizing planners of and participants in the IGY, preserving and making available items of historical significance from the IGY, and organizing commemorative activities and events.

The Education and Public Outreach component, which is the subject of this paper, is dedicated to bringing heliophysics education to all the people of the world.

3. IHY Education and Public Outreach Programme

"Demonstrate the beauty, relevance and significance of space and Earth science to the world" - this IHY basic objective is the focus of the Education and Public Outreach (EPO) Programme. IHY presents unique opportunities for expanding the education and awareness of space and Earth science. As a result, the education programme forms a cornerstone of IHY.

The programme aims to inspire the next generation of space and Earth scientists and explorers; and spread the knowledge of our solar system and the exciting process of scientific exploration to the people of the world. Encouraging young people to pursue careers in science and technology helps secure and, in the case of developing nations, build up human resources, which will improve the technological base and enhance the prospect for further development.

As hundreds of observatories and institutions are preparing for the scientific activities of IHY, the associated educational programmes are being linked into a worldwide network of schools, institutes, programmes and activities. In addition, the IHY programme was developed in conjunction with the United Nations. As a result, the IHY EPO programme is well situated to promote new international partnerships, bringing heliophysics education to all the people of the world. Our goal is to assist in increasing the visibility and accessibility of exemplary programmes and in the identification of formal or informal educational products beneficial to a given country. The programme aims to provide greater international exposure for existing programmes but also and foremost to make them accessible to regions where there are few EPO resources available. Furthermore, we aspire to leave a legacy of enhanced global access to resources and world-wide connectivity between those engaged in education and public outreach efforts in space and Earth science.

The organizational structure of the IHY EPO programme closely mirrors the overall organization of the IHY programme: a central coordinator, an advisory committee and educators serving as national coordinators for EPO (Table 1). The central coordinator is part of the IHY secretariat and serves as a contact point and facilitator for the IHY team members and their EPO activities. The EPO national coordinator is appointed by the IHY national coordinator and acts as the liaison between those working on education and public outreach in his/her country and the IHY secretariat, as well as to the coordinators of the other nations. The national coordinator for EPO is responsible for promoting local activities and inviting exemplary EPO programmes in their country to be a participating programme in the IHY.

To foster the partnerships and facilitate the access to exemplary programmes our website (`http://ihy2007.org/outreach/outreach.shtml`) lists the participating programmes with a links to their resources, which are available to all. They are asked to fill out a form where they are divided into the following categories: formal education (curriculum development, professional development, schools, colleges, universities); informal

Table 1. IHY national coordinators for EPO.

Country	Coordinator	Institute
BRAZIL	Adriana V.R. da Silva	MacKenzie University, Sao Paulo
BULGARIA	Penka Stoeva	Bulgarian Academy of Sciences
CZECH REPUBLIC	Rostislav Halas	Realne Gymnasium, Prostejov
FINLAND	Heikki Nevanlinna	Finnish Meteorological Inst., Helsinki
GERMANY	Frank Jansen	University of Greifswald, Greifswald
JAPAN	Shinichi Watari	Nat. Inst. of Info. & Com. Tech.
JORDAN	Hanna Sabat	Al al-Bayt University, Mafraq
MALAYSIA	Azreena Ahmad	National Space Agency, Putrajaya
MEXICO	Guadalupe Cordero	Instituto de Geofisica, Coyoacan
NIGERIA	Ayodele Faiyetole	English Obafemi Awolowo Univ.,Ile-Ife
NORWAY	Arve Aksnes	University of Bergen, Bergen
PERU	Maria-Luisa A. Hurtado	Univ. Nac. de San Marcos, Lima
SLOVAKIA	Karel Kudela	Slovak Academy of Sciences Kosice
SOUTH AFRICA	Lee-Anne McKinnell	Rhodes University, Grahamstown
SPAIN	Javier Rodriguez-Pacheco	Universidad de Alcala, Madrid
SWEDEN	Henrik Lundstedt	Swedish Inst. of Space Phys., Lund
UNITED KINGDOM	Lucie Green	University College London
UNITED STATES	Cherilynn Morrow	Space Science Institute, Boulder, CO
YEMEN	Nada Alhaddad	University of Sana'a
International Coordinator	M.C. Rabello-Soares	Stanford University, California, USA

education (museums, science centers, planetariums); mass media (news, video, journalism); clubs and community groups (e. g. youth groups, scouts, amateur astronomy clubs); evaluation and research; scientist in education (e.g. scientist who gives public talks); and last but not least translation. A searchable database where the participating programme data can be updated at any time by its leader will be available soon.

A good example of a participating programme is the Space Weather Monitor Project which is also part of the IHY/UNBSSI Observatory Development (see Scherrer *et al.* 2006). It will bring hands-on science to pre-college students in developing countries. It consists of inexpensive monitors used to track solar- and lightning-induced changes to the Earth's ionosphere. The project also provides classroom support materials in the six official languages of the United Nations.

Determining the need for multi-lingual adaptations of educational resources and facilitating their translation is an important aspect to accomplish our goal of globalizing heliophysics outreach. Participating programmes able to assist with the translation and adaptation are of great importance. Institutions and individuals are encouraged to help in this effort. Scientists and educators of developing countries play an important role not only in the selection of useful resources, but also in helping with the translation and dissemination of the EPO resources that they believe will improve science literacy in their countries. The available multilingual resources are accessible to all at our website. At the moment, among the languages that we have support to translate from and to English are: Arabic, Armenian, Azerbaijani, Chinese, Czech, French, Gaelic, German, Greek, Japanese, Polish, Portuguese, Romanian, Russian and Spanish. We hope to expand this further.

As IHY establishes a greater presence for research in space and Earth science in developing countries through the IHY/UNBSSI Observatory Development, it will provide opportunities for undergrad and grad students to participate actively in an international cutting-edge research project. In this way, IHY will foster the development of graduate and undergraduate programmes which, in turn, will encourage young people to become interested in the exciting field of heliophysics. Stronger research programmes mean stronger universities. To reinforce education at college level, the IHY has created the IHY Schools Programme to develop a series of schools in 2007 with the purpose of educating students in heliophysics.

3.1. *IHY Schools Programme*

The IHY Schools Programme is organized and operated by the IHY Schools Committee (ISC). The ISC consists of the following members: David Webb (Coordinator), Ilia Roussev, Nat Gopalswamy, M. Cristina Rabello-Soares (EPO Coordinator), Don Hassler, Cristina Mandrini, and Barbara Thompson. The ISC will initially invite and support students that are associated with two IHY thrusts: the Coordinated Investigation Programmes (CIPs) and the IHY/UNBSSI Observatory Development Programme.

At least four IHY schools are being planned in 2007. One will be in Boulder (Colorado, USA) to serve North America and is led by Don Hassler and David Webb. The second will be held at the ICTP in Trieste, Italy for the European/African region and is led by Nat Gopalswamy. The third will be in Latin America and is led by Cristina Mandrini. The last will be in the Asia-Pacific region and the leader will be determined soon. Additional schools under the IHY umbrella are being considered in individual countries.

The overall scope of the schools will be heliophysics, including its universal processes, Sun-Earth interactions as well as those at other planets, and the outer heliosphere. The schools will include seminars and hands-on data analysis sessions with databases acquired particularly through the CIPs and IHY/UNBSSI programmes, and collaborative efforts with other affiliated groups such as CAWSES, the European COST 724, NASA, NSF, AFOSR and the IAU. Each school will be encouraged to select local lecturers and students from their regions so as to reduce costs. The help of the appropriate national and regional IHY coordinators will be sought.

The ISC is developing a general curriculum that will be used as a model for all four schools. The curriculum will include lectures in the following universal physical processes in heliospace: 1) evolution and generation of magnetic structures and transients, 2) energy transfer and coupling processes, 3) flows and circulations, 4) boundaries and interfaces, and 5) synoptic studies of the 3-D coupled solar-planetary-heliospheric system.

4. Conclusions

The International Heliophysical Year in 2007–2008 presents a unique opportunity for expanding education on space and Earth science and to demonstrate the beauty, relevance and significance of space and Earth science to the world.

The developing countries have an important role in IHY not only in selecting useful resources, helping in the translation/adaptation and disseminating the resources to the teachers and general public, but also in providing different and innovative approaches and techniques in teaching.

Acknowledgement

MCRS is supported by NASA grant NNX06AD10G.

References

Davila, J.M., Poland, A.I. & Harrison, R.A. 2004, *Adv. Sp. Res.* **34**, 2453

Gopalswamy, N., Davila, J.M., Thompson, B. & Haubold, H.J. 2006, in: J. Hearnshaw & P. Martinez (eds.), *these proceedings*

Harrison, R.A, Breen, A., Bromage, B. & Davila, J.M. 2005, *Astronomy & Geophysics* 46, June 2005

Scherrer, D., Rabello-Soares, M.C. & Morrow, C. 2006, in: J. Hearnshaw & P. Martinez (eds.), *these proceedings*

United Nations Office for Outer Space Affairs (UNOOSA) 2006, *Putting the "I" in the IHY*: Comprehensive overview on the worldwide organization of the International Heliophysical Year 2007 (United Nations, Austria), V.05-89935, February 2006

Cristina Rabello-Soares

Astronomy for the developing world
IAU Special Session no. 5, 2006
J.B. Hearnshaw and P. Martinez, eds.

© 2007 International Astronomical Union
doi:10.1017/S1743921307007181

The United Nations Basic Space Science Initiative for IHY 2007

Nat Gopalswamy[1], Joseph Davila[1], Barbara Thompson[1] and Hans Haubold[2]

[1]NASA Goddard Space Flight Center, Greenbelt, MD 20878, USA
email: gopals@ssedmail.gsfc.nasa.gov

[2]United Nations Office of Outer Space Affairs, Vienna, Austria

Abstract. The United Nations Office for Outer Space Affairs and the International Heliophysical Year (IHY) community have joined hands to deploy arrays of small, inexpensive instruments such as magnetometers, radio telescopes, GPS receivers, all-sky cameras, and particle detectors around the world to provide global measurements of ionospheric, magnetospheric and heliospheric phenomena. The small instrument programme is envisioned as a partnership between instrument providers, and instrument hosts in developing countries as one of United Nations Basic Space Science (UNBSS) activity. The lead scientist will provide the instruments (or fabrication plans for instruments) in the array; the host country will provide manpower, facilities, and operational support to obtain data with the instrument, located typically at a local university. This paper provides an overview of the IHY/UNBSS programme, its achievements and future plans.

Keywords. Sun: atmosphere, Sun: coronal mass ejections (CMEs), solar-terrestrial relations, interplanetary medium, solar wind, solar system: general, acceleration of particles, plasmas, shock waves, magnetic field

1. Introduction

The International Heliophysical Year (IHY) will commence in 2007, marking the fiftieth anniversary of the International Geophysical Year (IGY, 1957–58). The IGY resulted in an unprecedented level of understanding of geospace and saw the beginning of the Space Age. Like the IGY, the objective of the IHY is to discover the physical mechanisms that link Earth and the heliosphere to solar activity. The IHY will focus on global effects but at a much greater physical scale (from Geophysics to Heliophysics) that encompasses the entire solar system and its interaction with the local interstellar medium (Davila *et al.* 2005).

The IHY activities are centred around four key elements: Science (coordinated investigation programmes or CIPs conducted as campaigns to investigate specific scientific questions), Observatory Development (an activity to deploy small instruments in developing countries), Public Outreach (to communicate the beauty, relevance and significance of the space science to the general public and students), and the IGY Gold Programme (to identify and honor all those scientists who worked for the IGY programme). The CIP is aiming at a large number of scientific campaigns in 2007 and 2008, and hundreds of observatories have expressed interest in participating in these campaigns. Thousands of scientists will analyze the data from these IHY CIP campaigns. The Observatory Development activity will provide additional data for the IHY campaigns and new opportunities for research and education in developing countries. This activity will be carried out as a partnership between IHY and United Nations.

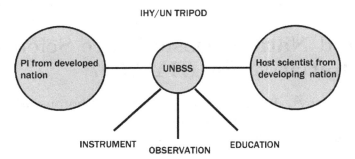

Figure 1. A schematic illustration of the IHY/UNBSS Tripod concept

2. United Nations Basic Space Science (UNBSS) Initiative

The United Nations Office for Outer Space Affairs (UNOOSA) implements the United Nations Programme on Space Applications and works to improve the use of space science and technology for the economic and social development of all nations, in particular developing countries. Under this programme, UNOOSA conducts training courses, workshops, seminars and other activities on applications and capacity building in subjects such as remote sensing, communications, satellite meteorology, search and rescue, basic space science, satellite navigation and space law. The subject of Basic Space Science includes: fundamental physics, astronomy and astrophysics, solar-terrestrial interaction and its influence on terrestrial climate, planetary and atmospheric studies, and origin of life and exo-biology. The IGY 1957 was one of the driving events to establish the United Nations Committee on the Peaceful Uses of Outer Space (UNCOPUOS). The UN Basic Space Science (UNBSS) programme is devised according to the requirements of the Member States of UNCOPUOS.

Recognizing the large overlap between the goals of IHY and the UNBSS activities, a partnership was established in 2004 during the UNBSS meeting in China. IHY and UN representatives met in October 2004 at NASA's Goddard Space Flight Center (GSFC) and decided to dedicate UNBSS resources and activities through 2009 to the IHY observatory development programme. The ultimate aim of this collaboration is to deploy small instruments that will provide an opportunity for the scientists in developing countries to effectively participate in IHY activities and promote space science education and research in their home countries.

2.1. *The UNBSS Tripod*

The basic framework of the UNBSS activity can be described by the Tripod concept, as illustrated in Figure 1. The three legs of the Tripod are (i) Instrument, (ii) Observation, and (iii) Education (Al-Naimy *et al.* 2004). During the period 1991-2004, the instrument leg was astronomical telescopes. From 2005 onwards, it represents various IHY instruments deployed in developing countries. The observation and education legs remain the same: data acquisition from IHY instruments and training in data analysis and instrumentation at the university level. Under the Tripod programme, scientists from developed countries or those who are willing and able, donate instruments to developing countries. These instruments will be used for scientific research and for university level education for young people. These deployments will serve as nuclei for a sustained development of scientific activities in the host countries. The data acquired from these instruments will also augment IHY data bases developed from IHY campaigns.

The instrument deployment projects need to adhere to the following basic principles so that the joint goals of the IHY and UNBSS programmes will be readily accomplished:
A) Quality Science: The projects must produce scientifically significant and publishable results pertaining to the objectives of the IHY activities.
B) Host Countries: The projects that can be readily carried out in developing nations (many of which are near the equator) need to be identified.
C) Cost/Technical Compatibility: The costs and technical requirements of the projects must be compatible with the resources available in the participating nations.
D) Legacy Potential: The projects must lead to a beneficial long-term relationship for the participants in developing nations.
E) Educational Component: The instrument deployment, observation and data acquisition, and analysis should ideally involve students, especially at the university level.

2.2. *Instrument concepts*

A detailed description of the instrument concepts proposed and accepted in 2005 can be found in the 2005 Seminars of the United Nations programme on Space Applications (Gopalswamy, 2005). The current instrument concepts can be grouped into the following classes: (i) Solar Telescope Networks, (ii) Ionospheric Networks, (iii) Magnetometer Networks, and (iv) Particle Detector Networks. These four groups together cover a substantial number of themes listed in section 1. In this section we provide a brief outline of the concepts and new initiatives considered in 2006.

2.2.1. *Solar telescope networks*

The solar telescope network consists of radio telescopes that detect radio bursts associated with solar eruptions. Of particular importance are the bursts produced by shocks and electron beams produced at the Sun, which can be remotely sensed by the radio telescopes. The telescopes will be deployed at several locations in the world, so that the Sun can be monitored continuously. There are currently two projects:

(*a*) The Compound Astronomical Low-cost Low-frequency Instrument for Spectroscopy and Transportable Observatory (CALLISTO) is a dual-channel frequency-agile receiver based on commercially available consumer electronics (PI: Arnold Benz, ETH-Zentrum, Switzerland). The complete spectrometer is very compact, very cheap and easy to replicate for deploying in many locations. Arrangements are being made to deploy one in India at the Radio Astronomy Center in Ooty. This network, in addition to the existing spectrometers at Hiraiso in Japan, ARTEMIS in Greece and Culgoora in Australia will form an excellent radio network for IHY science. Other locations in Mexico and Costa Rica are also being explored for a 2007 deployment.

(*b*) A related radio instrument is the Bruny Island Radio Spectrometer operating in Tasmania, which can be deployed complementary to CALLISTO (PI: R. MacDowall, NASA/GSFC). The frequency of operation is just above the ionospheric cutoff, which depends on the latitude. Currently the electromagnetic characteristics of a site in Gauribidanur, India are being studied for installation of such a spectrometer to work in conjunction with CALLISTO in Ooty.

2.2.2. *Ionospheric networks*

The ionospheric layer is most important for radio communications and broadcasting. Mapping the ionospheric properties around the globe is important for all kinds of communications. Deployment of several ionospheric networks is underway, as described below.

(*a*) The Atmospheric Weather Educational System for Observation and Modeling of Effects (AWESOME) instrument is an ionospheric monitor that can be operated by students around the world (PI: U. Inan, Stanford University). The monitors detect solar flares and other ionospheric disturbances. AWESOME monitors will be deployed in many African and Asian countries, so that the current data obtained in the western hemisphere can be combined with other data. This effort will provide a basis for comparison to facilitate global extrapolations and conclusions.

(*b*) Africa GPS is an effort to link many Global Positioning System (GPS) networks in Africa and the effort is coordinated by Tim Fuller-Rowell (NOAA-Boulder) and Christine Amory-Mazaudier (CETP, France). The overarching plan is to increase the number of real-time dual-frequency GPS stations worldwide for the study of ionospheric variability. Of particular interest is the response of the ionospheric total electron content (TEC) during geomagnetic storms over the African sector. This programme is particularly compatible with magnetometry.

(*c*) Scintillation Network Decision Aid (SCINDA) is a real-time, data driven, communication outage forecast and alert system (PI: K. Groves, AFRL). Its purpose is to aid in the specification and prediction of communications degradation due to ionospheric scintillation in the equatorial region of Earth. Scintillation affects radio signal frequencies of up to a few GHz and seriously degrades and disrupts satellite-based navigation and communication systems. SCINDA consists of a set of ground-based sensors and quasi-empirical models, developed to provide real-time alerts and short-term (< 1 hour) forecasts of scintillation impacts on UHF satellite communication and L-Band GPS signals in the Earth's equatorial regions. SCINDA will be deployed near Earth's magnetic equator (within 20 degrees on either side). The current thrust is in African countries, where there is a clear dearth of ionospheric data.

(*d*) The Remote Equatorial Nighttime Observatory for Ionospheric Regions (RENOIR, PI: Jonathan Makela, University of Illinois at Urbana-Champaign) is a suite of instruments dedicated to studying the equatorial/low-latitude ionosphere/thermosphere system, its response to storms and the irregularities that can be present on a daily basis. Through the construction and deployment of a RENOIR station, it is possible to achieve a better understanding of the variability in the nighttime ionosphere and the effects this variability has on critical satellite navigation and communication systems.

(*e*) The South America VLF Network (SAVNET, PI: Jean-Pierre Raulin, MacKenzie University, Brazil) is for monitoring the solar activity on long and short time scales and studying ionospheric perturbations over the South Atlantic Magnetic Anomaly region. The network will also be used for studying Earth's atmosphere. The basic data output is composed of the phase and amplitude measurements of VLF signals.

2.2.3. *Magnetometer networks*

A magnetometer network is a relatively low-cost method for monitoring solar-terrestrial interactions. A multi-continental network would provide an excellent basis for meso- and global-scale monitoring of magnetospheric-ionospheric disturbances and provide scientific targets for mid- and low-latitudes and opportunities for developing countries to host instruments and participate in the science investigations. Current projects are:

(*a*) The Magnetic Data Acquisition System (MAGDAS) is being deployed for space weather studies during 2005 to 2008 (PI: K. Yumoto, Kyushu University, Japan). The MAGDAS data will be used to map the ionospheric equivalent current pattern every day. The current and electric fields at all latitudes are coupled, although those at high, middle and low latitudes are often considered separately. By using the MAGDAS ionospheric current pattern, the global electromagnetic coupling processes at all latitudes will be clarified. MAGDAS will utilize the Circum-Pan Pacific Magnetometer Network involving several countries around the globe (Australia, Indonesia, Japan, Philippines, Russian Federation, United States and Taiwan Province of China).

(*b*) The Canadian Array for Real-time Investigations of Magnetic Activity (CARISMA, PI: Ian Mann, University of Alberta, Canada). CARISMA is the magnetometer element of the Canadian Geospace Monitoring (CGSM) project. Each proposed IHY magnetometer observatory shall consist of magnetometer station pairs separated meridionally by approximately 200 km. Other requirements are: a 2 × 3-component fluxgate magnetometer, data logger, GPS timing and power source.

2.2.4. *Particle detector networks*

Particle detectors have a wide range of applications: they can detect energetic particles from the Sun, galactic and extra-galactic sources and from the heliosphere. They can also indirectly observe large magnetic structures such as magnetic clouds and shocks from the Sun through the well-known process of Forbush decrease of cosmic ray intensity. The energetic particles also interact with Earth's atmosphere and produce air-showers (secondary particles). Some large solar energetic particle events interact with Earth's atmosphere leading to ozone depletion (Jackman *et al.* 2005).

(*a*) The SEVAN worldwide particle detector network (PI: Ashot Chilingarian, Aragats Space Environmental Center, Alikhanian Physics Institute, Armenia) is a combined neutron-muon detecting system. A flexible 32-bit microcontroller-based data acquisition electronics will utilize the correlation information from cosmic ray secondary fluxes, including environmental parameters (temperature, pressure and magnetic field). The high precision time synchronization of the remote installations via GPS receivers are crucial ingredients of the new detector. It is proposed to deploy such detectors in neighboring countries such as Bulgaria and Croatia.

(*b*) The muon detector network collaboration (PI:. Kazuoki Munakata, Shinshu University, Japan) consists of nine institutes from seven countries (Armenia, Australia, Brazil, Germany, Japan, Kuwait and the United States). Many of the countries are already operating muon detectors and some have recently installed them. The muon detector network can identify the precursory decrease of cosmic ray intensity that takes place more than one day prior to the Earth-arrival of shock driven by coronal mass ejection. This is an important forecasting tool for predicting space weather attributed to energetic solar eruptions.

2.2.5. *Recent Accomplishments*

AWESOME: The first deployment of AWESOME under the IHY/UNBSS programme commenced in October 2005 with the installation in Tunisia. Instruments were delivered to Algeria and Morocco in summer 2006. More deployments are planned in Libya, Egypt and South Africa. A large network of AWESOME monitors is expected to be in place during 2007–2008.

MAGDAS: IHY-Japan has made significant progress towards the completion of its 51-magnetometer MAGDAS global network with a new installation site on MacQuarie Island, a sub-Antarctic island between Tasmania and Antarctica. Recent deployments were made in Ethiopia, Nigeria, and Ivory Coast (August 2006). The next installation will be in Malaysia.

RENOIR: The RENOIR ionospheric observing station programme has received support for development, and will be making plans for instrument host sites later this year.

SAVNET: The South America VLF NETwork (SAVNET) has been recently approved by the São Paulo state funding agency FAPESP in Brazil for a duration of two years.

The deployment of the SAVNET VLF receiver chain will begin in 2006 with the target of being operational in 2007.

SCINDA: The SCINDA scintillation network is expected to double the size of their equatorial network. An instrumenters meeting was held in July 2006 in Cape Verde in preparation for new deployments. Deployments in Cape Verde and Nigeria have just been completed.

2.3. *Recent developments*

2.3.1. *The Flare Monitoring Telescope*

After the 2005 IHY/UNBSS workshop, a new instrument concept was proposed by Japan. This is an H-alpha telescope to be donated by Japan (PI: S. Ueno, Japan) and hosted in Peru. This will be a good complement to the radio telescope network, discussed in section 2.2.1. The H-alpha instrument consists of six telescopes in the telescope dome: 3 for observations in the H-alpha line center, blue wing, and red wing; one with the occulting disk for prominence observations, one for the continuum, and the last one has an optical guider for accurate tracking of the Sun.

2.3.2. *Data projects*

In addition to instrument deployments, a new element will be introduced during the 2006 workshop. The idea is to replace the Instrument leg of the IHY/UNBSS Tripod with a data base. Accessing and manipulating data from such data bases will be equally rewarding, similar to acquiring data from instruments. One of the examples is the Solar Anomalous and Magnetospheric Particle Explorer (SAMPEX) data base (effort leader S. Kanekal, University of Colorado). SAMPEX is the first in NASA's relatively low-budget, fast-track series of Small Explorer class of spacecraft, launched on July 3, 1992, to provide cosmic ray fluxes at the polar cap and radiation belts fluxes. The SAMPEX mission ended in July 2004, leaving behind a 12-year continuous record of observations. By providing the data with analyzing tools, scientists will be able to study Earth's radiation belts.

2.3.3. *Gnu Data Language*

There is a plan to develop the Gnu Data Language (GDL), which is a free, UNIX-based software that will be available for processing image and time series data. This will enable many scientists from developing countries to access, display and analyze IHY data.

3. UNBSS workshops

Implementation of the BSS Tripod concept is being done using the annual UNBSS workshops. Such workshops have been conducted since 1991 for enabling astronomical telescope deployments. From 2005 onwards, the UNBSS workshops have been devoted to IHY Observatory Development activity. Several things happen during the UNBSS workshops: (1) Scientists from developing and developed countries meet face-to-face to discuss collaborative projects under the UNBSS programme, (2) Scientific instrument host groups provide descriptions of the sites for instrument deployment and the facilities available for hosting the instrument, (3) Potential providers of scientific instruments describe their instruments and the key requirements in terms of infrastructure for a successful deployment and continued operation, (4) Progress reports after the previous workshop are presented and discussed, and (5) Several participants provide the necessary scientific background through a series of tutorial talks.

The first IHY/UNBSS Workshop on Basic Space Science, sponsored by UN, ESA, and NASA, was held in Abu Dhabi and Al-Ain, United Arab Emirates during 20–23

November, 2005. Workshop participants represented 44 countries, including a significant portion of North Africa, the IHY-West Asia region, as well as leadership from the remaining six IHY international regions. There were special sessions on IHY instruments and host institutions, as well as IHY science, global scientific initiatives, education programmes, astrophysical research in Arab nations, and the 2005 World Year of Physics. Special discussions also included planning for the IHY-Africa initiative, education and outreach activities, and the establishment of a working group on Infrastructure and Sustainability Issues in Global Space Sciences. Overall the conference accomplished more than expected, and the future workshops will build on this success.

The Second IHY/UNBSS Workshop was held from November 27 – December 1, 2006 in Bangalore, India and was sponsored by UN, NASA and several institutions in India. After several sessions on background science topics, presentations by the instrument donors on the current progress and future plans for the deployment projects were made. Like the 2005 workshop in UAE, this meeting focussed on identifying instrument host and deployment sites for IHY instrumentation.

4. Conclusions

The IHY activities recognize the importance of global efforts with participation from as many countries as possible and as many observatories (from ground and space) as possible. All four elements of IHY are built upon this global cooperation. The new instruments deployed by the IHY/UNBSS programme will also participate in the CIPs. Scientists and students from developing countries will also participate in the observations and data analysis, providing valuable training and education. The IHY schools programme will also help build a solid scientific background for young people throughout the world. The synergy between IHY and UNBSS activities is expected to make great progress during the IHY years and beyond.

Acknowledgements

We thank D.F. Webb for a careful reading of the manuscript. This work was supported by the NASA/LWS programme.

References

Al-Naimiy, H.M.K., Celebre, C.P., Chamcham, K., de Alwis, H.S.P., de Carias, M.C.P., Haubold, H.J., Troche Boggino, A.E. 2004, in: W. Wamsteker, R. Albrecht, & H.J. Haubold (eds.), *Developing Basic Space Science World-Wide: A Decade of UN/ESA Workshops*, Dordrecht: Kluwer p. 31

Davila, J.M., Thompson, B.J., Gopalswamy, N. 2005, *Seminars of the United Nations programmeme on Space Applications*, Vienna: UNOOSA, 2005, p. 37

Gopalswamy, N. 2005, *Seminars of the United Nations programmeme on Space Applications*, Vienna: UNOOSA, 2005, p. 47

Jackman, C. H., DeLand, M.T., Labow, G. J., Fleming, E. L., Weisenstein, D. K., Ko, M. K. W., Sinnhuber, M., Russell, J.M. 2005, *J. Geophys. Res.* 110, A09S27

Hans Haubold

Cristina Rabello-Soares and Keith Arnaud

Astronomy for the developing world
IAU Special Session no. 5, 2006
J.B. Hearnshaw and P. Martinez, eds.

© 2007 International Astronomical Union
doi:10.1017/S1743921307007193

The Space Weather Monitor Project: bringing hands-on science to students of the developing world for the IHY2007

Deborah Scherrer[1], M. Cristina Rabello-Soares[1] and Cherilynn Morrow[2]

[1]Stanford University, Solar Physics, HEPL-4085, Stanford, CA 94305-4085, USA,
dscherrer@solar.stanford.edu

[2]Space Science Institute, 4750 Walnut Street, Suite 205, Boulder, Colorado 80301, USA,
morrow@SpaceScience.org

Abstract. Stanford's Solar Center, Electrical Engineering Department, and local educators have developed inexpensive Space Weather Monitors that students around the world can use to track solar- and lightning-induced changes to the Earth's ionosphere. Through the United Nations Basic Space Science Initiative (UNBSSI) and the International Heliophysical Year (IHY) Education and Public Outreach programme, our Monitors are being deployed to 191 countries. In partnership with Chabot Space and Science Center, we are designing and developing classroom and educator support materials to accompany the distribution. Materials will be culturally sensitive and will be translated into the six official languages of the United Nations (Arabic, Chinese, English, French, Russian, and Spanish). Monitors will be provided free of charge to developing nations and can be set up anywhere there is access to power.

Keywords. education, Sun, solar-terrestrial relations, Earth, miscellaneous

1. Introduction

Earth's ionosphere reacts strongly to the intense X-ray and ultraviolet radiation released by the Sun during solar events and by lightning during thunderstorms. Imagine students being able to track these sudden ionospheric disturbances by using a receiver to monitor the signal strength from distant VLF transmitters and noting unusual changes as the waves bounce off the ionosphere. Stanford's Solar Center, Electrical Engineering Department, and local educators have developed inexpensive Space Weather Monitors that students around the world can use to track changes to the Earth's ionosphere. Two versions of the monitors exist – a low-cost version nicknamed SID designed to detect solar flares; and a more sensitive version (AWESOME) that provides both solar and nighttime research-quality data.

Through the United Nations Basic Space Science Initiative (UNBSSI) and the IHY International Education and Public Outreach programme, we expect to deploy, without charge, our monitors to high school-equivalents and universities in developing nations of the world for IHY 2007. Monitors are inexpensive (US$200 for the student version; US$2400 for the research quality). The monitors come preassembled, but students buy in by designing and building their own antenna, costing little and taking a few hours to assemble. Participants also provide a simple PC to record the data and, if possible, an internet connection to share their data with the rest of the worldwide team. The monitors can be set up anywhere there is access to power.

The IHY is ideally suited to support a global VLF receiver venture. This project directly relates to the goals of IHY by advancing our understanding of the fundamental heliophysical processes that govern the Sun, Earth and heliosphere and demonstrating the beauty, relevance and significance of space and Earth science to the world. As a result of IHY, AWESOME (research-quality) receivers have recently been set up successfully in Ireland, Tunisia, India, Algeria, and Morocco. Seventy of the inexpensive SIDs have already been distributed to the USA and a dozen more countries.

Our project enables researchers, educators, and students to produce scientific data from their own instrument and access data from other worldwide team members. It provides tools for doing research, both independently and collaboratively. A 2002 decadal survey by the USA National Research Council noted that "understanding and monitoring the fundamental processes responsible for solar-terrestrial coupling are vital to being able to fully explain the influence of the Sun on the near-Earth environment". Abundant evidence shows that high school and introductory college science teaching is improved when the students engage in doing real science on real data.

2. Students collect and analyze scientific data

Our monitors capture signal strength from ELF/VLF frequencies between about 30 Hz – 50 kHz, transmitted by nations to communicate with their submarines. SID data are easy to read and understand, thus readily accessible to introductory college and high school groups. In addition to solar flare phenomena, the graphs host a wealth of details about the Earth's ionosphere and how it changes during the day/night cycle, from season to season, and how it responds to lightning storms and other ionospheric events.

Data recorded by our instruments resemble seismograph data (Figure 1). Students receive their SID data as a signal strength value and a timestamp. The data collected are easily read and graphed either by Excel or a gnu-based plot programme provided. Solar events show up as spikes in the signal strength. The more sensitive AWESOME monitors allow students to track nighttime lightning phenomena as well.

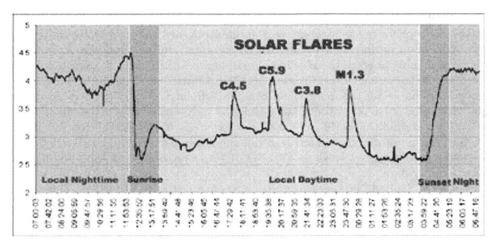

Figure 1. Sample SID data.

Stanford hosts a centralized data repository for both SID and AWESOME data. (http://sid.stanford.edu/database-browser/) Data are useful for solar researchers, for ionospheric researchers, and for educators and students, even those not hosting monitors. Observations are highly appropriate for creative use in student research. We suggest to students (http://solar-center.stanford.edu/SID/educators/) a collection of potential research projects relating to the Sun and space weather that will encourage their participation in hands-on research and collaboration, giving them experience in a field of vital worldwide interest. Research topics relate to:

- Sunrise and sunset ionization phenomena;
- Tracking solar storms back to the Sun;
- Understanding ionization effects;
- Nighttime research phenomena;
- Using ionization changes as earthquake predictors.

Student researchers with access to the internet can compare their data with that from NOAA's GOES satellites to help identify solar flares. Students using a monitor proto-type have already identified flares that GOES "missed" due to human error. Students can track events back to specific active regions on the Sun and can view corresponding images from solar observatories such as SOHO/MDI, SOHO/EIT, TRACE, and eventually SDO. Students can also compare their data with solar-induced effects on radio communications, high-flying aircraft, satellites, even pigeon racing. As an advanced research project, they might attempt to predict possible flare sites using MDI's farside imagery. (See `http://soi.stanford.edu/data/full_farside`). Students using prototype monitors have also attempted to correlate their solar events with environmental changes on Earth. In most of their research, students benefit by talking with other students and comparing data from other sites. The programme provides student access to Scientist Mentors to assist in undertaking and understanding their research.

3. Educational components

In partnership with Chabot Space and Science Center of Oakland, California, USA we are designing, developing, testing, and assessing classroom and educator support materials to accompany the distribution. Curriculum materials, designed and tested by Master Teachers, will be inquiry-based and professionally assessed. Online and DVD/video teacher training will be available to support teachers around the world. Extensive background and accompanying materials will be provided, both on the web and in hard-copy form. A centralized database and communications hub have already been set up by Stanford for communication amongst teachers and students worldwide.

This project is closely aligned with the USA National Academy of Science/National Research Council's National Science Education Standards (National Research Council, 1996). The Standards emphasize science as inquiry. Unifying concepts and processes parallel those in the Standards, and include focus on the fundamental aspects of solar activity and their effects on the Earth, on obtaining data, taking measurements, on observing and understanding change and constancy, and learning to interpret and explain results from data. The pairing of the instrument with scientific inquiry improves understanding of technological design and the functions technology can play in science. As the Standards encourage, the project facilitates teachers and students working directly with scientists and real data, providing them unique opportunities to experience science as a human endeavor and giving them direct insight into the nature of scientific knowledge. The activities developed for the Teachers Guides reflect "best practices" in education as experienced and interpreted by Master Teachers and include specific information about alignment with the National Science Standards. Although we understand that the US National Science Standards do not apply to other nations, we feel the concepts and guiding principles are strong and valid models for designing science education throughout the world.

To the extent possible, materials will be reviewed for cultural appropriateness and a subset will be translated into the six official languages of the United Nations (Arabic, Chinese, English, French, Russian, and Spanish).

Blindness is a serious problem in many developing nations, and often blind students are "left behind" in the education process. We are looking for funding for a special version of the education materials to be adapted for the blind, translated into Braille, and made available to any country upon request. SID data will be sonified to support equal access by blind students. Curricula support materials such as NASA's/Noreen Grise's *Touch the Sun* book (Grice, 2005) will also be included in this package.

The programme includes teacher and student access to Scientists Mentors for discussions about problems, data, and research. Scientist Mentors provide direct connections between students, teachers, and the international research community. Loosely modeled on JPL's Solar System Ambassador programme (http://www2.jpl.nasa.gov/ambassador), we are already experimenting with soliciting and using mentors – motivated individuals with scientific skill and experience in solar physics, geophysics, ionospheric physics, or radio. Mentor training, also modeled after the JPL programme, will be provided through teleconferences, the web, and supplied materials. Once our full Scientist Mentor programme is in place, we will publish and make available to sites a database (via web, email, or direct mailing) of Mentors' expertise, skills, interests, languages, and contact information. SID sites will be encouraged to build a relationship with a Scientist Mentor who has the skills necessary to solve their problems or answer their questions. Mentors will also serve as role models for Science-Technology-Engineering-Math (STEM) careers. Part of mentor training will provide information on encouraging students to consider STEM careers, no matter where in the world they might live.

4. Conclusions

The International Heliophysical Year aims to explore the response of the terrestrial atmosphere and magnetosphere to external drivers. The future of exploration lies in global understanding of the ways that the Sun affects Earth's environment. We propose a combined educational and research programme aimed at developing nations and featuring a network of sensors distributed to a worldwide team who will perform quantitative comparisons of ionospheric disturbances. The network consists of state-of-the-art ELF/VLF receivers sensitive to a broad range of ionospheric phenomena and student-appropriate low cost space-weather sensors sensitive to solar-flare-induced sudden ionospheric disturbances. Educational materials accompany the distribution and access to Scientist Mentors assures adequate student support. For more information about the Space Weather Monitor programme, see http://solar-center.stanford.edu/SID/.

Acknowledgements

This programme was originally developed and funded by NSF's Center for Integrated Space Weather Modeling. Additional funding was provided by NASA grant NAG5-3077, supporting the MDI instrument onboard the SOlar and Heliospheric Observatory (SOHO). SOHO is a project of international cooperation between ESA and NASA.

References

Grice, N. 2005, *Touch the Sun – A NASA Braille Book*. Joseph Henry Press (an imprint of the National Academies Press), Washington, DC

National Research Council 1996, *National Science Education Standards*. National Academies Press, Washington, DC

Section 9:

The virtual observatory
and developing countries

Astronomy for the developing world
IAU Special Session no. 5, 2006
J.B. Hearnshaw and P. Martinez, eds.

© 2007 International Astronomical Union
doi:10.1017/S1743921307007211

Virtual observatories and developing countries

Ajit Kembhavi

Inter-University Centre for Astronomy and Astrophysics, Pune, India
email:akk@iucaa.ernet.in

Abstract. I will discuss in this article the emerging concept of virtual observatories, the efforts being made in various countries to set up these structures, and the relevance of the concept to astronomy in developing countries.

Keywords. Virtual observatories, astronomical data, International Virtual Observatory Alliance

1. Introduction

A virtual observatory is a platform for launching astronomical investigations: it provides access to huge data banks, software systems with user-friendly interfaces for data processing, analysis and visualization, and even access to computers on which the work can be carried out. Virtual observatories the world over are seamlessly networked, and their resources can be accessed over the internet by astronomers regardless of their location, expertise and the level of access to their own advanced computing facilities. Due to their nature, virtual observatories can make an immense impact on the way astronomy is done in the developing world. I will consider some of the facilities that virtual observatories provide, discuss their possible use by astronomers, and how even small groups in the developing world can contribute to the setting up of virtual observatories, and benefit from their use.

2. Astronomical data

Astronomers carry out their observations using a variety of telescopes, based on the ground or on space platforms. They also use a variety of different detectors like photographic plates, CCD cameras, radio receivers, X-ray detectors, etc. With any of these instruments, there are two basic kinds of observing strategies which are followed: (1) observations of specific targets which are of interest to a specific research project and (2) observations which survey large portions of the sky and which can be used in a variety of scientific projects over the years.

Over the last two decades there has been great progress in telescope and detector technology, and it has been possible to build many large telescopes and increasingly sensitive detectors. The large installations are extremely expensive, and the trend has been to build telescopes and detectors through collaborative efforts and to make them available to a wide community. Astronomers all over the world can therefore use advanced facilities to which they may earlier not have had access. The data obtained using these facilities is generally archived and eventually made available to the entire community, regardless of who obtained it in the first place.

2.1. *Data volumes*

Data volumes generated in ongoing and planned modern surveys can range from several hundreds of gigabytes to many tens of terabytes. It is expected that some of the surveys which will be initiated over the next few years will generate terabytes of data per day. Storing, retrieving and scientifically using these vast databases is a formidable task. It requires the joint effort of astronomers and computer professionals to adapt existing hardware and software technology, and to develop it further to meet the challenging task of making the data accessible to all potential users.

At the present time devices which can usefully store many terabytes of data are expensive and difficult to manage, and it is expected that this situation will prevail for some time to come. It is therefore not practical to store all available data in many locations in the world for it to be locally available to astronomers. Moreover, maintaining mirror copies requires regular data transfers, which cannot always be done incrementally using the available bandwidth. Maintaining large data volumes also requires constant attention by computer professionals, and not all places would have access to such expertise. It is therefore necessary to store data in strategic locations and to make just what is needed available using the internet as well as other means for data transfer. This of course brings forth the issue of providing engines and interfaces to enable users to obtain and combine data from a number of locations.

2.2. *Data variety*

Data obtained in different parts of the electromagnetic spectrum, like the optical, radio or X-ray, requires vastly different kinds of processing before it is brought to a scientifically usable form. The data are also stored using quite different hardware and software systems, and techniques have to be developed for bringing together the different structures. Even for data in the same region of the spectrum, different observers use different notations, conventions and units, and comparing data from different sources can be an exacting task. The difficulty here becomes more pronounced when much of the processing is to be carried out with computers, avoiding human intervention. The solution is to provide extensive universal standard descriptors for the data which make automated analysis feasible.

3. Virtual observatory

A virtual observatory (VO) seeks to facilitate the tasks mentioned above, but it has to go beyond simply making large amounts of data available: it has to provide query tools for the required data to be accessed from the vast store, for analysis and visualization of the data, and for data mining, which will enable new scientific discoveries to be made. The queries needed to generate the data required by a user, and the subsequent analysis, can require computing facilities which may not be available at the user's establishment. A VO would seek to provide computing resources as well, either on its own site, or through a grid linking computers located at different sites.

Attempts to establish such structures have been made in the past, with moderate success, by different observatories and institutes. But the huge increase in the volume of data now available, and the need to carry out research simultaneously in many different parts of the electromagnetic spectrum, has made it necessary to make collaborative efforts, much in the manner of joint work undertaken to develop major new ground and space based telescopes. Using the internet, even a remotely located user can access the facilities offered by a VO, and can use the data just as well as anyone else. This is a development of far reaching consequence, particularly for astronomers in developing countries.

Astronomical data are available in the form of catalogues, spectra and images. The VO will enable astronomers to use these different kinds of data simultaneously, irrespective of their location and basic nature, for a full multi-wavelength analysis. The vast quantities of data will make it possible to look for very rare objects, patterns and relationships which remained totally inaccessible when only very limited data were available. Searching for these rare features will require the development of highly sophisticated data mining techniques for the search to be completed in a reasonable time. The features found will have to be subject to analysis, and to be compared with the results of numerical simulations. The VO seeks to provide hardware and software platforms on which all these operations can be carried out.

3.1. *VO projects in the world*

A virtual observatory, as the name implies, is a comprehensive concept which embodies computer hardware and software, data, and human expertise for providing services. A VO can occupy a small space in a single university department or research institution, or can be distributed over several locations. In fact all the VOs in the world can be considered to be meshed together to be parts of a single VO, providing data and related services to anyone who may need them.

There are several VO projects in operation in different countries in the world. Some projects are large, have huge budgets, and many people working on them. Other projects have just a small number of people engaged on very specific programmes, and work with modest resources; they have nevertheless managed to provide products which have proved to be very useful to astronomers all over the world. The bigger projects include the National Virtual Observatory (NVO) of the USA, the European Virtual Observatory (EURO-VO) which brings together many European countries, some of which also have their own individual VO projects (including those in France, Germany, Italy and Spain), and ASTROGRID, which is a VO project based in Britain. The smaller projects include those based in Armenia, Australia, Canada, China, Hungary, India, Japan, Korea and Russia. Links to various VO sites can be found on the Virtual Observatory - India (VO-I) homepage (`http://vo.iucaa.ernet.in/~voi/`) and the IVOA homepage (`http://www.ivoa.net`).

4. The International Virtual Observatory Alliance (IVOA)

As its name suggests, this is an alliance of VO projects based in various countries. It provides a forum at which astronomers, computer professionals and others who are engaged in VO activities in different countries can come together to share their experiences and resources, set up collaborations, and most importantly, develop common standards and infrastructure for data exchange and inter-operability. The VO concept involves sharing of data and resources, and the IVOA is most effective in bringing this about through continuous discussion and collaboration.

The IVOA has defined six major programmes which have to be undertaken to make progress towards building up of virtual Observatories. These are (1) REGISTRIES: These collect metadata about data resources and information services into a queryable database. The registry is distributed. A variety of industry standards is being investigated. (2) DATA MODELS: This initiative aims to define the common elements of astronomical data structures and to provide a framework to describe their relationships. (3) UNIFORM CONTENT DESCRIPTORS: These will provide the common language for metadata definitions for the VO. (4) DATA ACCESS LAYER: This provides standardized access mechanisms to distributed data objects. Initial prototypes are a Cone Search Protocol and a simple Image Access Protocol. (5) VO QUERY LANGUAGE: This will provide a standard query language which will go beyond the limitations of SQL. (6) VOTable: This is an XML mark-up standard for astronomical tables. Over the last few years, much progress has been made in these directions.

The IVOA has a carefully set up procedure for producing standards through a cycle of Working Drafts, Proposed Recommendations, and finally Recommendations to the international community as represented by the International Astronomical Union (IAU). The IVOA has several Working Groups and Interest Groups in various aspects of the VO, and these groups have produced an overall architectural plan for an operational VO that identifies the critical areas for current and future development of standards and technologies. Various VO groups meet at Inter-operability Meetings organized in different countries, and the community also regularly participates in important astronomical conferences to familiarize the community of the work done, and to demonstrate through real science applications the standards and tools which have been developed.

5. Some VO projects

The developmental work undertaken by different national VO projects depends on the needs of the community they serve, and the facilities and expertise available at the particular VO. A few examples are given below:

5.1. *Hungarian Virtual Observatory (HVO)*

The HVO (http://hvo.elte.hu/en/) has been developing various VO services, and archives of large databases. An important HVO project has been the Spectrum Service, through which spectra of galaxies and other astronomical objects are accessible through Web services, and several manipulations and transformations can be made on them. All of these services are very useful for astronomers. The HVO is creating the first dynamical synthetic spectrum service to generate spectra on the fly, using inputs from the calling service. In another HVO project, photometric red shifts have been obtained for over 100 million objects, and these have been made available through Data Release 5 of the Sloane Digital Sky Survey (SDSS), in close collaboration with the SDSS group at Johns Hopkins University.

5.2. *Russian Virtual Observatory (RVO)*

The RVO (http://www.inasan.rssi.ru/eng/rvo/) provides Russian astronomers effective access to international resources, and integrates Russian astronomical resources into the international VO structure. It also aims at providing observational resources to its user base when the required data are not in the archives, and to develop electronic educational resources. Several Russian astronomical observatories have unique collections of astronomical photographic plates obtained since the beginning of the past century. An important project for the RVO is to make these data available as a digital archive.

The RVO has undertaken various scientific projects to exploit the VO resources. These include creating a three-dimensional map of interstellar extinction in the Galaxy, in collaboration with NVO; compilation of the fundamental stellar parameters and and evolutionary status of close binary systems, in collaboration with the Observatory of Besançon in France; investigations of open clusters (in collaboration with German and Indian astronomers), and work on the MIGALE project together with French scientists.

5.3. *China - Virtual Observatory (China-VO)*

China-VO (http://www.china-vo.org/) is a consortium initiated by the National Astronomical Observatory of China (NAOC) and the Large Sky Area Multi-Object Fibre Spectroscopic Telescope (LAMOST) project, and has several partner institutions.

LAMOST is a meridian reflecting Schmidt telescope, with a large focal plane which can accommodate up to 4000 fibres, and can accumulate up to 10,000 spectra per night of objects as faint as 20.5 magnitude. China-VO will develop tools to process this huge volume of spectroscopic data automatically, and to make it available online as a VO-compatible archive.

China-VO has also been developing various services like VOfilter to transform tabular data files into OpenDocument format, to bridge the VO with current desktop applications; SkyMouse, which is an intelligent information collector which uses online astronomical resources, including VO services, and traditional web applications; and VO-IMPAT, which is an interactive imaging tool that allows the users to visualize digitized images of any part of the sky and interactively access related data and information from the Beijing Astronomical Data Center (BADC). VO-IMPAT is available as a stand alone version, which is to be installed on the user's machine.

5.4. *German Astrophysical Virtual Observatory (GAVO)*

The GAVO (http://www.g-vo.org/portal/) project aims at providing fast and easy access to astronomical data archives and related documentation, as well as a capability to use highly sophisticated software tools for new studies. In GAVO's pilot phase, work is concentrating on four main areas: archive technology and publication, data mining and knowledge discovery in federated astronomical archives, theory in the virtual observatory, and Grid-computing. A special area of interest to the GAVO is theory in the virtual observatory. This involves the publication of theoretical datasets, obtained through extensive numerical simulations, in ways similar to the publication of observational data, and the creation of services appropriate for the use of the theoretical data. The goal is to create an environment in which theoretical results can be used for the interpretation of observations, and observations can be used to constrain theoretical models. GAVO is pursuing a number of concrete projects and through collaborations is exploring techniques for publishing theoretical datasets. GAVO is also actively involved in the IVOA theory interest group, which aims at channelling the requirements from the theory community into the IVOA standards process.

6. Virtual Observatory - India (VO-I)

This project is based at the Inter-University Centre for Astronomy and Astrophysics located in the city of Pune in India, and its novel feature is that it is a collaboration between an astronomical institute and a major computer software company, Persistent Systems Pvt. Ltd. (PSPL), Pune, which has expertise in data management related products. The project, which is funded by the Ministry of Communications and Information Technology, as well as by the two participating organizations, was begun in early 2002 for an initial period of three years. VO-I brings together astronomers from IUCAA and other institutions and University departments in India, and software experts from PSPL.

VO-I has successfully completed a series of projects which include: (1) Development of a parser for data in VOTable format, which consists of a library of programs in C++ format which helps the user handle data in VOTable files. Versions are available which act on data in the non-streaming as well as streaming format; (2) Writers for converting data in other formats to the VOTable format; (3) A Web-based FITS file manager and (4)

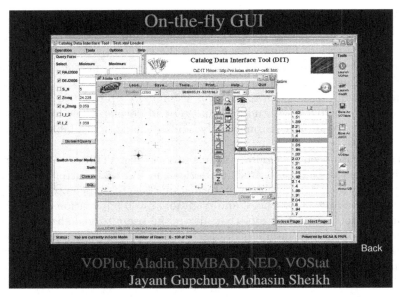

Figure 1. An example of using VOCat to study a catalogue. A specific source is selected, and the corresponding image of that source, shown as an inset, is obtained using ALADIN.

VOCat, which is a generic user interface for large data sets, and (5) VOPlot and VOStat which are major packages described below. As a part of VO-I activity, many useful large databases like the Sloane Digital Sky Survey (DR3), 2MASS, CHANDRA, etc. have been installed at IUCAA on RAID arrays, and access to these through appropriate user interfaces is being provided. An example of the use of VOCat is shown in Figure 1.

6.1. *VOPlot*

The VOPlot tool was developed in collaboration between VO-I and CDS, France, and is a menu and button driven tool for graphically visualizing data available in the form of catalogues. It is available as a web-based version fully integrated with VIZIER, which is an on-line catalogue service provided by CDS. VOPlot is also available in a stand alone version to be installed on the user's machine. VOPlot uses data in the VOTable format. It can be used to display the distribution, as a histogram, of any data field, or to plot two data fields against each other. Simple transformations can be applied to the extracted data. The visualized portion can be just a dynamically selected subset of extracted data. Some simple statistical information about the selected data can be obtained.

VOPlot goes far beyond being just a graphics tool. Each plotted point in a graph is active, and all the data related to it are available for further processing. If the data includes sky coordinates of the listed sources, it is possible to obtain the distribution of the sources in the sky in different projections. Selecting a number of points on a plot leads to the selection of the same points on the sky plot, which proves to be very useful in investigating outliers in a plot. An important feature of VOPlot is that it enables the user to pass from specific data points to images of the corresponding objects using ALADIN, which is an interactive software sky atlas developed by CDS. VOPlot is under continuous development, and is being used by the international astronomical community as a very useful new tool compatible with VO objectives. An example of the use of VOPlot is shown in Figure 2.

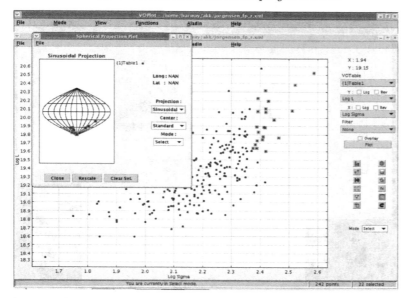

Figure 2. An example of a plot of galaxy properties made using VOPlot. The positions of the clusters from which the galaxies are drawn is shown as an inset.

Packages which are related to VOPlot, and have specialized capabilities, are VOPlot3D and VOMegaplot VOPLot3D is used in making three-dimensional plots, which can be viewed from different dynamically selected lines of sight. This allows three-dimensional correlations, like the fundamental plane of elliptical galaxies to be spotted easily. An example of the use of VOPlot3D is shown in Figure 3. VOMegaplot is used for plotting points numbering in the millions. This process would generally take substantial time to execute, and any further dynamical interaction with the data through the plot would be very slow. VOMegaplot gets over this limitation by creating a number of index files from the large data set, which are automatically used in all further interactions with that particular data set.

6.2. *VOStat*

This is a VO-compatible web-based statistical tool which was originally developed in collaboration between VO groups at Penn State University and the California Institute of Technology. A comprehensive new version of VOStat has been developed by VO-I, in collaboration with the original developers, as well as with statisticians from the University of Calcutta and the Indian Statistical Institute at Calcutta. VOStat uses the publicly available statistical package called *R*, and through an elegant interface provides access to a large number of statistical tests, which can be easily applied to data sets which are present with the user, or on remote data bases. VOStat will be made available on various sites, and will be integrated with other tools like VOPlot, so that data visualization and statistical analysis can be carried out simultaneously.

7. Science with the VO

The main aim of a VO is to facilitate astronomical research in such a manner that projects which depend on very large data sets and multi-wavelength studies can be easily undertaken. Mining large datasets will enable the discovery of rare objects, events and interesting new processes which remain obscured in the noise in small data sets.

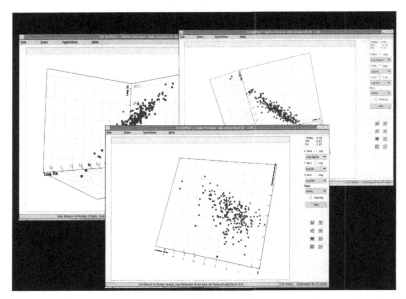

Figure 3. An example of a three-dimensional plot of galaxy properties made using VOPlot3D. Views along different lines of sight are shown.

Multi-wavelength studies will allow the form of the radiation spectrum to be better established for different classes of objects, and for correlation studies to be made across different regions of the spectrum. Most interestingly, with large samples it will be possible to examine how various properties change as a function of the types of objects being considered. For example, a sample of thousands of galaxies can be used to examine whether bulges of all types of galaxies show the same scatter from the fundamental and photometric planes.

Many scientific projects based on various VO facilities and tools are in progress at at the present time, and such projects are growing at a rapid rate. Many interesting results still remain unpublished at the time of writing, but excellent summaries of some of the work can be found in the presentations made at the Special Session on *The virtual observatory in action: new science, new technology, and next generation facilities* at the IAU XXVI General Assembly in Prague in 2006, which are available at http://www.ivoa.net/pub/VOScienceIAUPrague/programme/index.html.

The range of topics discussed in the special session include VO studies of (1) SDSS AGN with X-ray emission from ROSAT pointed observations, (2) the origin of soft X-ray emission in obscured AGN, (3) determination of radio spectra from catalogues and identification of GHz peaked sources, (4) the environment around QSOs with redshift of 1.3, (5) galaxy formation and evolution using multi-wavelength, multi-resolution imaging, and (6) super star clusters in nearby galaxies, (7) mapping galactic spiral arm structure, (8) discovery and characterization of brown dwarfs, (9) spectral classification of cool stars using high resolution spectra, (10) parameters for stellar population synthesis from SDSS galaxy spectra using evolution strategies, (11) photometric identification of quasars and AGN from the Sloane survey, (12) multi-wavelength properties of a sample of Texas Radio Survey steep spectrum sources and (13) near-IR properties of Spitzer selected sources. The investigations described here could all have been carried out using traditional methods, but that would have involved manually combining data from different sources after time-consuming searches through the literature, extensive programming to collate the data and to analyse it, and the use of various systems for graphical representation and statistical analysis. Use of the facilities provided by the VO makes all this very quick, and leads in fact to work which would have been very difficult to do without the new tools, and to results which otherwise would have remained obscure for a long time. The range of topics reported shows that people working in every discipline would benefit from using the VO, and that soon it will become an indispensable tool. It is very easy to gain access to these tools by navigating through various web-sites and it is also easy to learn to use these tools proficiently.

8. Virtual observatories and the developing world

Virtual observatory facilities are open to everyone, and even limited internet access is sufficient to enable any astronomer to use data and tools provided through various VO portals. Astronomers in the developing world, who have only limited facilities in their own centres can therefore participate in front-line research, and can use the data in the same manner that their colleagues in the most advanced centres do. In addition to enabling research, VOs also act as great teaching facilitators; projects at every level can be easily set up using their resources. VOs also provide excellent public outreach platforms for spreading awareness about astronomy in a very exciting way to young students as well as to the public.

A very important aspect of VOs is that astronomers everywhere can contribute actively to their development. It is clear from the examples of some of the national projects described above that widely used VO tools have been developed by small groups of people working with limited resources. Many developing countries are now developing world class expertise in software technology, and collaboration between software specialists and astronomers in such countries can enable significant contributions to be made. Due to the highly collaborative nature of the enterprise, groups of interested persons can contribute in the manner most suited to their expertise, and in turn benefit from work of other groups, and from the data generated by advanced observing facilities the world over. Exciting new opportunities have therefore become available for astronomers everywhere, and they can both benefit from all the developments and actively contribute to sustaining the virtual observatory movement.

9. Further information

Due to the nature of this rapidly evolving field, the best resources for information and listing of the latest technical advances scientific work done using VO facilities are the web-sites of the various VO projects listed above, the links that they provide, and the papers that are regularly posted on them. Many VO sites also provide forums for discussion and users can join specialized mailing lists.

10. Acknowledgements

The author wishes to thank the Ministry of Communications and Information Technology of the Government of India for supporting the Virtual Observatory-India projects, and colleagues at IUCAA and Persistent Systems Pvt. Ltd, Pune, with whose help all the work was done. The author also wishes to thank the International Astronomical Union for providing partial support to him for attending the XXVI IAU General Assembly in Prague.

Alan Batten

Nat Gopalswamy

Astronomy for the developing world
IAU Special Session no. 5, 2006
J.B. Hearnshaw and P. Martinez, eds.

© 2007 International Astronomical Union
doi:10.1017/S1743921307007223

Improving the ease of use and efficiency of software tools

Ganghua Lin, Jiangtao Su and Yuanyong Deng

National Astronomical Observatory, Chinese Academy of Sciences,
Datun Rd. 20A, Chaoyang District, Beijing, 100012, P. R. China,
email: lgh@bao.ac.cn

Abstract. For developing countries it is very important to derive maximum use of data obtained from their own telescopes. This is not only related to maximizing science returns on capital investment, but also to maximizing science output. In this paper we describe how we are utilizing software tools to realize this goal. This paper discusses the design and main features of our software tools, and planned future developments. The primary vehicle for general data interpretation is through various interactive techniques of data visualization. Our software employs an object oriented approach which facilitates data processing for experienced users as well as being easier to learn for novice users. This leads to greatly increased efficiency in every phase of data analysis. For developing countries the kind of software we are developing and the virtual observatory concept holds out the hope of advancing capability and efficiency in scientific research.

Keywords. data analysis, software, object oriented programming, solar physics, virtual observatories

1. The Multi-Channel Solar Telescope

The Multi-Channel Solar Telescope (MCST) is a ground-based video magnetograph that can measure the solar 2-dimensional magnetic field and velocity field using different spectral lines. It is consists of a local magnetic field, full-disc vector magnetograph, full-disc H-alpha telescope, etc. It is one of the few integrated-function solar telescopes in the world. The MCST is used to investigate and predict solar magnetic activity and solar-terrestrial interaction effects. Users of its data include solar physicists, the Solar Activity Prediction Centre of Chinese Academy Sciences (CAS), the Space Environment Prediction Centre of CAS, university students and the public.

2. Frequently encountered problems

The following problems usually cost the user plenty of time: programming data analysis codes or codes to translate between various data formats; data calibration; image manipulation, etc. These operations are often carried out using software developed by other users and supported on a best-effort basis. If these problems can be solved in an integrated way, then usability of the data will be greatly enhanced. The easier it is to utilize the data, the higher will be the processing efficiency and resulting data processing gains. Researchers can then pay their attention to solving scientific problems rather than solving data processing problems.

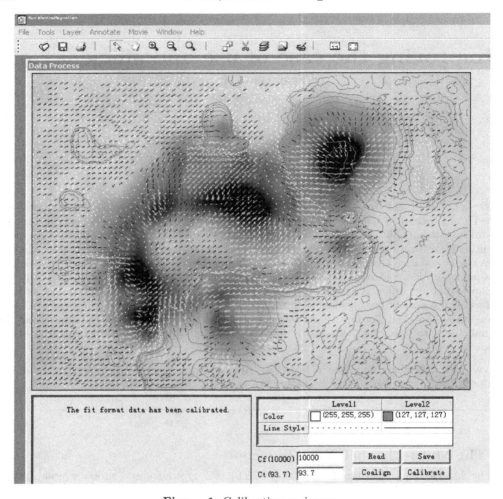

Figure 1. Calibrating an image.

3. Software tools

3.1. *Functionality designed into our software*

We have devised a robust platform for improving the efficiency of data processing. In developing the system we focused on the frequently encountered problems mentioned above and developed tools to:

- analyse data from different instruments;
- transform efficiently raw or processed data between different formats;
- calibrate data;
- co-align images;
- adjust parameters dynamically;
- annotate graphics;
- produce images with arbitrary numbers of rows and columns;
- save images in any common format for presentation or publication;
- produce graphs, contour plots, histograms and animations;
- process data as single files or in batch mode.

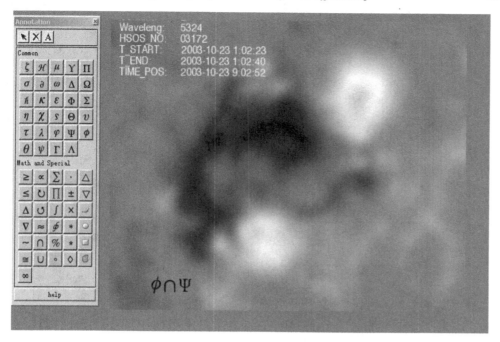

Figure 2. Annotating an image.

These software tools lead to a number of important advantages from the user perspective.

(*a*) Users no longer have to know or care how many bits per pixel images have, or how many rows and columns images have when combining data from different instruments. These data attributes are now all transparent to the user.

(*b*) Likewise the availability of data in various formats no longer presents a problem. It is now easy to translate between the old local data format and FITS, PS, JPEG or GIF.

(*c*) Data calibration is best handled locally by experts who apply all the necessary instrumental corrections so that users can concentrate on using the data. The calibration processing of our vector magnetograms is an example of this. This requires several steps such as: changing the size of the stored data array to the size of the CCD; reduction of bias noise; co-alignment of L, Q, and U data; computing components of magnetic field along the of line of sight and the horizontal azimuthal angle of the horizontal component; and eliminating an uncertainty of 180 degrees.

(*d*) Co-alignment of images is used in making vector magnetograms and movies.

(*e*) Adjusting parameters dynamically becomes fast and simple with object oriented software.

(*f*) We have gathered symbols and characters that are frequently used and placed them in a palette for easy use (Fig. 2). These include commonly used mathematical symbols, arrows, lines, various shapes (circle, triangle, rectangle, etc.). The user can easily change the position, size, colour, etc. of these symbols.

(*g*) Likewise the attributes of graphical objects, such as line type, thickness and colour may be readily changed to create expressive images.

(*h*) Output results of graphics processing are now easily produced to user-specified image sizes.

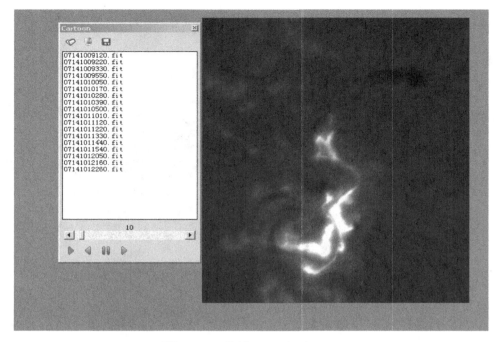

Figure 3. Calibrating an image.

(*i*) Files can be processed individually or in batches, thus facilitating processing of large data sets.

3.2. *Principal software functions*

The principal functions of the software are:

- Reading data in FITS format and in our old local format;
- Making and displaying graphics: graphs, images, contour plots, histograms, and movies;
- Calibrating data;
- Adjusting dynamically various elements in graphics such as line style, axis titles, colours and so on (Fig. 1);
- Adjusting dynamically parameters during the data processing (e.g. Fig. 1);
- Zooming in or out of an image;
- Cropping an image or cutting from an image;
- Annotating graphics, as shown in Fig. 2;
- Writing images in FITS and several other common image formats, and writing movies in MPEG and AVI formats;
- Deriving information from images, locally and globally (Fig. 4);
- Converting from the old local format into FITS, and vice versa; also converting from FITS into PS, JPEG, GIF, etc.;
- Online help for all user functions.

3.3. *Features*

The main features of the software are:

- Higher data processing throughput;
- The algorithms allow dynamic adjustments to key parameters, as shown in Fig. 1;.

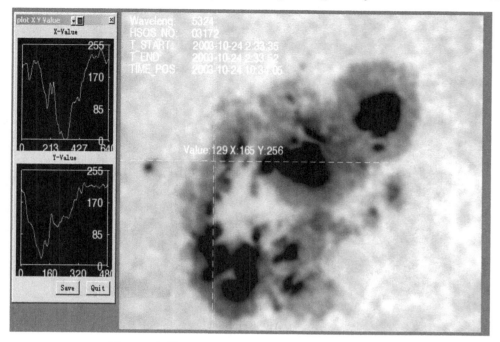

Figure 4. Transects of intensity values in an image.

- The software is menu-driven with a graphical user interface;
- The software can be used in a flexible manner to allow easy viewing and fine tuning of various parameters;
- The software is independent of other proprietary software, such as IDL.
- The software is built in a modular fashion and may be easily extended by adding more modules.

4. Future developments

Thus far we have discovered several software problems and limitations which are being addressed. Future plans for further development of the system include the addition of a log of software changes and new data products. We plan to add more functions, such as roll-correction, stretching, translation, solar rotation-compensation, etc. We also plan to introduce on-line database access and on-line data processing capabilities.

Ajit Kembhavi

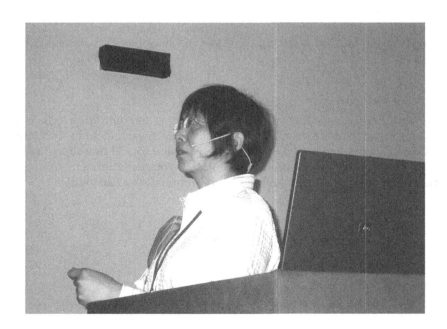

Ganghua Lin

Astronomy for the developing world
IAU Special Session no. 5, 2006
J.B. Hearnshaw and P. Martinez, eds.

© 2007 International Astronomical Union
doi:10.1017/S1743921307007235

A series of technologies that advance the goal of data sharing

Ganghua Lin

National Astronomical Observatory, Chinese Academy of Sciences,
Datun Rd. 20A, Chaoyang District, Beijing, 100012, P. R. China,
email: lgh@bao.ac.cn

Abstract. One of the purposes of a Virtual Observatory is to facilitate data sharing. Data products from the Solar Multi-Channel Telescope and the Solar Radio Telescope at Huairou Solar Observing Station, Beijing, are not only used for solar research but also for solar activity and space environment predictions. To provide these services, we have exploited a number of technologies, which we discuss in this article. These include the setting up of a WWW server, a local area network, network security facility, data processing software, etc. We discuss the implementation of a Virtual Solar Observatory (VSO) and show how it meets various user requirements for unified international metadata. We also discuss future plans for further development of the system.

Keywords. Virtual solar observatories, astronomical data processing, solar physics

1. The virtual solar observatory concept

The Solar Multi-Channel Telescope at Huairou Solar Observation Station, Beijing comprises a local magnetic field, full-disc vector magnetograph, a full-disc H-alpha telescope, etc. The resulting data are not only used for scientific research but also for solar activity predictions and space environment predictions. Direct users of the data products include solar physicists, the Solar Activity Prediction Centre (SAPC) of the Chinese Academy Sciences (CAS), the Space Environment Prediction Centre (SEPC) of CAS, university students and the public. To provide these services requires real-time or quasi real-time data, databases and database management software. We are therefore exploiting a number of technologies to provide these services.

2. ICT technologies in the VSO

2.1. *Setting up WWW servers*

Our web page is a shop window for the VSO. Our website provides an introduction to our telescopes, our data and metadata sets that can be downloaded, simple data processing programs, etc. We also publish a Solar News column so that more users can learn about and use the data on our site. To ensure easy access, our data are available in FITS format. We have also made available translation software to transform our older local data into FITS files. Every day, the latest data and vector magnetic field image are added to the web site. The rapidly increasing volume of data and the website software infrastructure are backed up regularly, thereby allowing us to recover promptly from failures to maximize the up-time of our website.

2.2. *Building up the network*

The Huairou Solar Observation Station (HSOS) is located on a small island near the north bank of the Huairou Reservoir, 6.5 km west of the county seat of Huairou and 60 km north of Beijing. At the time the website was set up, there was no internet connection to the station, so the only solution was to transmit the data by telephone line to the WWW server at headquarters. We used a Linux platform for the server because of its relative immunity to viruses and because of its relative stability. We used the telephone connection with 64 kbps modems at each end for about three years. Because of the need for transmitting data in real time, we started a project to build up a fibre-based Local Area Network (LAN) connecting the various domes and and buildings in the Huairou Solar Observing Station. The LAN has a transmitting speed of 100 Mbps, and can even go up to 1000 Mbps when necessary. The Wide Area Network was recently upgraded to 10 Mbps and is enough to satisfy current needs.

2.3. *Network security*

Unfortunately the network not only brings us high-speed connections, but also viruses and the threat of hacker invasions and network attacks. In HSOS there are various telescopes, instruments, data processing and data transmitting systems, all of which need to be protected. The various safety measures employed are shown in Fig. 1. Firstly, a network virus server was installed to protect every computer in HSOS. The server controls the upgrading and scanning of the client ends. The second network security element is a firewall to prevent invasion from outside the network. The third element is the invasion detection system, which complements the firewall and warns the network administrator of abnormal information transfers. The last element is a leak scan system which can find leaks in a computer operating system and report them to the administrator for corrective action.

2.4. *Data processing software*

Users of the data products are divided into a few groups: people with programming experience who are able to process data; people who know about solar physics but know less about the data and data processing methods; and lastly beginners who have neither programming nor data analysis skills. There is much time spent and much duplicated effort by various users to write software for manipulating and processing data.

An integrated data processing system is clearly advantageous as it reduces duplication of effort and improves data processing efficiency. Beginners can start processing data with a simple mouse click and thus obtain the results they need without having to learn all the specialized programming and data manipulation skills previously required.

The software makes use of a Graphical User Interface (GUI) to allow the user to:

- read files in different formats (FITS or local), either as single files or in batch mode;
- convert from FITS to different image formats, such as PS, GIF and JPEG;
- write data in the several common image formats;
- make contour plots;
- calibrate data;
- making vector magnetic images;
- make a movie from image files and write it out in AVI or MPEG format;
- edit and annotate images with text and graphics of various colors, sizes and shapes.

During image processing users can adjust parameters dynamically until they obtain satisfactory results. Users can enter parameters such as position coordinates, magnetic field strength, tilt angles, etc. at any point. The software will be extended to include roll correction, stretching, translation, solar rotation compensation, image co-alignment, etc.

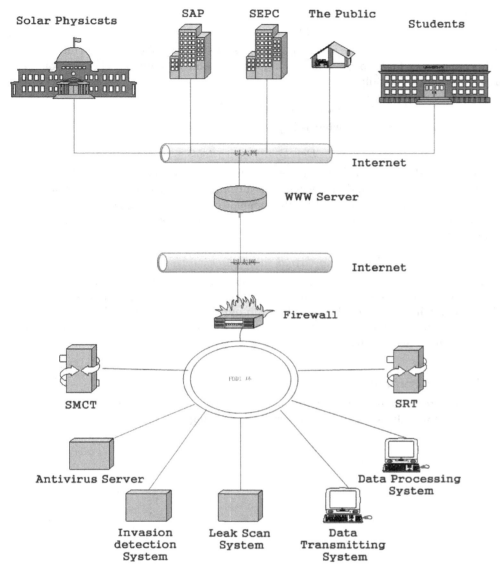

Figure 1. Schematic representation of the system and technologies used to supply data and services to various user groups.

We employ an object oriented approach to programming because of advances of object oriented programming in compiler technology. These advances make the software easy to maintain and adapt to other uses. For instance, with object-oriented programming, once a graph is drawn, all its attributes like color, size, text and markers can be modified interactively with a few mouse clicks. Three-dimensional graphs can be rotated with a mouse movement to be able to visualize the data from different view points. Several curves with different attributes can be superimposed in one graph. In this way a foundation for developing further functionality is established.

3. Further effort

3.1. *Uploading data in real time*

Having the assurance of secure network access, we have the ability to transmit data in real time and at a high speed. We are able to respond to requests of users for data promptly and automatically.

3.2. *Automatic updates of web pages*

This function helps users learn what is happening on the surface of Sun in near real time.

3.3. *Developing query tools*

When searching through data sets, the availability of query tools greatly enhances efficiency. Examples of this would be querying data from solar active regions at low and high latitudes, or studying rotation-related phenomena. Hence designing good query tools is one of our focus areas. Connecting databases with the data processing software is another direction of our efforts.

3.4. *Developing virtual solar observatories (VSO)*

The VSO is a new kind of research tool for solar physics that will enable progress in many significant solar research problems. Data from new observing instruments will be published online in the VSO. The availability of our data in the VSO is likely to produce new research projects. We will link the VSO web site to our web site so that more users can learn about it and use it.

4. Acknowledgements

I would like to acknowledge support and help provided by my colleagues and by the employees of HSOS during the work described in this paper.

Subject Index

Author Index